AN INTRODUCTION TO
CONTROL SYSTEMS

T0320616

ADVANCED SERIES IN ELECTRICAL AND COMPUTER ENGINEERING

Editor: W. K. Chen

Advanced Series in Electrical and Computer Engineering – Vol. 8

AN INTRODUCTION TO
CONTROL SYSTEMS

(Second Edition)

KEVIN WARWICK
Department of Cybernetics, University of Reading

World Scientific
Singapore • New Jersey • London • Hong Kong

Published by

World Scientific Publishing Co Pte Ltd
P O Box 128, Farrer Road, Singapore 912805
USA office: Suite 1B, 1060 Main Street, River Edge, NJ 07661
UK office: 57 Shelton Street, Covent Garden, London WC2H 9HE

British Library Cataloguing-in-Publication Data
A catalogue record for this book is available from the British Library.

First printed in 1989 by Prentice Hall International (UK) Ltd.
— A division of Simon and Schuster International Group

AN INTRODUCTION TO CONTROL SYSTEMS

ISBN 981-02-1563-0

Printed in Singapore by Uto-Print

To the Staff and Students of the University of Reading,
Cybernetics Department

Contents

Preface

Automatic control encroaches upon many of the defined engineering and mathematics university courses in terms of both theory and practice. In common with most engineering disciplines it is a topic which is constantly changing in order to keep pace with modern day application requirements. In the 1960s control engineers were predominantly concerned with analog design procedures, the analog computer being a fundamental building block. Things have changed considerably since that period.

Many new techniques have risen to prominence in the area of control systems during the last decade. Significantly, the increasing influence of microcomputers has led to a much greater need for an emphasis to be placed on digital aspects of control. Many control ideologies that were once thought to be unrealizable, except in a few extreme cases, because of hardware limitations, have now become everyday realities in practical industrial applications due to the transference of complex algorithms to a software base. With the advent of VLSI design in association with special-purpose machines, higher speed and greater memory capacity computers are becoming available, allowing further control schemes to be considered.

The undergraduate students of engineering taking an introductory control course as part of their overall degree must necessarily become aware of the developments in digital control which have now become standard practice in many industrial applications and are fundamental to any further degree courses or research in control. This book treats digital control as one of the basic ingredients to an introductory control course, including the more regularly found topics with equal weight.

Organization of the book is as follows. A general overview of the field called control systems is introduced in Chapter 1. This includes a brief history of control and points out many of the standard definitions and terms used. In Chapter 2 it is shown how mathematical models can be formed in order that they may be representative of a system. Basic handling rules for these models are given and the model's relationship with everyday engineering tools and apparatus is stressed.

Differential equations are examined in Chapter 3 and in Chapter 4 system stability and performance, important aspects of control systems design, are considered. These two chapters, in effect, detail the necessary mathematical tools used in the analysis of a system and give an indication as to how the system responds to a particular set of conditions. Chapter 4 concentrates mainly on the Routh–Hurwitz criterion in terms of stability and root-locus techniques for performance by investigating the effect of parameter variations on the roots of the system characteristic equation.

Chapter 5 looks into the frequency response methods due to Bode and Nyquist. Details are given with emphasis placed on how to design compensation networks using the methods described. The same approach is then taken in the discussion based on the Nichols chart which is used in obtaining the system closed-loop frequency response. It is considered that Chapters 3, 4 and 5 cover what is now termed classical control techniques; the rest of the book is therefore devoted to more recently devised methods.

Design philosophies and system descriptions based on the state space are brought together in Chapters 6 and 7. The use and variety of state equations and their relationship to system transfer functions is found in Chapter 6, whereas Chapter 7 is devoted more to system control using state feedback, subject to this being possible. Realization of the state variables when these are not measurable or physically obtainable is also considered.

Digital control uses the z operator and is closely linked with computer control via sampled-data systems. Computers, however, are also used for control design in terms of system simulations and modeling. Chapters 8 and 9 include the important topics of digital and computer control, computer-aided control system design being highlighted. The final chapter looks in depth at the most widely used control scheme, the PID controller, and includes details of its implementation in both continuous and discrete time.

The book is intended to be used in relation to an introductory course on control systems, for which students need a mathematical background such that Laplace transforms, solutions to differential equations and complex numbers have been taught.

This text has been put together with the help of several people. I would particularly like to thank John Westcott, John Finch, Mohammed Farsi and Keith Godfrey for their technical appraisal, and numerous former students who have provided valuable feedback on the material presented. I would also like to acknowledge the assistance of Sukie Clarke and Karen Smart in preparing the text and correcting my tewible English. Finally, I am immeasurably grateful for the support I have received from my family during the writing of this book, including practical help from my daughter Madeline: I hope the reader concludes that it has been worth it.

Preface to the Second Edition

One of the reviewer's comments on the first edition was that the section on Analog Computers was rather dated. In the second edition, apart from a brief mention, this section has been completely removed, to be replaced by a completely new section on Fuzzy Logic Controllers, which is very much of interest both from an academic and industrial viewpoint.

I am delighted that World Scientific Publishing Co. have decided to publish this second edition, following the first edition selling out very quickly. My thanks go to Liz Lucas for all her help in putting together this edition, her husband Rodney for helping with the figures and Barbara Griffin at World Scientific for her patience.

About the Author

Kevin Warwick has been Professor of Cybernetics and Head of the Department of Cybernetics at the University of Reading, UK, since 1988. He previously held appointments at Oxford University, Imperial College, London, Newcastle University and Warwick University, as well as being employed by British Telecom for six years. Kevin is a Chartered Engineer, a Fellow of the IEE and an Honorary Editor of the IEE Proceedings on Control Theory and Applications.

Professor Warwick has received Doctor of Science degrees for his research work on Computer Control both from Imperial College London and the Czech Academy of Sciences, Prague. He has also published well over 200 papers in the broad areas of his research interests, including intelligent systems, neural networks and robotics. He has written or edited, a number of books, including *Virtual Reality in Engineering* and *Applied Artificial Intelligence*.

1

Introduction

1.1   General introduction

When first encountering the subject of control systems, it should in no way be considered that the topic is a completely new field with which one has had no prior contact whatsoever. In everyday life we encounter control systems in operation and actually perform many controlling actions ourselves. Take, for example, picking up and opening this book, which involves the movement of our hands in a coordinated fashion. This movement is made in response to signals arising due to information obtained from our senses – in this case sight in particular. It is worth mentioning however that the selection of this book, as opposed to other books written by different authors, is an example of the human being behaving as an intelligent control system exhibiting, in this case, signs of good taste and common sense – both of which are difficult properties to quantify and study.

Certain properties are common between many different fields, examples being distance, height, speed, flow rate, voltage, etc., and it is the mere fact of this commonality that has given rise to the field of control systems. In its most general sense a system can be virtually any part of life one cares to consider, although it is more usual for a system to be regarded as something to which the concepts of cause and effect apply. A control system can then be thought of as a system for which we manipulate the cause element in order to arrive at a more desirable effect (if at all possible). In terms of engineering it therefore follows that the study of control systems is multi-disciplinary and is applicable equally well in the fields of chemical, mechanical, electrical, electronic, marine, nuclear, etc., engineering.

Although lying within distinctly different fields, possibly different branches of engineering, systems often exhibit characteristics which are of a similar, if not identical, nature. This is usually witnessed in terms of a system's response to certain stimuli, and although the physical properties of the stimuli themselves can take a different form, the response itself can be characterized by essentially the same information, irrespective of the field in which it lies. Consider, for example, heating a pot of water: this exhibits the same type of exponential response as that witnessed if either a capacitor is charged up or a

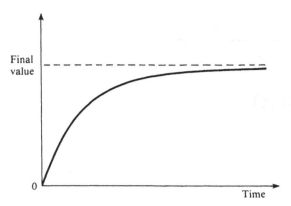

Fig. 1.1   Exponential response of a system

spring is compressed, see Fig. 1.1. The stimulus is, say, quantity of gas (if the water is gas heated), voltage and force respectively; however, the response, whether it be in terms of temperature, charge or stored energy, is still characterized as in Fig. 1.1. The study of control systems in terms of a multi-disciplinary basis therefore means that the results of performance and design tests in one discipline can readily be reposed and re-evaluated within an alternative discipline.

There are essentially two main features in the analysis of a control system. Firstly system modeling, which means expressing the physical system under examination in terms of a model, or models, which can be readily dealt with and understood. A model can often take the form of an appropriate mathematical description and must be satisfactory in the way in which it expresses the characteristics of the original system. Modeling a system is the means by which our picture of the system is taken from its own discipline into the common control systems arena. The second feature of control systems analysis is the design stage, in which a suitable control strategy is both selected and implemented in order to achieve a desired system performance. The system model, previously obtained, is therefore made great use of in the design stage.

The design of a control system only makes sense if we have some objective, in terms of system performance, which we are aiming to achieve. We are therefore trying to alter the present performance of a system in order to meet our objectives better, by means of an appropriate design, whether this is to be in terms of a modification to the system itself or in terms of a separate controller block. Actual performance objectives are very much dependent on the discipline in which the system exists, e.g. achieving a particular level of water or a speed of rotation, although the objectives can be stricter, e.g. requiring that a level of water does not vary by more than 2% from a nominal mean value.

An underlying theme in the study of automatic control systems is the assumption that any required controlling/corrective action is carried out automatically by means of actuators such as electromechanical devices. The concept of a human operative reading a value from a meter and applying what is deemed an appropriate response by pulling a lever or twisting a dial is therefore not really part of the subject matter covered in this

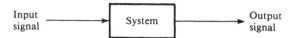

Fig. 1.2   Basic system schematic

book. The human being as an element in a control system is briefly considered later in this first chapter, merely to serve as an example. It is in fact becoming more often the case that a digital computer is employed to turn the measured (meter) value into an appropriate responsive action, the main advantage of such a technique being the speed with which a computer can evaluate a set of what could be complicated equations. It must be remembered however that interfacing is required, firstly to feed the measured value into the computer and secondly to convert the calculated response from the computer into a signal which is suitable for a physical actuator.

In terms of the system under control, an actuator is used to apply a signal as input to the system whereas a suitable measurement device is employed to witness the response in the form of an output from the system. Schematically, a system can therefore be depicted as a block which operates on an input signal in order to provide an output signal. The characteristics of that system are then contained within the block, as shown in Fig. 1.2.

1.2   A concise history of control

To put forward a chronologically ordered set of events in order to show the logical development of a technical subject is certain to be fraught with wrong conclusions and misleading evidence. The study of control systems is certainly no exception in this context, a particular problem being posed by much of the development of the mathematical tools which now form the basis of control systems analysis. It is also a difficult task to point the finger at someone way back in time and accuse them of starting the whole thing. Indeed some historical accounts refer to times well before the birth of Christ for relatively early examples.

As far as automatic control (no human element) is concerned, a historical example cited in many texts is James Watt's fly-ball governor. The exact date of the invention/ development seems to vary, dependent on the particular text one is looking at, with dates ranging from 1767 to 1788 being quoted. Although the actual year itself is not particularly important, it serves a primary lesson in control systems analysis not to rely completely on any measured value. James Watt's fly-ball governor is shown in Fig. 1.3, where the control objective is to ensure that the speed of rotation is approximately constant. As the fly-balls rotate so they determine, via the valve, how much steam is supplied; the faster the rotation – the less steam is supplied. The rate of steam supplied then governs, via the piston and flywheel, the speed of rotation of the fly-balls. Although tight limits of operation, in terms of speed variation, can be obtained with such a device,

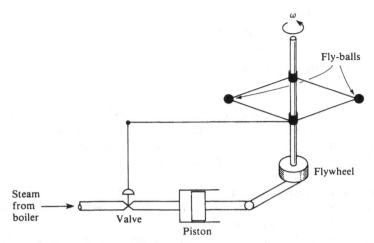

Fig. 1.3   James Watt's fly-ball governor

there are unfortunately several negative features, not the least of these being the tendency of the speed to oscillate about a mean (desired) speed value.

Around 1868, many years after Watt's governor, J. C. Maxwell developed a theoretical framework for such as governors by means of a differential equation analysis relating to performance of the overall system, thereby explaining in mathematical terms reasons for oscillations within the system. It was gradually found that Maxwell's governor equations were more widely applicable and could be used to describe phenomena in other systems, an example being piloting (steering) of ships. A common feature with the systems was the employment of *information feedback* in order to achieve a controlling action.

Rapid development in the field of automatic control took place in the 1920s and 30s when connections were drawn with the closely related areas of communications and electronics. In the first instance (1920s) this was due to analysts such as Heaviside who developed the use of mathematical tools, provided over a century before by Laplace and Fourier, in the study of communication systems, in particular forming links with the decibel as a logarithmic unit of measurement. In the early 1930s Harry Nyquist, a physicist who had studied noise extensively, turned his attention to the problem of stability in repeater amplifiers. He successfully tackled the problem by making use of standard function theory, thereby stressing the importance of the phase, as well as the gain, characteristics of the amplifier. In 1934 a paper appeared by Hazen entitled 'Theory of servomechanisms' and this appears to be the first use of the term 'servomechanism', which has become widely used as a name to describe many types of feedback control system.

The Second World War provided an ideal breeding ground for further developments in automatic control, particularly due to finance made available for improvements in the military domain. Examples of military projects of that time are radar tracking, anti-aircraft gun control and autopilots for aircraft; each of which requires tight performance achievements in terms of both speed and accuracy of response.

Since that time mechanization in many industries, e.g. manufacturing, process and power generation, has provided a stimulus for more recent developments. Originally, frequency response techniques such as Bode's approach and Laplace transform methods were prominent, along with the root locus method proposed by Evans in the late 1950s. (*Note:* This was also known in the UK at one time as Westcott's method of π lines.) However, in the 1960s the influence of space flight was felt, with optimization and state–space techniques gaining in prominence. Digital control also became widespread due to computers, which were particularly relevant in the process control industries in which many variables must be measured and controlled, with a computer completing the feedback loop.

The 1970s saw further progress on the computer side with the introduction of microprocessors, thus allowing for the implementation of relatively complicated control techniques on one-off systems at low cost. The use of robot manipulators not simply for automating production lines but as intelligent workstations also considerably changed the requirements made of a control system in terms of speed and complexity. The need for high-speed control devices has in the 1980s been a contributing factor too and has made great use of hardware techniques, such as parallel processors, whereas at the same time ideas from the field of artificial intelligence have been employed in an attempt to cope with increased complexity needs. Finally the low cost and ease of availability of personal computers has meant that many control systems are designed and simulated from a software basis. Implementation, which may itself make use of a computer, is then only carried out when good control is assured.

1.3   Open-loop control

An open-loop control system is one in which the control input to the system is not affected in any way by the output of the system. It is also necessary however that the system itself is not varied in any way in response to the system output.

Such a definition indicates that open-loop systems are in general relatively simple and therefore often inexpensive. An excellent example is an automatic electric toaster in which the control is provided by means of a timer which dictates the length of time for which the bread will be toasted. The output from the toasting system is then the brownness or quality of the toast, the assumption being that once the timer has been set the operator can only wait to examine the end product.

Clearly the response of an open-loop system is dependent on the characteristics of the system itself in terms of the relationship between the system input and output signals. It is apparent therefore that if the system characteristics change at some time then both the response accuracy and repeatability can be severely impaired. In almost all cases however the open-loop system will present no problems insofar as stability is concerned,

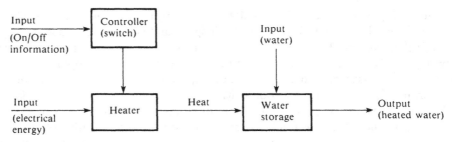

Fig. 1.4   Water heating device

i.e. if an input is applied the output will not shoot off to infinity – it is not much use as an open-loop system if this is the case.

Another example of an open-loop system is the water heater shown in Fig. 1.4, in which the controller is merely an on/off switch which determines when the heater is supplying heat in order to provide heated water at a certain specified temperature.

A problem with this open-loop system is that although today the water may be provided at a nice temperature at the output, tomorrow this might not be the case. Reasons for this could be a change in ambient conditions, a change in the temperature or amount of water input to the storage device or a drop in the voltage used to supply the heater. Merely leaving the system as it is and hoping for the best is clearly not satisfactory in the majority of cases. It would be much more sensible to measure the temperature of the heated water such that the on/off information can be varied appropriately in order to keep the output at approximately the temperature desired. We have now closed the loop, by providing feedback from the system output to an input, hence the system is no longer open-loop, but rather is closed-loop.

1.4   Closed-loop control

In a closed-loop system the control input is affected by the system output. By using output information to affect in some way the control input of the system, feedback is being applied to that system.

It is often the case that the signal fed back from the system output is compared with a reference input signal, the result of this comparison (the difference) then being used to obtain the control or actuating system input. Such a closed-loop system is shown in Fig. 1.5, where the error = reference input – system output.

Very often the reference input is directly related to the desired value of system output, and where this is a steady value with respect to time it is called a set point input.

By means of the negative feedback loop shown in Fig. 1.5 (negative because the system output is subtracted from the reference input) the accuracy of the system output in relation to a desired value can be much improved when compared to the response of an open-loop system. This is simply because the purpose of the controller will most likely be

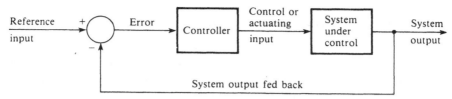

Fig. 1.5   Closed-loop system

to minimize the error between the actual system output and the desired (reference input) value.

A disadvantage of feedback is related to the fly-ball governor of Fig. 1.3, where oscillations can occur in the system output, the speed of rotation, which would not occur if the system were connected in open-loop mode. The oscillations are due to the attempt to get the error signal to as low a magnitude as possible, even if this means swinging the output first one way and then the other.

The open-loop heating system of Fig. 1.4 can be converted into a closed-loop system by measuring the temperature of the heated water (output) and feeding this measurement back such that it affects the controller by means of modifying the switch on/off information. It is then apparent that when variations in performance occur due to system modifications or a change in ambient conditions, the effect on the system output will be much reduced because of the feedback arrangement, i.e. with feedback the system is less sensitive to variations in conditions. Let us assume that we require the water temperature to be 50 °C. With the system in open-loop, a variation in system characteristics would merely cause the temperature to drift away from the required value. With the system in closed-loop however, any variation from 50 °C will be seen as an error when the actual measured temperature is compared with the required value, and the controller can then modify the control input in order to reduce the error to zero.

1.4.1   Effects of feedback

Open-loop systems rely entirely on calibration in order to perform with good accuracy; any system variations or effects caused by outside influences can seriously degrade this accuracy. Although only common in relatively simple systems, an important property is found in that an increase in system gain (amplification) does not affect the stability of an open-loop system (oscillations cannot be induced in this way).

Once feedback is applied, the system is in closed-loop. Closed-loop systems can achieve much greater accuracy than open-loop systems, although they rely entirely on the accuracy of the comparison between desired and actual output values and therefore on the accuracy of the measured output value. Effects of system variations or outside influences are much reduced, as are effects due to disturbances or nonlinear behavior. Unfortunately the advantages obtained when feedback is employed are at the expense of system stability (oscillations can often be induced by increasing system gain).

1.5   Some examples of control systems

Many simple control systems in everyday life include a human being, not only in the feedback loop but also to provide the actuating signals and even as the control system itself. Consider drinking a cup of coffee, which involves lifting a cup through a desired height. Feedback is provided via touch and sight, with a comparison being made between where the cup actually is at any time and where we want it to be, i.e. by our mouth – this is the error signal. An actuating signal is then provided in terms of lifting the cup with our hand at a rate which is dependent on how large the error signal is at any time. Performance is measured in terms of how quickly we can lift the cup – lifting it too quickly or going past the desired end position will result in coffee being spilt and possibly several broken teeth, conversely lifting it too slowly will result in severe boredom and even death through lack of liquid intake. This is really an example of a *biological control system*, everything within the system, except the cup and coffee, being part of the human being.

Note that outside influences such as windy weather, a lead weight in the bottom of the cup or the temperature of the coffee have not been considered. Although all of these would affect the performance in question, they were considered to be outside the scope of the system definition in the first instance. This is an important initial point because when modeling a control system the more characteristics that are included in the model, the more complex that model becomes. A trade-off therefore has to be made in terms of model complexity and the ability of that model to account for all eventualities. It would not be sensible to include the effect of high winds when modeling the act of drinking a cup of coffee because the vast majority of cups of coffee are not drunk in the presence of high winds.

Often the human being merely forms part of an overall control system, examples being driving an automobile and cooking, where the human can merely form the feedback loop – measuring the actual output, comparing this with the desired value and then modifying the control input accordingly.

Many control systems have been automated by replacing the human feedback element with more accurate equipment which also responds more rapidly. Examples of this are *ship steering* and *flight control*, for which pilots are replaced by autopilots whereby information from sensors, such as roll detectors and gyros respectively, is used to ensure constant velocity and constant heading. Any change in either of these requirements can be fed in merely as a change in set-point value. Taken further, both *ship docking* and *aircraft landing* can also be automatically carried out, by means of computational procedures either operating *in situ* or from a remote station. These latter examples serve as particular reminders that many different characteristics must be taken into account when modeling a system; neglecting the effect of rough weather or the presence of other craft could be fatal in this instance.

Other transit systems also employ automatic control techniques; certain automatic electric trains fall into this category along with autonomous guided vehicles (AGVs) which are not restricted to operation along certain lines, such as rails. With both visual (camera) and distance (laser and sonar) information it is now possible for these vehicles to 'drive' themselves and it should not be long before greater use is made of such techniques in automobiles. The military interest in such vehicles is important for their development, as indeed it is for future developments in gun positioning, radar tracking and missile control, all of which involve rapid tracking of a target, with minimum error.

In manufacturing, *automated production lines* are not only supervised by an overall automatic controller, but each individual workstation can consist of a robot manipulator which either merely performs a set routine repeatedly with good accuracy and speed or which modifies its actions due to visual and/or tactile information feedback. On a similar basis the control of *machine tools* is gradually increasing in complexity from simple profile copying mechanisms, through numerical control for exact coordination, to machine tools which can modify their procedure in response to varying conditions such as differing workpiece requirements.

Power generation is also an important area in the application of automatic control systems. *Nuclear reactor control* is concerned with the rate at which fuel is fed into the reactor and it is in areas such as this that the effects of hazardous environment must also be taken into account. Certainly the large number of variables which need to be measured in order to provide a useful picture of the system presents many problems in terms of accuracy and reliability. Indeed other systems concerned with power generation and distribution share many instrumentation difficulties. Stricter controls imposed due to *environmental pollution* have meant not only increased efficiency requirements in the conversion schemes themselves but also the use of automatic control devices to control effluence.

Further down on the electrical power scale, *voltage stabilizers* are employed to retain the output voltage from a source as close as possible to a reference voltage value. This is done by comparing the actual voltage output with the reference, any difference (error) then being used to vary the resistance provided by a transistor which is connected in series with the source.

Uniformity in the thickness of such as paper, steel and glass is based on the control of the motor speed of rollers and drawing machines. Rollers are often not perfectly cylindrical and, combining this with possible large variations in the product quality (e.g. pulp obtained from a different factory), controlling output thickness within tight limits is not a simple task. Letting the thickness increase merely wastes money whereas letting it decrease produces an inferior product, hence tight limits are necessary. The control of *motors* both in terms of position and speed is however a much wider field than mentioned here, the complexity of control being dependent on the accuracy requirements made.

In terms of financial outlay, large amounts have been spent in providing accurate automatic systems for *process control*. In oil and chemical industries a small percentage improvement in reducing output variations (e.g. flow rate, concentration) in many cases results in millions of dollars being saved. Process plants are usually characterized as fairly slowly varying systems with only occasional changes in reference input. Controllers are required in order to minimize the effect of disturbances on the output signal. *Biochemical control* and *biomedical control* are in most cases closely related to process control systems, with the former including such examples as drug production and fermentation, and the latter respiratory monitoring and pain relief.

Automatic control has become increasingly more important in *agricultural engineering*, with intelligent tractors now able to cope with wheel slip and to provide a constant ploughing level despite surface variations. Improved automobile *engine performance* and efficiency has become more and more dependent on complex control schemes, with power generated by means of the engine itself being used to provide energy for the electronic/microcomputer-based control circuitry.

Obvious advantages can be obtained by successfully modeling *economic systems* in order to obtain suitable automatic control schemes. Apart from the most simple cases however it is extremely difficult to account for many of the spurious events that can occur in practice and which have considerable effect on the economic system under control, an example being the price of shares on the stock market. *Social systems* are similar to economic systems in terms of modeling difficulties and also suffer from the fact that very often there are both a large number of outputs and also different measures of performance which are often contradictory.

Hopefully a good idea of the wide variety of system types in which automatic control schemes operate has been given. When one is new to the subject, examples encountered are often related to relatively simple applications with which the reader may be familiar, e.g. electromechanical systems, electronic amplifiers, chemical plant, mechanical systems and electrical machines. It should be remembered that many of the simple techniques considered remain as simple techniques even when applied to the most complicated plant.

1.6   Definitions of standard terminology

As with most subject areas, terminology is used in the study of control systems which aptly describes phenomena within the field. It can, however, be seen as rather vague and in many cases confusing from the outsider's point of view. A list of definitions is therefore given here in order to help remove any barriers which do exist (see Fig. 1.6).

1. Lower case letters refer to signals, e.g. voltage, speed; and are functions of time, $u = u(t)$.
2. Capital letters denote signal magnitudes, as in the case of $u(t) = U \cos \omega t$, or otherwise Laplace transformed quantities, $U = U(s)$. Where $s = j\omega$, this is indicated by $U(j\omega)$.
   Note: s is the Laplace operator and $\omega = 2\pi f$ where f is frequency.
3. The system under control is also known as the *plant* or *process*, G.
4. The *reference input*, v, also known as the set-point or desired output, is an external signal applied in order to indicate a desired steady value for the plant output.
5. The *system output*, y, also known as the controlled output, is the signal obtained from the plant which we wish to measure and control.

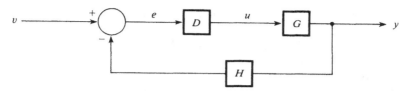

Fig. 1.6 Closed-loop system with feedback element *H*

6. The *error* signal, *e*, is the difference between the desired system output and the actual system output (when $H = 1$).
   Note: See 8.

7. The *controller*, *D*, is the element which ensures that the appropriate control signal is applied to the plant. In many cases it takes the error signal as its input and provides an actuating signal as its output.

8. The *feedback element H* provides a multiplying factor on the output *y* before a comparison is made with the reference input *v*. When $H \neq 1$ the error *e* is the error between *v* and *Hy*, i.e. it is no longer the error between *v* and *y*.
   Note: Often $H = 1$, although *H* can represent the characteristics of the measurement device in which case most likely $H \neq 1$.

9. The *feedback signal* is the signal produced by the operation of *H* on the output *y*.

10. The *control input*, *u*, also known as the actuating signal, control action or control signal, is applied to the plant *G* and is provided by the controller *D* operating on the error *e*.

11. The *forward path* is the path from the error signal *e* to the output *y*, and includes *D* and *G*.

12. The *feedback path* is the path from the output *y*, through *H*.

13. A *disturbance*, or noise (not shown in Fig. 1.6), is a signal which enters the system at a point other than the reference input and has the effect of undermining the normal system operation.

14. A *nonlinear* system is one in which the principles of superposition do not apply, e.g. amplifier saturation at the extremes, or hysteresis effects. Almost all except the most simple systems are nonlinear in practice, to an extent at least. The vast majority of systems can however be dealt with by approximating the system with a linear model, at least over a specific range.

15. A *time-invariant* system is one in which the characteristics of that system do not vary with respect to time. Most systems do vary slowly with respect to time, e.g. ageing, however over a short period they can be considered to be time-invariant.

16. A *continuous-time* system is one in which the signals are all functions of time *t*, in an analog sense.

17. A *discrete-time* system is a system such as a digital system or a sampled data system in which the signals, which consist of pulses, only have values at distinct time instants. The operator *z* is used to define a discrete-time signal such that $z^3 y(t) = y(t + 3)$ means the value of signal $y(t)$ at a point in time three periods in the future, where a period *T* (sample period) is defined separately for each system.

18. A *transducer* converts one form of energy (signal) into another, e.g. pressure to voltage.
19. *Negative feedback* is obtained when $e = v - Hy$.
20. *Positive feedback* is obtained when $e = v + Hy$.
    Note: This is not shown in Fig. 1.6.
21. *A regulator* is a control system in which the control objective is to minimize the variations in output signal, such variations being caused by disturbances, about a set-point mean value.
    Note: A regulator differs from a servomechanism in which the main purpose is to track a changeable reference input.
22. A *multivariable* system is one which consists of several inputs and several outputs. In this text only single variable systems are considered.

1.7   Summary

In this first chapter the purpose has been to gently introduce the subject area of control systems, by providing a brief historical account of developments leading to its present state and also by putting forward the idea of feedback control in terms of some elementary forms. To do this it has been necessary to employ representative diagrams to show connections between signals and systems. These diagrams are described in much greater detail in the following chapter.

Many examples of the application of control systems were given in an attempt to show the multi-disciplinary nature of the subject. Once the fundamentals of a particular control technique have been studied in one field, they are generally applicable in many other fields. Control engineers therefore come from many different backgrounds in terms of subjects studied and an even more diverse set exists in terms of areas in which a control engineer can show a certain amount of understanding.

In the chapters which follow many techniques are shown for the modeling, control and study of system behavior. The ideas put forward should not be seen as limiting or restrictive in any way, in terms of applications, but rather as an insight to the wide variety of possibilities.

Problems

1.1   A thermostat is often used in order to control the temperature of a hot water tank. If the water is heated from cold, explain, with the aid of a sketch of water temperature versus time, the principle of operation of such a device. Why is a simple open-loop system not used instead?

1.2   Explain the open-loop operation of traffic signals at a road crossing. How can improved traffic control be achieved by means of a closed-loop scheme?

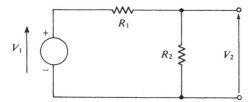

Fig. 1.7

1.3 Consider a missile guidance system which is based on an open-loop scheme, once the missile has been launched. How can a closed-loop technique be devised in order to track a target and what problems can this cause?

1.4 Consider the potential divider in Fig. 1.7. The output voltage is V_2 and the input voltage is V_1.
How can the system be both an open- and closed-loop system at the same time?

1.5 The population of rabbits in a particular field is unaffected by events external to the field, including visiting rabbits. The population consists of adult males, adult females and young rabbits.
Each year 20% of adult males and females die, while the number of newborn young which survive to the following year is twice the number of adult females alive throughout the present year.
Each year 40% of the young become adult males the following year and 40% of the young become adult females the following year. Meanwhile, the remaining 20% of the young die.
Find equations to describe each section of the population from one year to the next.

1.6 Consider the simplified economic system describing the price of a toy. Demand for the toy will decrease if the price of the toy increases, whereas supply of the toy increases if its price increases. By finding the difference between supply and demand, show how this can be used as an error signal to effect a controlling action on the price.

1.7 Foxes move into the field inhabited by the rabbits of Problem 1.5, and use the rabbits as a food source. F is the number of foxes at any one time and R is the number of rabbits. The rate of change of foxes is equal to a constant K_1 multiplying the error between R and a number $\bar{R}$, whereas the rate of change of rabbits is equal to a constant K_2 multiplying the error between a number $\bar{F}$ and F.
How does the system operate; what happens to F and R with respect to time?

1.8 Consider the end effector of a robot manipulator. Explain how, in many cases, this is operated in open-loop mode. How could a closed-loop system be constructed for the manipulator?

1.9 In a liquid level system, what problems could occur if an open-loop scheme is employed? Can all of these problems be removed by employment of a closed-loop technique? Are any problems brought about due to the closed-loop format itself?

1.10   Consider driving a vehicle without using any feedback, e.g. close your eyes and ears and drive (maybe you drive this way already!).

What advantages are obtained if a closed-loop vehicle driving scheme is used, which includes a human operator? (obvious answers). How could the vehicle driving scheme be configured without a human operator? Are there any particular advantages or disadvantages to such a technique?

Further reading

Bennett, S., *A history of control engineering, 1800–1930*, Peter Peregrinus Ltd, 1986.

Black, H. S., 'Inventing the negative feedback amplifier', *IEEE Spectrum*, 1977, pp. 55–60.

Bode, H. W., 'Feedback – the history of an idea', in *Selected papers on mathematical trends in control theory*, Dover, New York, 1964, pp. 106–23.

Danai, K. and Malkin, S. (eds.), *Control of manufacturing processes*, ASME, 1991.

Dorf, R. C., *Modern control systems*, 6th ed., Addison-Wesley, 1992.

Enos, M. J. (ed.), *Dynamics and control of mechanical systems*, Mathematical Soc., 1993.

Hazen, H. L., 'Theory of servomechanisms', *J. Franklin Institute*, **218**, 1934, pp. 543–80.

Leigh, J. R., *Control theory: a guided tour*, Peter Peregrinus Ltd., 1992.

Lighthill, M. J., *Fourier analysis and generalised functions*, Cambridge University Press, 1959.

Maskrey, R. H. and Thayer, W. J., 'A brief history of electrohydraulic servo mechanisms', *ASME J. of Dynamic Systems, Measurement and Control,* 1978, pp. 110–16.

Maxwell, J. C., 'On governors', *Proceedings of the Royal Society* (London), **16**, 1868, pp. 270–83.

Mayr, O., *Origins of feedback control*, MIT Press, 1971.

Minorsky, N., 'Directional stability of automatically steered bodies', *J. Am. Soc. Naval Eng.*, **34**, 1922, p. 280.

Minorsky, N., 'Control problems', *J. Franklin Inst.*, **232**, 1941, p. 451.

Nyquist, H., 'Regeneration theory', *Bell System Technical Journal*, **11**, 1932, pp. 126–47.

Popov, E. P., *The dynamics of automatic control systems*, Addison-Wesley, 1962.

Routh, E. J., 'Stability of a given state of motion', Adams Prize Essay, Macmillan, 1877.

Singh, M., *Encyclopedia of systems and control*, Pergamon Press, 1987.

Thaler, G. J., *Automatic control: classical linear theory*, Dowden, Huchinson and Ross Inc., Stroudsberg, Pa., 1974.

Tischler, M. B. (ed.), 'Aircraft flight control', *Int. J. Control* (Special Issue), **59**, No. 1, 1994.

Tzafestas, S. G. (ed.), *Applied control*, Marcel Dekker, 1993.

2

System representations

2.1 Introduction

In order to analyze a system in terms of controller design a basic assumption is made, as will be seen in the proceeding chapters, that we already have in our possession an accurate mathematical model of the system. It is not usually the case however that a mathematical model is employed as a starting point from which a physical system is constructed, the most likely occurrence being that a physical system is already in existence and it is required that a controller be obtained for this system.

Forming a mathematical model which represents the characteristics of a physical system is therefore crucially important as far as the further analysis of that system is concerned.

Completely describing, in a mathematical sense, the performance and operation of a physical system, will most likely result in a large number of equations, especially if it is attempted to account for all possible eventualities, no matter how simple the system might at first appear to be. Further, it must be remembered that once the equations have been obtained, they will be used for analysis purposes and hence a complicated collection of identities and equalities will not generally be widely acceptable, especially if this means an extremely costly design exercise due to the time taken because of the complexity. Conversely the model describing a physical system should not be over simple so that important properties of the system are not included, something that would lead to an incorrect analysis or an inadequate controller design. Therefore a certain amount of common sense and practical experience is necessary when forming a mathematical model of a physical system in order to decide which characteristics of the system are important within the confines of the particular set-up in which the system must operate.

The systems considered in this text are inherently dynamic in nature, which means that differential equations are an appropriate form in which their characteristics can be described, although for physical systems the equations will in general be nonlinear to a certain extent. In some cases the nonlinear characteristics are so important that they must be dealt with directly, and this can be quite a complicated procedure. However, it is possible in the vast majority of cases to linearize the equations in the first instance such that subsequent analysis is much simplified. It is important to remember, though, that any design carried out on the basis of a linearized model is only valid within the range of operation over which the linearization was carried out.

Once a physical system has been modeled by a finite number, which may be only one, of linear differential equations, it is possible to invoke the Laplace transform method with a consequent ease of analysis, by thinking of the system's operation in terms of a transfer function, as is done in Section 2.2 of this chapter. It is shown in the same section, by means of two examples, how linear approximations can be made, over a set operating range, to nonlinear system description equations. An alternative form of representing system characteristics in terms of a more visual approach is then discussed with the introduction of block diagrams in Section 2.3. Finally, both electrical and mechanical systems are considered in much greater depth in order to show the relationship between initial system networks and the transfer functions which are obtained as a result.

2.2   Transfer functions

In order to analyze a control system in any reasonable depth it is necessary to form a model of the system either by obtaining response information from a system which is already in existence or by combining the information from individual components to assess the overall system obtained by bringing those components together. In forming a model we need to supply an answer to the question 'What will be the system response to the particular stimulus?', i.e. for a given input our model must tell us what output will occur, and this must correspond with that of the actual system if the same input were applied. If we are modeling the system in a mathematical way, we must produce a function which operates on an input in such a way as to arrive at the correct output. If an input is applied to the function, it will transfer us to the corresponding output value – such a function is therefore called a *transfer function*.

In its simplest form a transfer function can be simply a multiplying operator; consider a chocolate bar machine for example. Let us assume that from the machine in question each chocolate bar costs 50 cents. If we insert 50 cents (the input) into an appropriate slot on the machine we will be rewarded with one chocolate bar (the output), so the transfer function of this system is simply $1/50$, as long as the input and output are scaled in cents and bars respectively. Unfortunately many systems are rather more complicated than the simple chocolate bar machine example, and hence their transfer function which relates input to output must take on a more complex form.

It is often the case that the relationship between system input, $u(t)$, and system output, $y(t)$, can be written in the form of a differential equation. This could mean, in general terms, that the input to output transfer function is found to be:

$$(D^n + a_{n-1} D^{n-1} + \cdots + a_1 D + a_0)y(t)$$
$$= (b_m D^m + b_{m-1} D^{m-1} + \cdots + b_1 D + b_0)u(t) \quad (2.2.1)$$

in which

$$Dy(t) = \frac{d}{dt} y(t), \qquad D^2 y(t) = \frac{d^2}{dt^2} y(t), \text{ etc.}$$

Also the coefficients $a_0, \ldots, a_{n-1}$ and $b_0, \ldots, b_m$ are real values which are defined by the system characteristics, whereas the differential orders n and m are linked directly with the system structure.

As long as the integers n and m are known, along with all of the coefficients $a_i: i = 0, \ldots, n-1$; and $b_i: i = 0, \ldots, m$; then for a given set of initial values at $t = t_0$, the output response for any $t > t_0$ can be calculated by finding the solution to the differential equation (2.2.1), with respect to a set of input and input derivative values for the same $t > t_0$. However, this is not a simple problem, especially for large values of n and m, and is therefore not the standard technique employed for the manipulation of transfer functions in order to find a solution to the output response. Having said that, for specific cases it may well be necessary to revert to a differential equation description (2.2.1) in order to find an exact solution, and many computer packages are available for such cases.

A much simpler approach to the analysis of a system transfer function is to take the Laplace transform of (2.2.1) and to assume that all initial signal values at time $t = t_0$ are zero. The effect of this is simply to replace the D operator in (2.2.1) by the Laplace operator s, and this results in an input to output transfer function:

$$(s^n + a_{n-1}s^{n-1} + \cdots + a_1 s + a_0) Y(s)$$
$$= (b_m s^m + b_{m-1}s^{m-1} + \cdots + b_1 s + b_0) U(s) \quad (2.2.2)$$

in which it must be noted that both the input and output signals are also transformed values.

Equation (2.2.2) can be written as

$$Y(s) = G(s)U(s) \qquad\qquad (2.2.3)$$

where it can be seen that $G(s)$, which itself is independent of both the input and output signals, forms the function which transfers the input $U(s)$ to the output $Y(s)$.

As far as the rest of this text is concerned a transfer function will be regarded, in its most general form, as the ratio of the Laplace operator polynomials with initial conditions zero, such that

$$G(s) = \frac{b_m s^m + b_{m-1}s^{m-1} + \cdots + b_1 s + b_0}{s^n + a_{n-1}s^{n-1} + \cdots + a_1 s + a_0} \qquad (2.2.4)$$

Also it will be considered that $G(s)$ is linear, where the property of linearity is defined in Section 2.2.3, and that the coefficients a_i and b_i are time-invariant.

2.2.1   Transfer functions – worked example

Consider the series RLC circuit shown in Fig. 2.1, assuming that no current, $i(t)$, flows until the switch S is closed at time $t = t_0 = 0$, and also that until $t = 0$ the capacitor, C, remains uncharged.

The problem is then to find the transfer function relating input voltage, $v_i(t)$, to output voltage, $v_o(t)$ in terms of the Laplace operator, for any $t \geqslant t_0$.

In order to find the solution, we can initially put to one side the output voltage and find the relationship between $v_i(t)$ and $i(t)$ if it is remembered that by the voltage law of Kirchhoff, the voltage input, $v_i(t)$ must be equal to the sum of the voltage drops across the inductor, capacitor and resistor, i.e.

$$v_i(t) = L\frac{di(t)}{dt} + \frac{1}{C}\int_0^t i(t')\,dt' + Ri(t) \tag{2.2.5}$$

where t' is the time, considered over the interval $0 \leqslant t' \leqslant t$.

In order to remove the integral term, both sides of (2.2.5) can be differentiated to achieve

$$\mathrm{D}v_i(t) = \left(L\,\mathrm{D}^2 + R\,\mathrm{D} + \frac{1}{C}\right)i(t) \tag{2.2.6}$$

and on taking Laplace transforms of both sides, this becomes

$$CsV_i(s) = (LCs^2 + RCs + 1)I(s)$$

or

$$I(s) = \frac{Cs}{LCs^2 + RCs + 1}\,V_i(s) \tag{2.2.7}$$

We have therefore obtained a transfer function which relates the input voltage signal transform $V_i(s)$ to the current transform $I(s)$. A further step is therefore required to find an equation (transfer function) which relates the current to the output voltage transform, $V_o(s)$, in terms of its transform.

The voltage across the capacitor C, which is also the output voltage, is given

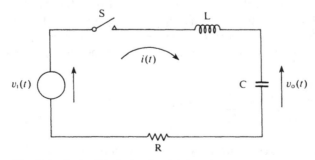

Fig. 2.1   Series RLC circuit with input $v_i(t)$

as

$$v_o(t) = \frac{1}{C} \int_0^t i(t') \, \mathrm{d}t'$$

(2.2.8)

or

$$\mathrm{D} V_o(t) = \frac{1}{C} i(t)$$

with a Laplace transform

$$V_o(s) = \frac{1}{Cs} I(s)$$

(2.2.9)

and on substitution of $I(s)$ from (2.2.7), the resultant transfer function from input to output voltage is thus

$$V_o(s) = \frac{1}{LCs^2 + RCs + 1} V_i(s)$$

(2.2.10)

In this example one of the basic properties of transfer functions has been shown, namely the multiplicative property exhibited by the combination of the transfer functions (2.2.7) and (2.2.9) in order to obtain the final form of (2.2.10).

2.2.2   The impulse response

Having discussed the use of a transfer function to describe the response of a system to an input signal, in this section it is shown how the Laplace transformed function of (2.2.4) is related to a time domain solution, in particular the response to an impulse input is investigated. Applying an impulse to a system is one of the oldest forms of analysis. As an example consider the chocolate bar machine mentioned earlier: if the machine was to fail, possibly by not presenting us with a chocolate bar after our money is inserted, then we apply an impulse as a means of fault diagnosis, i.e. we hit the machine as hard as we can. Strictly speaking an impulse is a signal which lasts for a time period equal to approximately zero, or if a signal lasts for a small time period, $\delta t'$, then as $\delta t' \to 0$ so the signal becomes an impulse. The output from a system which is obtained in response to an impulse applied at the input is then known as the *impulse response*.

If an impulse is applied as the input, $u(t)$, to a system, then the output, $y(t)$, will also be the impulse response, $g(t)$: where for all causal systems (all physical systems) $g(t) = 0$ for all $t < 0$, if the input is applied at $t = 0$. If any other input signal is applied, the output can be found, as will be shown here, in terms of the impulse response and the applied input.

Actually applying an impulse as introduced, with $\delta t' \to 0$, is not a realistic idea. In fact the approach taken is to define a unit impulse function as being one which lasts for a time period $\delta t'$ and has a magnitude equal to unity. The impulse response of a linear system is then generally taken to be the response of the system to a unit impulse function applied at the input. The unit impulse function is said to have an *impulse strength* equal to unity such that an impulse function of magnitude 5 and time period $\delta t'$ would have an impulse strength of 5. It is not difficult to see therefore that if the impulse response is the output of a linear system in response to a unit impulse function then the system output in response to a 5-unit impulse function will be $5 \times$ the impulse response. This basic technique can be extended to account for a general input, as follows. Consider the input signal, $u(t')$, shown in Fig. 2.2, in which the fixed time t is an integer number n of the time periods $\delta t'$ and the input is assumed to be zero for all $t' < 0$. The general time scale is given as t' and the time instant t is considered to be a certain point along this scale, e.g. $t = 3$ seconds. The continuous time signal $u(t')$ is in Fig. 2.2, for the purpose of our analysis, approximated by a series of pulses, such that as the pulse width $\delta t' \to 0$ so (a) the approximation $\to u(t')$ and (b) the pulses become impulses.

In terms of the pulse approximation to $u(t')$, the output $y(t)$ can be calculated as the impulse strength multiplied by the impulse response, i.e. for the pulse at $t' = 0$, the first pulse, so

$$y(t) = \lim_{\delta t' \to 0} \{ u(0) \cdot \delta t' \cdot g(t) | 0 \leqslant t' < 2\delta t' \}$$

Similarly, the output found in terms solely of the second pulse is:

$$y(t) = \lim_{\delta t' \to 0} \{ u(\delta t') \cdot \delta t' \cdot g(t - \delta t') | \delta t' \leqslant t' \leqslant 2\delta t' \}$$

where, for this second pulse the impulse strength is given as the pulse magnitude $u(\delta t')$ multiplied by its time period $\delta t'$. Also in respect of the output at time instant t, the impulse response will only have effect from time $\delta t'$ to time t, thus resulting in the $(t - \delta t')$ indexing.

But, for the signal shown in Fig. 2.2, the output at time t will in fact be equal to the summation of all of the output parts due to each of the input pulses up to t, i.e. the output due to the first pulse must be added to the output due to the second pulse, etc., in

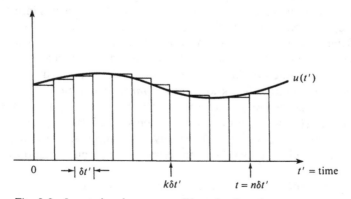

Fig. 2.2 Input signals as a set of impulse functions

order to find the output at time t due to the input from time $t' = 0$ to $t' = t$. In other words the actual output at time t is found to be:

$$y(t) = \lim_{\delta t' \to 0} \left\{ \sum_{k=0}^{n} u(k\, \delta t') \cdot \delta t' \cdot g(t - k\, \delta t') \right\} \qquad (2.2.11)$$

where $t = n\delta t'$.

In terms of an integral this becomes

$$y(t) = \int_0^t u(t')g(t - t')\, dt' \qquad (2.2.12)$$

which is known as the *convolution integral*, written as

$$y(t) = u(t)^* g(t) \qquad (2.2.13)$$

such that (2.2.12) and (2.2.13) mean exactly the same thing.

One property of the convolution integral is the interchangeability of the integrands, i.e. we also have

$$y(t) = g(t)^* u(t)$$

which means that

$$y(t) = \int_0^t g(t')u(t - t')\, dt' \qquad (2.2.14)$$

This does not imply that the input function and the impulse response change roles, it is merely another way of writing (2.2.12). It can in fact be observed quite simply by reference to the first two and last two terms in (2.2.11) which are:

$$y(t) = \lim_{\delta t' \to 0} \{ u(0) \cdot \delta t' \cdot g(t) + u(\delta t') \cdot \delta t' \cdot g(t - k\, \delta t') + \cdots$$

$$+ u(t - \delta t') \cdot \delta t' \cdot g(\delta t') + u(t) \cdot \delta t' \cdot g(0) \}$$

If now Laplace transforms are taken on both sides of (2.2.14) we have

$$Y(s) = G(s)U(s) \qquad (2.2.3)$$

in which $G(s)$ is the Laplace transform of $g(t)$ and $U(s)$ is the Laplace transform of the input signal $u(t)$.

So the transform relationship between system input and output has been obtained by starting from an impulse function input and impulse response, both continuous time signals. In particular the transfer function has been shown to be the Laplace transform of the impulse response.

A basic assumption was made earlier that the system $G(s)$ is linear and time-invariant, and the results of this section have been derived on that understanding. It is however the case for many systems that neither of these assumptions hold. The problem of approximating a nonlinear system with a linear model will be discussed in the next section. The means of dealing with a time-varying transfer function, however, are very much dependent on just how rapidly the transfer function is changing, and in what way, with respect to time. One thing to remember is that any controller design carried out which is based on a system transfer function depends primarily on the accuracy of the transfer function coefficients. If the system transfer function varies slowly with respect to time then it will most likely be necessary to redesign the controller every so often, but as the time variations become more rapid so a more frequent redesign may well be necessary. This then leads on to the idea of an adaptive controller which periodically updates its transfer function representation in order to carry out a controller redesign, a technique which is particularly suited to a computer control framework.

2.2.3   Linear approximations

All physical systems can exhibit nonlinear behavior; in most cases this is simply due to a system variable or signal being increased well beyond its normal scope of operation. Consider for example the series RLC circuit of Fig. 2.1, where with the switch closed, if the input voltage $v_i(t)$ is continually increased, after a while one of the components would malfunction, i.e. it would no longer operate in its normal mode. This may be due to overheating such that when the input voltage is reduced, normal operation may be resumed. However, the component might be permanently damaged and it may well not be possible to return to the initial state. We have therefore for a physical system that it will be in some sense nonlinear, but that it will very likely have a certain range of operation over which it can be considered to be linear. But what do we mean when we say a system is linear?

A system is said to be linear if it has the two following properties:

1. If an input $u_1(t)$ to a particular system produces an output $y_1(t)$ and if an input $u_2(t)$ to the same system produces an output $y_2(t)$, then in order for the system to be linear it is necessary that an input $u_1(t) + u_2(t)$ will produce an output $y_1(t) + y_2(t)$.
2. Homogeneity: If an input $u_1(t)$ to a particular system produces an output $y_1(t)$, then in order for the system to be linear it is necessary that an input $b_1 u_1(t)$, where b_1 is a constant multiplier will produce an output $b_1 y_1(t)$.

Both of these properties are necessary to define a linear system and when combined mean that an input $b_1 u_1(t) + b_2 u_2(t)$, where b_1 and b_2 are constant multipliers, will produce an output $b_1 y_1(t) + b_2 y_2(t)$.

In the previous section it was shown that a system input to output relationship could be given in terms of the convolution integral (2.2.12), i.e.

$$y(t) = \int_0^t g(t' - t')u(t') \, dt'$$

where it has been assumed that (a) $u(t') = 0$ for $t' < 0$ and (b) $g(t - t') = 0$ for $t < t'$.

Although the linear range of many electrical and mechanical systems, some examples of which are given in the following sections, can appear to be large in terms of component and signal values, for other system types it may well be necessary to consider the linear range as existing over only a small set of signal values, indeed fluid and heating systems fall into this category. An excellent example of a nonlinear system which can be considered to be linear over a small signal range is the transistor amplifier, which is usually analyzed in terms of several different linear models, depending on the frequency of the input signal. So for a low frequency signal one linear amplifier model is used, whereas for high frequencies a different model is employed, as is true for mid-frequencies. Two examples of linearly approximating mechanical systems are detailed in the next section.

2.2.4   Linear approximations – worked examples

The first example of approximating a nonlinear mechanical system by a linear model is the pendulum oscillator system, depicted in Fig. 2.3.

The equation of motion for the pendulum in terms of the displacement angle is

$$\frac{d^2\theta}{dt^2} = -\frac{g}{l}\sin\theta \qquad (2.2.15)$$

in which g is the acceleration due to gravity. Consider θ to be the system input and $d^2\theta/dt^2$ the system output. It is then apparent that the equation of motion is nonlinear as it breaks the linear property rules set out in the previous section because of the sine function. If the sine function can be approximated by a linear function then the equation of motion will be 'linearized', and indeed this is a possibility for small values of θ about the mid-position $\theta = 0$. The assumption can be made that for small values of θ we have $\theta \simeq \sin\theta$, which can be seen by truncating the Taylor series expansion to its first term only, as shown in (2.2.16):

$$\sin\theta = \theta - \frac{\theta^3}{3!} + \frac{\theta^5}{5!} - \frac{\theta^7}{7!} + \cdots \qquad (2.2.16)$$

It follows that by using the truncated Taylor series, the nonlinear equation

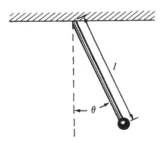

Fig. 2.3   Pendulum oscillator system

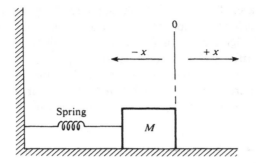

Fig. 2.4   Mass–spring system

(2.2.15) becomes the linear equation (2.2.17)

$$\frac{d^2\theta}{dt^2} = -\frac{g\theta}{l} \qquad (2.2.17)$$

for small values of θ.

So the nonlinear pendulum oscillator system has been approximated by a linear system over a small range for the input signal θ.

As a second example consider the mass–spring system of Fig 2.4.

The equation of motion for the mass–spring system in terms of the displacement x is:

$$M\frac{d^2x}{dt^2} = -f(x) \qquad (2.2.18)$$

where $f(x)$ is the force, as a function of displacement, applied by the spring, as shown in Fig. 2.5.

Although the general effect of the force $f(x)$ is to cause (2.2.18) to be a nonlinear equation, if the mass displacement is restricted to lie within the range

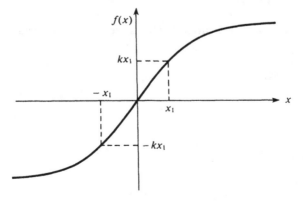

Fig. 2.5   Plot of spring force against displacement

$-x_1 \leqslant x \leqslant x_1$ then the term $f(x)$ in (2.2.18) can be replaced by kx where k is some constant value. So the linearized version of (2.2.18) is then

$$M\frac{d^2x}{dt^2} = -kx \qquad (2.2.19)$$

as long as the displacement lies in the range $-x_1 \leqslant x \leqslant x_1$.

Both of the examples given in this section have shown that, although the system under examination is inherently nonlinear, by considering the input signal to exist within a relatively small range the system response can be effectively described by means of a linear equation.

2.3   Block diagrams

In Section 2.2 the mathematical description of systems in terms of a transfer function representation was considered. While dealing with a particular system in its simplest form there is really no need for any other alternative representation. However, as soon as that system is connected up in some fashion to another system, or a controlling function is applied to the system, then the overall transfer function will become that due to the combined functions, dependent on just how the connections were made. Although it is perfectly acceptable to think of the system connections in terms of solely algebraic relationships, it rapidly becomes a difficult problem to relate various equations to the underlying physical systems that they represent. The result of this is that it is often best to think of the system concerned in terms of a graphical block diagram representation. When other systems and/or controllers are connected on to the original system it is then quite straightforward to consider each separate system in terms of its own block diagram.

When a set of systems are connected it is quite possible that the overall relationship produced could have been achieved either by connecting the same system up in a different way and/or by connecting some slightly different systems. It is one advantage of the block diagram form of system representation that, by means of block manipulation, alternative forms of connection can be considered. In this section the basic ingredients for block diagram system representation are introduced and subsequently methods for block manipulation are discussed.

2.3.1   Basic building blocks

The transfer function system description introduced in Section 2.2 is represented in block diagram form as Fig. 2.6. Hence the block diagram in Fig. 2.6 and the equation $Y(s) = G(s)U(s)$ mean exactly the same thing, they are just alternative representations.

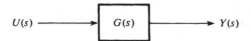

Fig. 2.6   Block diagram – basic transfer function

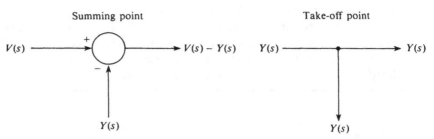

Fig. 2.7   Summing and takeoff points

It can be seen in Fig. 2.6 that the signals $U(s)$ and $Y(s)$ are shown in terms of a direction given by the arrows, the block containing the operator $G(s)$. The only other two basic elements of a block diagram representation are the summing point and the takeoff point, as shown in Fig. 2.7.

The summing point, which is sometimes shown with the symbol Σ inside the circle, can have any number of signals entering, each being given a $+$ or $-$ sign to indicate whether they are to be added or subtracted. There is only one resultant signal from a summing point, and if this signal must appear in a number of places then an appropriate number of takeoff points can be used.

2.3.2   Block diagram manipulation

If two blocks are connected together in cascade, they can be replaced by a single block, the function within which is a multiple of the two original block functions, as shown in Fig. 2.8 where the (s) indexing is dropped to aid explanation, i.e. Laplace functions are assumed.

If a takeoff point is introduced into the cascading example shown in Fig. 2.8, then the equivalence of the three representations shown in Fig. 2.9 can be examined.

In each of the cases shown in Fig. 2.9 we have that (a) $Y = G_1 G_2 V$ and (b) $E = G_1 V$, but a different takeoff point has been employed. In terms of an overall block diagram

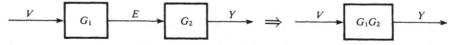

Fig. 2.8   Blocks in cascade

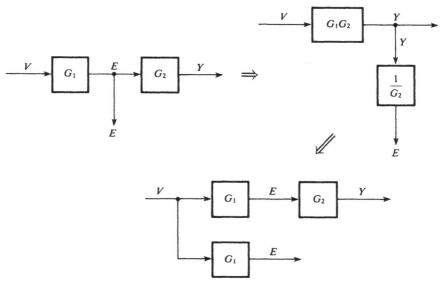

Fig. 2.9 Moving a takeoff point

any of the three representations can be used, the decision being dependent on whichever seems most appropriate for a particular purpose, whether this be in terms of an algebraic or a physical sense.

The same is true for a summing point, as shown in Fig. 2.10, where in each case the output is found to be $Y = G_1 G_2 V + G_1 G_3 E$.

The one remaining basic operation to be considered is the method of dealing with the use of feedback, i.e. when the output is fed back by means of a takeoff point, to be one of the inputs as well. A negative feedback loop is shown in Fig. 2.11, along with its equivalent single block representation.

From the original (left-hand) negative feedback block diagram we have two equations, one which describes the open-loop system transfer function

$$Y = GU$$

and the other which describes the system input as consisting of an applied input V summed with a signal fed back from the output

$$U = V - HY$$

On substitution for U into the first of these equations we have our final closed-loop transfer block given by

$$Y = GV - GHY$$

or

$$Y = \frac{G}{1 + GH} V \qquad (2.3.1)$$

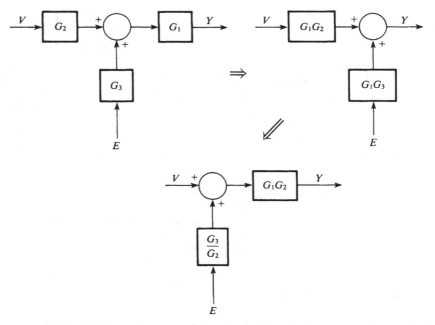

Fig. 2.10   Moving a summing point

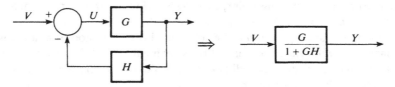

Fig. 2.11   Negative feedback loop

With the basic elements as described in the previous section and the block manipulations discussed in this section it is possible to analyze, modify and, where necessary, reduce even the most complicated control system either algebraically in terms of the component transfer function or in terms solely of block diagrams.

2.3.3   Block diagrams – worked examples

Two examples of block diagram manipulation are given in this section. The first is to show how the basic ideas of block manipulation can be employed one set at a time to deal with systems which initially appear quite complex. The second example is included as a reminder that the transfer functions contained within each block can often contain factors which may cancel out in a later part of the analysis.

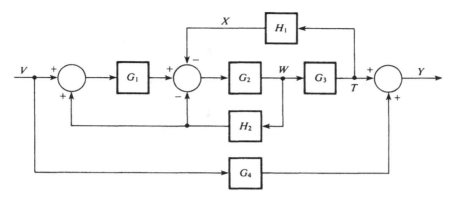

Fig. 2.12 Worked example, initial block diagram

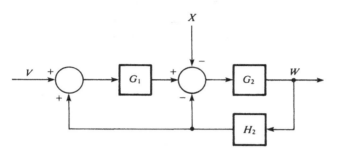

Fig. 2.13

Consider the block diagram shown in Fig. 2.12. It is required to reduce this diagram into a single block with input V and output Y.

As a first step towards diagram reduction the G_1 block can be moved onto the other side of the leftmost summing junction, so Fig. 2.13 becomes Fig. 2.14(a) or Fig. 2.14(b) and on reduction of the positive feedback loop we have Fig. 2.15.

Now we can investigate the negative feedback loop shown in Fig. 2.16 in which the G_3 block has been cascaded. The feedback loop can be reduced, as described in the previous section, to give the overall diagram as shown in Fig. 2.17.

All that remains is to cascade G_1 and then to add on G_4 in order to obtain the final block diagram as in Fig. 2.18.

For the second example in this section, consider the block diagram shown in Fig. 2.19, which is required to be redrawn in terms of a unity feedback loop, that is in terms of a system of the form introduced in Fig. 2.11 with H = unity; note that K is some constant gain value.

The $(s + 2)$ feedback numerator factor can be taken over to the other side of the summing junction to give Fig. 2.20.

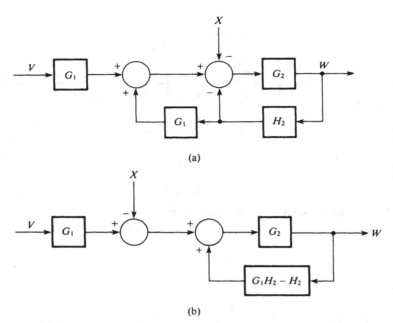

(a)

(b)

Fig. 2.14

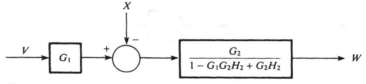

Fig. 2.15

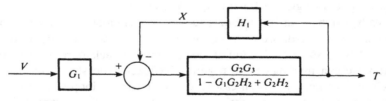

Fig. 2.16

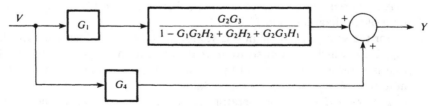

Fig. 2.17

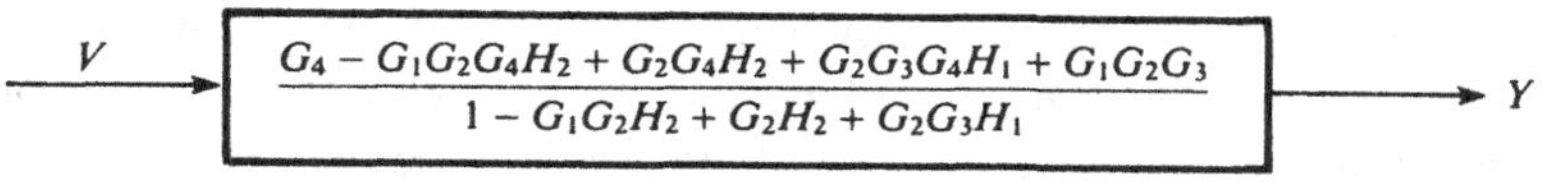

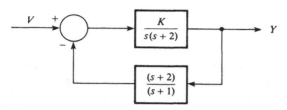

V
$\dfrac{G_4 - G_1G_2G_4H_2 + G_2G_4H_2 + G_2G_3G_4H_1 + G_1G_2G_3}{1 - G_1G_2H_2 + G_2H_2 + G_2G_3H_1}$
Y

Fig. 2.18 Worked example, final block diagram

V
+
−
$\dfrac{K}{s(s+2)}$
Y
$\dfrac{(s+2)}{(s+1)}$

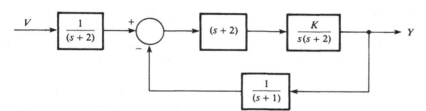

Fig. 2.19 Worked example, initial block diagram

V
$\dfrac{1}{(s+2)}$
+
−
$(s+2)$
$\dfrac{K}{s(s+2)}$
Y
$\dfrac{1}{(s+1)}$

Fig. 2.20

V
$\dfrac{1}{(s+2)}$
+
−
$\dfrac{K}{s}$
Y
$\dfrac{1}{(s+1)}$

Fig. 2.21

V
$\dfrac{(s+1)}{(s+2)}$
+
−
$\dfrac{K}{s(s+1)}$
Y

Fig. 2.22

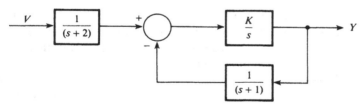

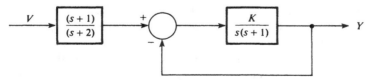

On cascading the $(s + 2)$ term now in the forward path, this becomes Fig. 2.21. On moving the $(s + 1)$ feedback denominator factor to the other side of the summing junction, the final unity feedback system is as shown in Fig. 2.22.

2.4   Physical system realizations

In this chapter methods of representing physical systems in terms of an algebraic sense by means of a transfer function or a graphical sense by means of a block diagram have been introduced. Once a system model has been obtained, further analysis and controller design work can then be carried out, employing the model instead of the actual system. Hence testing of a new control scheme can be carried out in terms of the model rather than the actual system – so if a closed-loop control system was to go wrong because of poor control action, the result would only be in terms of the model output, possibly just a number in a computer, rather than an actual output.

To accurately characterize, with a model, the exact nature of any real physical item would require a model which was at least as complicated as the item itself, any simplification meaning that at least some information was lost. However, in a complicated form the model may well be time-consuming and expensive, not only to obtain but also to comprehend and make use of, causing it to be unsuitable for analytical purposes. So the basic requirement is for a relatively simple model which encapsulates the main/important properties of a physical system. In terms of a straightforward input–output transfer function relationship simplicity is directly related to the order of the system, i.e. the number of coefficients required such that the model well describes the system's behavior.

One advantage in the study of control systems is that results obtained for say an electrical system are also true for a chemical system or an aeronautical system, etc. In each case the basic components may well be different as will the type of input and output signals; however the mathematical model employed in order to carry out the analysis could have been obtained from any kind of physical system. Once a mathematical model has been arrived at, the physical system from whence it came loses much of its significance for analysis purposes and only really matters when one checks to see how sensible or practically applicable the analysis results are. Indeed the physical system itself could easily consist of several different types, for instance it might be part electrical, part mechanical, etc.

In this section the text deals with obtaining transfer functions, as described in Section 2.2, in order to represent firstly electrical and secondly mechanical systems.

Methods are considered in which variables from one system type can be regarded as being analogous to those from another system type, showing how the study of control systems brings together the various engineering and physical disciplines by means of a common mathematical toolbox. Subsequently in Section 2.5 overall transfer functions of systems in which at least two physical systems are present are discussed in terms of a motor/generator problem.

2.4.1   Electrical networks

An electrical network has already been used as the worked example in Section 2.2.1, and in a sense this section merely acts as a backup for that previous work. The three basic elements of an electrical circuit are the resistor, the capacitor and the inductor, each of these forming a different relationship between the signal variables with which they are associated. Consider initially a resistor, Fig. 2.23, which relates the voltage across it to the current through it by means of the equation $v(t) = R \cdot i(t)$, or in terms of the Laplace operator this is

$$V(s) = R \cdot I(s) \tag{2.4.1}$$

This basic equation (2.4.1) can be viewed in two ways, either we apply the current (input) $I(s)$ and measure the voltage drop (output) $V(s)$ or vice versa. So although just one equation (2.4.1) is deemed to characterize the physical response of a resistor, hence relating current to voltage, it can also relate voltage to current. The block diagram representing a resistor can therefore be drawn as either Fig. 2.24 (a) or (b) dependent on which direction is appropriate for signal flow.

A capacitor, Fig. 2.25, can similarly be viewed in one of two ways as far as block diagram representation of its transfer function is concerned. We have that $i(t) = C \, dv(t)/dt$, or

$$sV(s) = \frac{1}{C} I(s) \tag{2.4.2}$$

and this means that either Fig. 2.26 (a) or (b) is suitable for representation purposes.

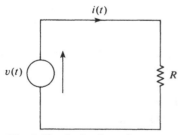

Fig. 2.23   A simple resistance

(a)   (b)

Fig. 2.24   Block diagram for the resistor

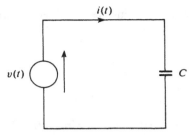

Fig. 2.25   A simple capacitor

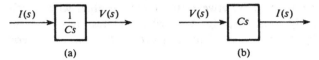

(a)   (b)

Fig. 2.26   Block diagrams for the capacitor

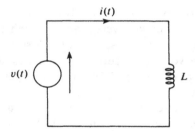

Fig. 2.27   A simple inductor

The final individual electrical element is the inductor, Fig. 2.27, which has a transfer function relating current to voltage of $v(t) = L \, di(t)/dt$, or

$$V(s) = LsI(s) \tag{2.4.3}$$

As far as block diagrams are concerned the inductance transfer function can be written as that in either Fig. 2.28 (a) or (b).

As an example of the use of block diagrams and transfer functions in the modeling and analysis of an electrical network, consider the series RC circuit with input $v_i(t)$ and output $v_o(t)$ shown in Fig. 2.29.

A block diagram for the RC circuit can be built up quite easily by recognizing that the voltage dropped across the resistance R is in fact $v_i(t) - v_o(t)$. The block diagram shown in Fig. 2.30 then follows directly.

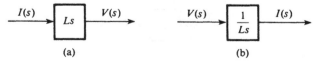

(a)   (b)

Fig. 2.28   Block diagram for the inductor

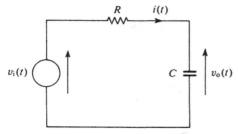

Fig. 2.29 Series RC circuit with input $v_i(t)$

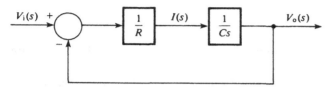

Fig. 2.30 Block diagram representation of the RC circuit

Using the block manipulation techniques described in Section 2.3, once the two forward path blocks have been cascaded, the final transfer function can be obtained by removing the negative feedback loop, resulting in

$$V_o(s) = \frac{1}{1 + RCs} V_i(s) \tag{2.4.4}$$

2.4.2 Mechanical systems

The elementary components and basic relationships involved in electrical circuits were described in the previous section. Mechanical systems will now be investigated, the overall intention being to characterize the essential physical relationships by means of equations similar to those used for the electrical analysis. The mechanical case is however a different entity in that there are two completely different types of mechanical action, translational motion in a straight line, and rotational motion about a fixed axis. The two types will be dealt with separately, the initial discussion being concerned with translational motion.

Translational

The fundamental signals within the study of translational mechanical systems are force $f(t)$ and velocity $w(t)$; in general the force is applied to a certain mass in order to move it at a specific velocity, although as will be shown shortly a more direct link is apparent via Newton's law of motion between force and rate of change of velocity, i.e. acceleration. Some of the applied force is however used up in order to overcome frictional effects, which are unfortunately nonlinear and rather difficult to model. However, by neglecting effects such as static friction, which opposes the commencement of motion, it is possible to model friction simply as viscous friction, which is a linear retarding force dependent on the relative velocity. The idealized viscous friction resistive component called a dashpot is shown in Fig. 2.31, and has the effect that the applied force $f(t)$ causes a velocity drop between piston and cylinder.

If B is the viscous frictional component then the transfer function of a dashpot is $f(t) = Bw(t)$ or

$$F(s) = BW(s) \qquad (2.4.5)$$

Hence two possible ways occur in which this can be described in block diagram format, as shown in Fig. 2.32.

A second element present in translational mechanical systems is a spring which in practice can constitute an actual spring, although more likely it is the capacity to store potential energy which is present in a cable or connecting rod. As was the case for a dashpot, the response of a spring is inherently nonlinear, although over a small range it can be approximated fairly closely by the simple linear relationship $df(t)/dt = Kw(t)$ or

$$sF(s) = KW(s) \qquad (2.4.6)$$

in which K is the spring stiffness, or K^{-1} is the compliance of the spring.

In terms of a block diagram, the response of a spring can be shown as in Fig. 2.33.

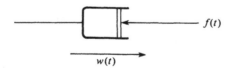

Fig. 2.31   Dashpot as a frictional element

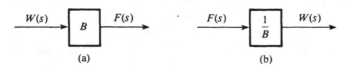

Fig. 2.32   Block diagram for the dashpot

Fig. 2.33   Block diagram for the response of a spring

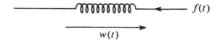

Fig. 2.34   Spring as an energy storage element

Perhaps the simplest way to consider the physical effect of a spring is that the application of a force produces a related displacement (Fig. 2.34), i.e. $f(t) = K \int w(t) \, dt$.

If one end of the spring is fixed to an immovable object, then the end of the spring attached to that object will have zero velocity whereas the other end will have a velocity which depends on the force applied, via (2.4.6).

The final element for study in the area of translational mechanical systems is mass M, where the weight of an object is equal to its mass multiplied by g, the free acceleration due to gravity (9.81 m/s^2). It was stated earlier in this section that one of the basic concepts of a translational mechanical system is the application of force to a mass in order to achieve an acceleration of that mass in terms of the relationship $f(t) = M \, dw(t)/dt$, or

$$F(s) = MsW(s) \tag{2.4.7}$$

When represented by a block diagram this relationship then takes on one of the two forms shown in Fig. 2.35.

Although it is quite possible for a mass to have different forces applied to it at different points, the mass is considered to be solid and therefore will have just one absolute velocity associated with it; this is shown in Fig. 2.36 as the mass having a velocity compared to the zero velocity of an immovable object.

As an example of the use of schematic diagrams and transfer functions in the modeling and analysis of a translational mechanical network, consider the mechanical system

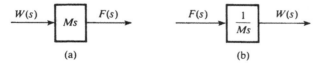

(a)   (b)

Fig. 2.35   Block diagram for a mass

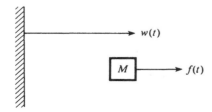

Fig. 2.36   Mass with absolute velocity

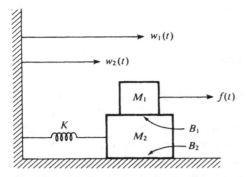

Fig. 2.37   Mechanical system with force input $f(t)$

shown in Fig. 2.37, where a force $f(t)$ is applied to a mass M_1 which sits on another mass M_2, there being a frictional resistance B_1 between the two masses. The system can also be described by means of element schematic diagrams as in Fig. 2.38.

Assuming that $F_1(s)$ is the resultant force on the dashpot with frictional resistance B_1 then from Fig. 2.38 the forces across mass M_1 balance via the equation

$$F(s) - F_1(s) = M_1 s W_1(s) \tag{2.4.8}$$

Further the velocity drop across the B_1 dashpot is described by

$$F_1(s) = B_1(W_1(s) - W_2(s)) \tag{2.4.9}$$

Also, balancing the forces across mass M_1 gives

$$F_1(s) - \left[B_2 + \frac{K}{s}\right] W_2(s) = M_2 s W_2(s) \tag{2.4.10}$$

where the resultant force on the B_2 dashpot is equal to $B_2 W_2(s)$ and the resultant force on the spring equal to $K W_2(s)/s$. By eliminating force $F_1(s)$ and velocity $W_2(s)$ from the three equations (2.4.8–10) we obtain the input impedance transfer function of the

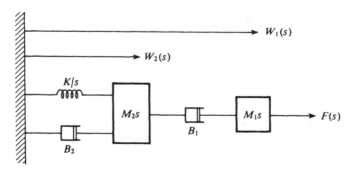

Fig. 2.38   Mechanical system – network schematic

mechanical system as:

$$\frac{F(s)}{W_1(s)} = \frac{M_1 s (M_2 s^2 + [B_1 + B_2] s + K) + B_1 (M_2 s^2 + B_2 s + K)}{M_2 s^2 + [B_1 + B_2] s + K} \qquad (2.4.11)$$

The above final transfer function was obtained by algebraic manipulations on the three previous equations; it could however have equally well been obtained by means of a block diagram approach.

Rotational

The second type of mechanical system considered is that concerned with rotational motion, in which the fundamental variables are torque $q(t)$ and angular velocity $\omega(t)$. As with translational mechanical systems there are three basic idealized elements and these are briefly introduced here.

Resistance to rotational motion is due largely to friction and is modeled by the damper, shown in Fig. 2.39.

In general the resistance equation is given as:

$$Q(s) = B\Omega(s) \qquad (2.4.12)$$

However, the two different diagrams in Fig. 2.39 indicate (a) one shaft, therefore one speed but a difference in torque along the shaft; and (b) two separate shafts, therefore a different speed on each shaft but with the same torque transmitted through the element. In many cases with a damper of type (b), the left-hand shaft of the damper is shown fixed, indicating a simple braking system.

The compliance of a shaft, when a torque is applied, is depicted by the torsional spring shown in Fig. 2.40.

The compliance equation in which K is the torsional spring constant, is then

$$sQ(s) = K\Omega(s) \qquad (2.4.13)$$

(a)   (b)

Fig. 2.39 Damper depicting resistance to rotational motion

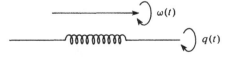

Fig. 2.40 Torsional spring

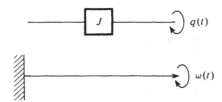

Fig. 2.41   Inertia of rotating bodies

The final element in the study of rotational systems is the inertia J of a body, which is a measure of that body's ability to store kinetic energy, given that that body has an absolute velocity associated with it, as shown in Fig. 2.41.

The inertia equation is

$$Q(s) = Js\,\Omega(s) \tag{2.4.14}$$

in which J is dependent on the density of the body and its rotational axis, for example its radius.

Consider as an example the rotational mechanical network shown in Fig. 2.42.

A block diagram for this system is given in Fig. 2.43, where a torque drop occurs over each of the dampers and an angular velocity drop over the torsional spring of constant K.

By block manipulation or algebraic transfer function manipulation, the input impedance for this network is found to be:

$$\frac{Q(s)}{\Omega(s)} = \frac{B_1(Js^2 + B_2s + K) + K(Js + B_2)}{Js^2 + B_2s + K} \tag{2.4.15}$$

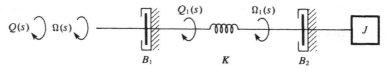

Fig. 2.42   Rotational mechanical network − example

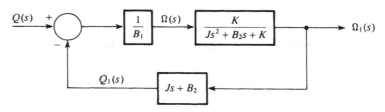

Fig. 2.43   Block diagram for the rotational mechanical example

2.4.3 Analogous circuits

In Section 2.4 thus far the relationships of basic electrical and mechanical systems with transfer functions and block diagrams have been discussed. It is readily observable that many similarities exist between electrical and mechanical systems in terms of the equations considered and this makes possible the investigation of a mechanical system by means of an electrical analogy or vice versa. One big advantage of employing the mechanical to electrical analogy is that the mechanical network under consideration may well be large, expensive and potentially dangerous whereas its electrical analog will most likely be a cheap bench-top exercise with the inherent safety that that affords.

In forming an analogous circuit, an analogy is drawn not only between the different network elements concerned but also between the fundamental signals. There are in fact two possibilities: either a link can be formed between force in a translational mechanical system and voltage in an electrical system or, as will be considered here, force and current, the latter connection being the most widely encountered. Once the initial link has been made the remaining connections between the different network types follow automatically, as shown in Table 2.1.

As an example of forming an electrical analog from a translational mechanical system, consider the network shown in Fig. 2.44.

The input to this mechanical network is in the form of a force $f(t)$ which, in the analog, is directly mapped to become current $i(t)$. Also the masses M_1 and M_2 have absolute velocities associated with them; these are mapped as capacitances with absolute voltages across them $v_1(t)$ and $v_2(t)$ respectively. The dashpot B_1 also has a velocity drop $w_2(t)$ across it, this therefore becomes a resistance with a voltage drop $v_2(t)$ across it. The dashpot B_2 however has a velocity drop $w_2(t) - w_1(t)$ across it and this becomes a resistance with a voltage drop $v_2(t) - v_1(t)$ across it. The overall electrical circuit diagram is then as shown in Fig. 2.45.

The input impedance of the network considered can be found by means of either the mechanical or the electrical set-up to be:

$$\frac{W_2(s)}{F(s)} = \frac{M_1 s^2 + B_2 s + K}{(M_1 s^2 + B_2 s + K) \cdot (M_2 s + B_1 + B_2) - B_2^2 s} \tag{2.4.16}$$

from the balance equations:

$$F(s) = M_2 s W_2(s) + B_1 W_2(s) + B_2[W_2(s) - W_1(s)]$$

and

$$B_2[W_2(s) - W_1(s)] = M_1 s W_1(s) + \frac{K}{s} W_1(s)$$

Table 2.1 Analogies, force–current

Electrical	Mechanical (translational)	Mechanical (rotational)
Current, $I(s)$	Force, $F(s)$	Torque, $Q(s)$
Voltage, $V(s)$	Velocity, $W(s)$	Angular velocity, $\Omega(s)$
Conductance, $\dfrac{1}{R}$	Dashpot, B	Damper, B
Inductance, L	Spring compliance, $\dfrac{1}{K}$	Torsional compliance, $\dfrac{1}{K}$
Capacitance, C	Mass, M	Inertia, J

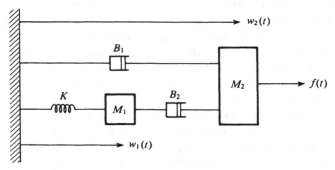

Fig. 2.44 Mechanical network, analog example

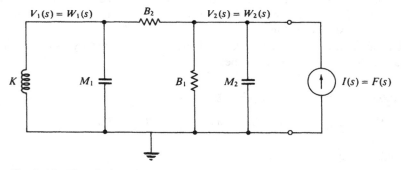

Fig. 2.45 Electrical analog of the mechanical network

2.5   Electromechanical devices

Having considered separately mechanical and electrical networks, and discussed their similarities in terms of analog relationships, an important aspect also arises in devices which involve both types of energy, usually when energy conversion takes place. Such electromechanical devices can exhibit very nonlinear characteristics, an example being the two-phase induction motor which requires an AC electrical input in order to provide its nonlinear torque–speed output relationship. Although many years ago AC motors proliferated, nowadays they are found only occasionally, usually when a special feature such as high noise rejection is required, by far the most commonly encountered motor being of the direct current type. Due to design improvements DC motors are made with very low inertia rotors and hence their time constant of operation is generally small, making them ideal for use in high-speed environments such as disk drives and printers.

In this section the discussion is concerned primarily with the modeling of DC motors, but a detailed analysis of such motors is considered to be beyond the scope of this text. A basic assumption is made that the motor is of the more usual separately excited field type which means that the field and armature windings are separate, thereby resulting in a wide control signal range. The series field motor, which is not discussed further here, is in fact only used for special low-speed, high-torque applications.

For completeness, the reverse process of electrical generation from a mechanical input will also be considered briefly from a control point of view.

2.5.1   DC motors

The DC motor is used to convert an electrical input signal into a mechanical output signal. There are in fact two electrical inputs to a separately excited DC motor, as shown in Fig. 2.46, one to the armature circuit and one to the field circuit.

The armature circuit consists of a resistance R_a in series with an inductance L_a and an idealized back e.m.f. $v_e(t)$ when motoring operation takes place. The field circuit however is simply a resistance R_f in series with an inductance L_f. The mechanical output circuit which contains an inertia J and frictional resistance B, has a generated torque $q_e(t)$ and a resultant load torque $q_l(t)$.

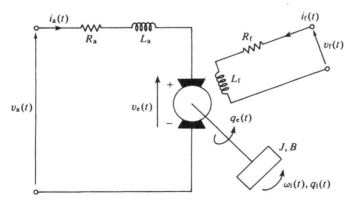

Fig. 2.46   Separately excited DC servomotor

On the assumption that various nonlinear characteristics such as saturation are to be neglected, the air gap flux $\phi(t)$ is considered to be proportional to the field current, $i_f(t)$. Two basic physical equations then describe the electromechanical characteristics of the motor, and these are:

$$q_e(t) = K_m i_a(t) i_f(t) \tag{2.5.1}$$

$$v_e(t) = K_m \omega_1(t) i_f(t) \tag{2.5.2}$$

where K_m is the electromechanical coupling constant. Because they both involve products of two variables the two basic equations (2.5.1) and (2.5.2) are inherently nonlinear, although it is quite straightforward to employ each of them in a linear fashion by holding one of the terms constant. So, in the case of (2.5.1) we can either hold the field current $i_f(t)$ constant ($= I_f$), which presents us with a linear relationship between $i_a(t)$ and $q_e(t)$, or we can hold the armature current $i_a(t)$ constant ($= I_a$), which presents us with a linear relationship between $i_f(t)$ and $q_e(t)$. Both of these cases, which are termed armature control and field control respectively, are discussed in this section, starting with an armature controlled motor.

An *armature controlled* motor is one in which the field current is held constant, such that

$$Q_e(s) = K_{mf} I_a(s) \tag{2.5.3}$$

where

$$K_{mf} = K_m I_f = \text{constant} \tag{2.5.4}$$

The armature current is related to the input armature voltage by means of the armature circuit equation

$$I_a(s) = \frac{V_a(s) - V_e(s)}{R_a + L_a s} \tag{2.5.5}$$

Also the load torque is made up simply of the torque provided by the motor $Q_e(s)$; however in some cases a torque due to external disturbances $Q_d(s)$ must be overcome before any movement is possible. The disturbance torque is included here for completeness although in many cases it will not be present and hence can be neglected from any analysis. The load torque is therefore given by

$$Q_1(s) = Q_e(s) - Q_d(s) \tag{2.5.6}$$

The mechanical output circuit means that load speed $\Omega_1(s)$ is obtained from the load torque by means of the equation

$$\Omega_1(s) = \frac{Q_1(s)}{Js + B} \tag{2.5.7}$$

A block diagram for the complete armature control motor is shown in Fig. 2.47.

The load position can be obtained directly from the load speed by employing the

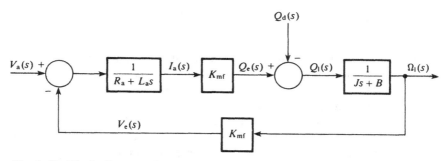

Fig. 2.47 Block diagram of an armature controlled DC motor

equation

$$\Theta_1(s) = \frac{1}{s} \cdot \Omega_1(s) \tag{2.5.8}$$

where $\Theta_1(s)$ is the output position transform.

A *field controlled* motor is one in which the armature current is held constant such that

$$Q_e(s) = K_{ma} I_f(s) \tag{2.5.9}$$

where

$$K_{ma} = K_m I_a = \text{constant} \tag{2.5.10}$$

The field current is related to the input field voltage by means of the field circuit equation

$$I_f(s) = \frac{V_f(s)}{R_f + L_f s} \tag{2.5.11}$$

which results in a transfer function from input voltage $V_f(s)$ to output speed of

$$\frac{\Omega_1(s)}{V_f(s)} = \frac{K_{ma}}{(Js + B)(L_f s + R_f)} \tag{2.5.12}$$

A block diagram of the field controlled DC motor is shown in Fig. 2.48.

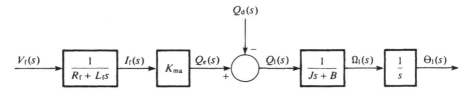

Fig. 2.48 Block diagram of a field controlled DC motor

2.5.2 Generator

In order to carry out a relatively simple analysis the motor diagram Fig. 2.46 and basic equations (2.5.1) and (2.5.2) are employed again to discuss the operation of a generator, one slight alteration being that the motor output speed ω_l and torque q_l are now replaced by the generator input speed ω_m and torque q_m respectively. It must be emphasized from a practical point of view, however, that DC generators do not commonly occur, the usual method of obtaining a direct current being to rectify a generated alternating current, as this procedure is both much cheaper and more robust.

A DC generator can be dealt with in roughly the same way as that used for the DC motor, the basic mode of operation is though essentially different and can be described as follows. Mechanical energy is supplied by means of the rotor and this is converted to electrical energy which appears in the form of a generated armature current. In the case of the DC generator the armature circuit is thus the output with $v_e(t)$ the generated voltage as shown in Fig. 2.49.

So, if R_b and L_b are the load resistance and inductance respectively, we have for the armature output circuit that

$$V_e(s) = [(R_a + R_b) + (L_a + L_b)s]I_a(s) \tag{2.5.13}$$

and

$$V_a(s) = [R_b + L_b s]I_a(s) \tag{2.5.14}$$

By referring to equation (2.5.2), there are two possibilities for the linear control of a DC generator; either the speed ω_m can be held constant and a varying field current i_f applied, or the field current can be held constant while varying the rotor speed.

A *field controlled* generator is one in which the rotor speed is held constant, such that, from (2.5.2):

$$V_e(s) = K_{m\omega}I_f(s) \tag{2.5.15}$$

where

$$K_{m\omega} = K_m\Omega_m = \text{constant} \tag{2.5.16}$$

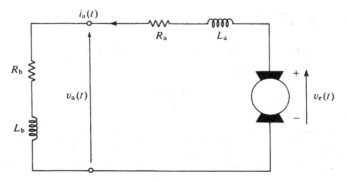

Fig. 2.49 Armature output circuit for a DC generator

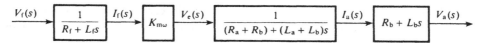

Fig. 2.50 Block diagram of a field controlled DC generator

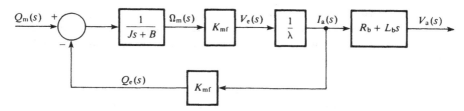

Fig. 2.51 Block diagram of a speed controlled DC generator
$$\lambda = (R_a + R_b) + (L_a + L_b)s$$

The field current is related to the input field voltage, see Fig. 2.46, by means of the field circuit equation (2.5.11), and hence an overall block diagram can be built up for the field controlled generator as shown in Fig. 2.50.

A *speed controlled* generator is one in which the field current is held constant, such that

$$V_e(s) = K_{mf}\Omega_m(s) \tag{2.5.17}$$

where K_{mf} was defined in (2.5.4)

It is fairly straightforward to obtain a closed-loop block diagram of the speed controlled DC generator by referring to $q_m(t)$ as the applied input torque and neglecting any torque due to external disturbances, indeed this is shown in Fig. 2.51.

It must be pointed out that in this look at electromechanical systems in terms of DC motors and generators, energy is converted between electrical and rotational mechanical systems; rules of the same kind apply and a similar analysis can be carried out for the conversion of energy between electrical and translational mechanical systems.

2.6 Summary

Step number one in the analysis of, or the design of a controller for, a physical system is a consideration of the relationship governing the variables, both signals and parameters, within the system in order to obtain a mathematical model which adequately describes their interactions. The overall system may well in fact consist of a number of different system types such as fluid, thermodynamic, mechanical and electrical, such that certain of the variables may well relate one physical type to another.

Since the systems to be controlled are dynamic, differential equations are the most usual form employed for describing operating relationships. However, in practice it is often the case that the system is quite complex, resulting in at the least a set of nonlinear differential equations. It is therefore necessary to make certain assumptions with respect to system operation, such as range of signal values possible, in order to linearize the system equations and ease further analysis. Once a set of linear differential equations have been obtained, the Laplace transform can be made use of, as was shown in Section 2.2, in order to define a system transfer function.

The transfer function approach to system representation highlights the importance of labeling one signal as the system input, i.e. the signal which is applied, and another signal as the system output, i.e. the signal which results. This technique also leads on to a graphical form of system representation, namely block diagrams, as was shown in Section 2.3. Fundamental analysis of the system variables can then be carried out either by means of an algebraic transfer function description or by means of the equivalent block diagrams, or indeed a combination of the two approaches can be used.

In Section 2.4 it was shown how system transfer functions can be obtained from electrical and mechanical system types, and how once a transfer function has been obtained it can in fact be regarded in the form of any desired appropriate physical type. This leads to the idea of analogous circuits in which a physical system of one type is considered within the framework of a physical system of another type, the two systems being related by a common transfer function. Although in this chapter the links between mechanical and electrical systems were dwelt upon, similar analogies can be made with other physical system types.

Problems

2.1   Write down the transfer function which relates input $V(s)$ to output $Y(s)$ in Fig. 2.52.

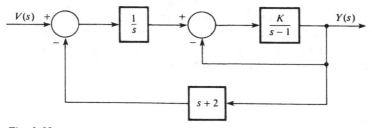

Fig. 2.52

2.2   The two-input/two-output system shown (Fig. 2.53) has inputs $V_1(s)$, $V_2(s)$ and outputs $Y_1(s)$, $Y_2(s)$.

Find an expression for:

(a) The output $Y_1(s)$ as a function of the inputs;
(b) The output $Y_2(s)$ as a function of the inputs.

What is the effect of making $G_1 G_4 = (G_2 G_3)^{-1}$? Assuming that $G_1 G_4 \neq (G_2 G_3)^{-1}$, find the value of G_s such that $Y(s) = G_1 V_1(s)$, where $Y(s) = Y_1(s) + G_s Y_2(s)$.

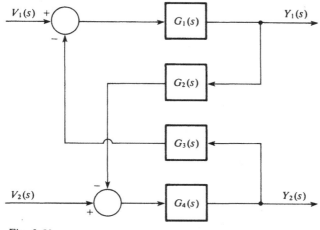

Fig. 2.53

2.3   Find the transfer function for the electrical circuit shown in Fig. 2.54.

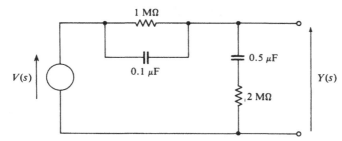

Fig. 2.54

2.4   Find the transfer function which relates input to output in Fig. 2.55.

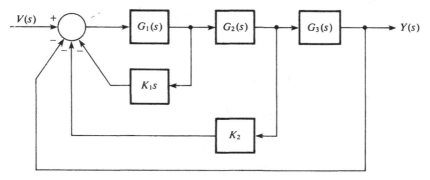

Fig. 2.55

2.5   Find the transfer function which relates input to output in Fig. 2.56.

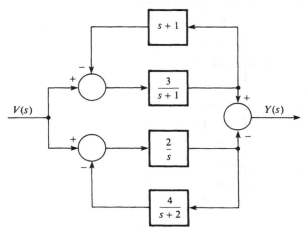

Fig. 2.56

2.6   Find the transfer function for the electrical circuit shown in Fig. 2.57.

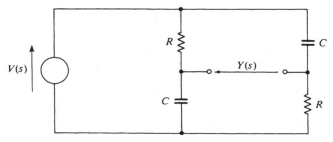

Fig. 2.57

2.7   Two tanks of water are connected by a pipe (Fig. 2.58). Water flows into tank 1 at a rate v and this tank, which is of unit cross-sectional area, has a head of water equal to height r. Water flows out of tank 1, and into tank 2, at a rate

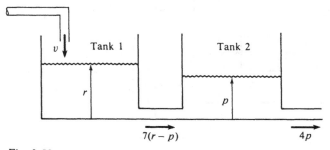

Fig. 2.58

$7(r - p)$, where p is the head of water in the unit cross-sectional area tank 2. Water flows out of tank 2 at a rate $4p$.

Show that the equations which govern the system are:

Tank 1: $\dot{r} = v - 7r + 7p$
Tank 2: $\dot{p} = 7r - 11p$

Until time $t = 0$ the tanks are empty. From that time onwards, $v = 1$. Find values for r and p from time $t = 0$ onwards and show that they tend to steady values.

2.8 Consider the system in Fig. 2.59 where $K \gg 1$.

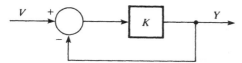

Fig. 2.59

What is the effect on the input transfer function if K is reduced to $K/4$? What would have happened if K was part of an open-loop system, i.e. if no unit feedback loop was present?

2.9 Find the relationship between the input and output voltages in Fig. 2.60.

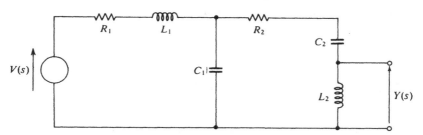

Fig. 2.60

2.10 Find the relationships between the input and the outputs $Y_1(s)$, $Y_2(s)$ and $Y_3(s)$ in Fig. 2.61. Show the solution in the form of a block diagram.

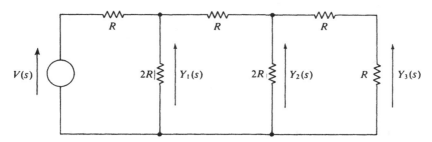

Fig. 2.61

2.11   The input to a system is designated as $V(s)$, whereas the output is $X(s)$. That output is in turn the input to a second system whose output is $Y(s)$.

If the systems are characterized by the equations

(a) $(D^2 + 7D + 12)x(t) = 3(D + 1)v(t)$

and   (b) $(D^2 + 4D + 3)y(t) = 5(D + 4)x(t)$

find the overall transfer function which relates input $V(s)$ to output $Y(s)$.

2.12   Determine the transfer function relating the input applied force $F(s)$ to the resultant velocity $W(s)$ in Fig. 2.62.

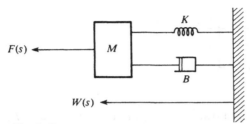

Fig. 2.62

2.13   Determine the transfer function relating the input applied force $F(s)$ to the resultant velocity $W(s)$ in Fig. 2.63. Thereby, construct an electrical circuit analog, using the force–voltage analogy.

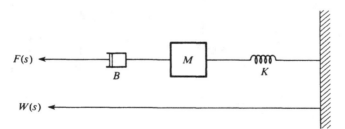

Fig. 2.63

2.14   Find the transfer function relating the output angular velocity $\Omega(s)$ to the input torque $Q(s)$ in Fig. 2.64.

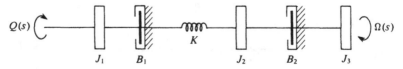

Fig. 2.64

2.15   Find a rotational mechanical system which is an analog of the electrical system shown, by using a torque–current analogy in Fig. 2.65.

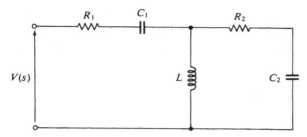

Fig. 2.65

2.16 An armature current controlled motor drives a load with inertia J kgm^2 through a shaft with compliance $1/K$ rad/Nm. Find the transfer function which relates the input armature current to output load angular velocity. The coupling constant $K_{mf} = K_m I_f =$ constant, and the rotor has a resistance B_r Nm/rad/s and inertia J_r kgm^2. Assume the effect of external disturbances $Q_d = 0$.

2.17 An armature voltage controlled motor, with armature resistance R_a Ω and inductance L_a H, drives a load with resistance B Nm/rad/s and inertia J kgm^2. The rotor has a resistance B_r Nm/rad/s and inertia J_r kgm^2 and $K_{mf} = K_m I_f =$ constant.

Find an electrical analog for this circuit, in terms of the transfer function relating armature current to armature voltage (the input impedance).

2.18 An armature voltage controlled motor, with armature resistance $R_a = 5$ Ω and negligible inductance, drives a load with inertia $J = 4$ kgm^2 and negligible resistance. Given that $R_r = J_r = 0$ and $K_{mf} = 3$ Nm/A, show that the transfer function which relates input armature voltage to output load angular velocity is given by:

$$G(s) = \frac{K}{s + \alpha}$$

where

$$K = \frac{K_{mf}}{JR_a} \quad \text{and} \quad \alpha = KK_{mf}$$

2.19 A field voltage controlled generator, with field resistance R_f Ω and field inductance L_f Ω, drives a purely capacitive load C F. If the armature resistance is R_a Ω and the armature inductance L_a H, find the transfer function which relates input field voltage to output load voltage when $K_{m\omega} = K_m \Omega_m =$ constant.

2.20 The generator output of Problem 2.19 is now connected to a series resistance R Ω, capacitance C F load. The field voltage is found as the output from an amplifier, with gain K, which has as its input the error between desired load voltage and actual load voltage. Find the transfer function which relates the desired to the actual load voltages.

2.21   If $V_1(s)$ and $V_2(s)$ are absolute velocities associated with masses M_1 and M_2 respectively, obtain the transfer function relating $V_1(s)$ to $F(s)$ (Fig. 2.66). Using the force–current analogy, construct an analogous electrical circuit.

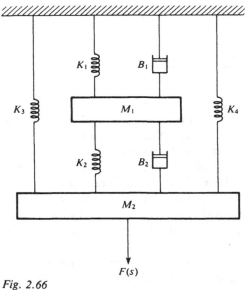

Fig. 2.66

Further reading

Bacon, D. H. and Stephens, R. C., *Mechanical technology*, Newnes-Butterworths, 1977.

Beachley, N. H. and Harrison, H. L., *Introduction to dynamic systems analysis*, Harper and Row, 1978.

Bender, E. A., *An introduction to mathematical modelling*, John Wiley, 1978.

Close, C. M. and Frederick, D. K., *Modeling and analysis of dynamic systems*, Houghton Mifflin, Boston, 1978.

DiStefano, J. J., Stubberud, A. R. and Williams, I. J., *Feedback and control systems*, McGraw-Hill (Schaum), 1976.

Doebelin, E. O., *System modeling and response*, Wiley, New York, 1980.

Draper, A., *Electrical circuits*, Longman, 1972.

Dym, C. L., *Principles of mathematical modelling*, Academic Press, 1984.

Edwards, J. D., *Electrical machines*, Macmillan, 1986.

Gould, L. A., *Chemical process control*, Addison-Wesley, 1969.

Houts, R. C., *Signal analysis in linear systems*, W.B. Saunders, 1991.

Hughes, E., *Electrical technology*, Longman, 1975.

Johnson, A., *Process dynamics, estimation and control*, Peter Peregrinus Ltd, 1985.

Kamen, E., *Introduction to signals and systems*, 2nd ed., Macmillan, 1990.

Karnopp, D. C. and Rosenberg, R. C., *System dynamics*, Wiley, 1975.

MacFarlane, A. G. J., *Dynamical system models*, Harrap, 1980.

Murphy, J., *Power control of AC Motors*, Pergamon Press, 1990.

Nash, P., *Systems modelling and optimisation*, Peter Peregrinus Ltd, 1981.

Nicholson, H. (ed.), *Modelling of dynamical systems – Vol. 1*, Peter Peregrinus Ltd, 1980.

Nicholson, H. (ed.), *Modelling of dynamical systems – Vol. 2*, Peter Peregrinus Ltd, 1981.

Raven, F. H., *Automatic control engineering*, 2nd ed., McGraw-Hill, 1990.

Shearer, J., *Dynamic modelling and control of engineering systems*, Macmillan, 1990.

Shinners, S. M., *Modern control system theory and application*, Addison-Wesley, 1978.

Truxal, J. G., *Automatic feedback control system synthesis*, McGraw-Hill, 1955.

Van Dixhoorn, J. J. and Evans, F. J. (eds.), *Physical structure in systems theory*, Academic Press, 1974.

Wellstead, P. E., *Introduction to physical system modelling*, Academic Press, 1979.

Ziegler, B., *Theory of modelling and simulation*, John Wiley, 1976.

3

System response using
differential equations

3.1   Introduction

In Chapter 2 various ways were considered in which a system to be controlled could be modeled in terms of a representative transfer function relating the system's input to its output. The Laplace operator was used to great effect, and it can be seen through the remainder of this text that this operator is a major mathematical tool in the formulation and design of continuous-time control systems.

The application of a controller to any system will result in an increase in overall complexity and, perhaps more obviously, will usually require some financial outlay. It is important therefore to decide, in the first instance, whether or not it is necessary to employ a controller in a particular case, the actual controller design is then of only secondary importance. So the initial question to be answered is, 'Does the performance of the present system satisfy our requirements?' But this raises a further problem in that we must have a clear definition of what we mean by performance. As an example, for a car, we could be interested in speed, reliability, comfort and/or fuel consumption in terms of one criterion in particular or more likely in some combination of all of them.

In this chapter some of the fundamental performance criteria used for system analysis are introduced by considering the output response of a system in terms of its dependence on one of a standard set of basic test input signals. Not only is it much simpler and cheaper to apply a fairly straightforward test input, the subsequent analysis is generally easier than that required for a complex test input and it is most likely the case that the input signals which are in practice applied to that particular system can be thought of as combinations of the basic forms of input.

In assessing the performance of a system, analysis can be concentrated on either the initial reaction of the system output, following a particular test input, or the reaction observed once things have settled down to a steady form of response. It is noticeable that the output response is viewed in terms of its time domain features, and although it may not be initially obvious, it is shown here that s-domain system representations are an ideal form and that no inverse transformation is necessary when investigating time domain specifications. The total output response witnessed is therefore considered to consist of two parts (a) the initial reaction, and (b) the steady reaction, which is known as the steady-state response and is essentially that part of the response which is left when the transient response dies down to zero. In the detailed study of the steady-state response of a system, carried out in Section 3.2, it is shown how the observed output is dependent on the input signal applied and in Section 3.4 the possibility of an error existing in the steady-state between the actual output and the desired output/reference input is shown to be directly related to the application of specific input signals to certain types of systems.

All physical systems, when they are asked to respond, do not do so instantaneously by jumping to a final steady value. In some cases it might be that the response is rather sluggish and slowly reaches its final value, whereas in other cases it is possible for the output to very rapidly overshoot its final value and then to slowly home in on it by decreasing its overshoots as time goes on. These different transient response possibilities are considered in Section 3.3.

A further, but vital, contribution of this chapter is to introduce some of the positive aspects gained with the application of feedback to form a closed-loop control system, such as reduction of the effect of noise on the output signal and reduction in output sensitivity to plant variations. Some of the main points made in the chapter are then brought together in the extended servomechanism example of Section 3.5, in which the use of feedback to improve both transient and steady-state system performance is discussed and this results in a basic controller design problem.

3.2   Steady-state response

Because the total response of a system is made up of two parts, the transient and steady-state responses, it is possible to think of the steady-state response as being the response which remains when the transient response has died down, or more simply as the transient response subtracted from the total response. Indeed this latter definition is that employed to investigate response errors in Section 3.4.

It is wrong to think that the steady-state response of a system refers in every case to a steady system output value although, as is shown in this section this might be the case, it is not necessarily so. If a time invariant signal is applied at the input then a different steady output response would be expected when compared with the response obtained by applying a sinusoidal input to the same system. So the steady-state response of a system can only be considered by making reference to the type of input applied, and as the transient response will by definition tend to zero at time $t \rightarrow \infty$, so the steady-state response to an input will be that apparent as $t \rightarrow \infty$.

A variety of standard input signals are available for employment in order to determine the response characteristics of a system, a common element between these inputs being their relatively simple implementation. In most cases the system, when in normal operation, may well have to deal with a multitude of different input signals or conversely only have to face one specific type of input. Whatever the case it is usual to investigate the system response to only one or two of the standard input forms, because in general this will provide sufficient information for further analysis.

In the section which follows directly, the most commonly encountered standard input test signals are described, with the more popular types discussed in greater detail. Subsequently the steady-state response of systems to certain of the inputs is considered with examples being given of some low order, fairly straightforward system transfer functions.

3.2.1   Control test inputs

The input to some systems can, during normal operation, be considered as random or at least pseudo-random, one example being systems whose input is affected by a large random disturbance. It may then be necessary to apply a pseudo-random input in order to obtain detailed information as to the characteristics of the system; the same is true for many complicated systems which are inherently of high order. However for essentially simple systems, and in order to obtain important fundamental information on more complicated systems, the test signals applied as inputs are generally of a fairly simple standard form.

Standardizing test input signals is important when assessing the relative merits of several systems by comparing the output responses to a common input signal. Perhaps the simplest of all inputs to employ is a *step input* which can be applied, for example, by switching a voltage in an electrical circuit or a sudden force which is retained in a mechanical network. If it is assumed that the step is of unit magnitude and is applied at $t = 0$, then this is shown in Fig. 3.1.

The unit step input is a primary test signal and can be described by

$$u(t) = \begin{cases} 1: & t > 0 \\ 0: & t < 0 \end{cases} \qquad (3.2.1)$$

where the input magnitude is shown to be unity after time $t = 0$. In practice it will most likely be sensible to apply a step input of some magnitude other than unity, this being

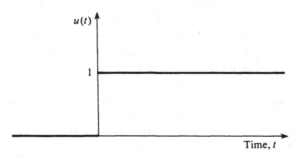

Fig. 3.1   Unit step input signal

particularly true for a low gain system when the presence of noise on the output could be problematic unless a high magnitude step is used. It can also be noticed that the step input is undefined for $t = 0$, the assumption being that at $t = 0$ the input changes from 0 to 1, an idealistic view, since a finite but very small time is actually necessary to carry out such a transition. In general though the response of systems considered in this text, governed by their time constants, will be much slower than the step transition time, which can therefore be neglected. For the study of very high speed systems, logic gates for example, a more complicated description of the step input would, however, most likely be necessary.

The Laplace transform of the ideal unit step input (3.2.1) is given by

$$U(s) = 1/s \tag{3.2.2}$$

and it is worth noting that a 4-unit step input, for example, has a Laplace transform $U(s) = 4/s$.

A second test signal is the *impulse function*. This was considered to an extent in the previous chapter, and is a simple derivative of the step input; it therefore has a Laplace transform of

$$U(s) = 1 \tag{3.2.3}$$

when applied as a control input signal.

In its strictest form an impulse function lasts for a time interval δt as $\delta t \to 0$, and is of infinite magnitude. As the practical application of such a signal poses obvious problems, it is usual to approximate the impulse with a signal of unit magnitude which lasts for a very short time period δt, and this is in fact discussed at length in Chapter 2.

If the integral is taken of the step input this results in a function which increases linearly with respect to time from a zero value at time $t = 0$. The *ramp function*, which can also be regarded as a velocity function, is shown in terms of a unit magnitude ramp in Fig. 3.2.

The unit ramp input can be written as

$$u(t) = \begin{cases} t: & t > 0 \\ 0: & t < 0 \end{cases} \tag{3.2.4}$$

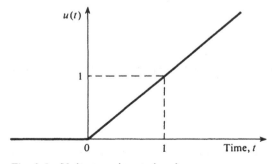

Fig. 3.2   Unit ramp input signal

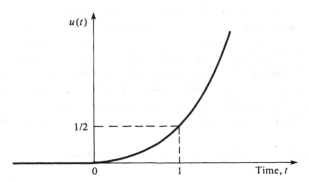

Fig. 3.3 Unit parabolic input signal

or in terms of the Laplace transform

$$U(s) = 1/s^2 \tag{3.2.5}$$

By integrating the ramp function, hence obtaining an acceleration function, a square law signal known as a *parabolic function*, is achieved. This is shown in terms of a unit parabolic signal in Fig. 3.3.

The unit parabolic input can be written as

$$u(t) = \begin{cases} t^2/2: & t > 0 \\ 0: & t \leqslant 0 \end{cases} \tag{3.2.6}$$

where the division by 2 in (3.2.6) is included in order to think of the function as a step acceleration term with respect to the original step input (3.2.1). This can also be considered as

$$U(s) = 1/s^3 \tag{3.2.7}$$

by means of the Laplace transform.

It is apparent that by combining the step, ramp and parabolic inputs to form one single input signal, for all $t > 0$, this can be written as

$$u(t) = 1 + t + \frac{t^2}{2}$$

It is quite possible, however, to consider further integrated terms in order to arrive at an overall input

$$u(t) = 1 + t + \frac{t^2}{2} + \cdots + \frac{t^p}{p!}$$

although a more general expression is obtained by considering each separate term to have a magnitude equal to a constant d_i, resulting in a power series or *polynomial function*

$$u(t) = d_0 + d_1 t + d_2 \frac{t^2}{2!} + \cdots + d_p \frac{t^p}{p!} \tag{3.2.8}$$

for $t > 0$; $u(t) = 0$ for $t < 0$.

This polynomial function can though be written more concisely as

$$u(t) = \sum_{i=0}^{p} d_i t^i / i! : \quad t > 0$$

where d_i are scalar constant values.

The final standard test signal introduced here is the *sinusoidal function* which can be written in unit magnitude form as

$$u(t) = \begin{cases} \cos \omega t : & t > 0 \\ 0 : & t \leqslant 0 \end{cases} \tag{3.2.9}$$

This type of input signal is considered in the following section in terms of the steady-state response which results from its application.

3.2.2 Sinusoidal and polynomial inputs

Analyzing the steady-state response of a system to a sinusoidal input is not in general possible by means of the Laplace transform final value theorem. This theorem requires that there exists a limiting value to which a signal converges as time tends to infinity, and for a sine wave no such value exists.

For the sinusoidal input defined in (3.2.9) we have that for all $t > 0$, $u(t) = \cos \omega t$, remembering that if the input is of magnitude U, in terms of a phasor diagram this can be represented as in Fig. 3.4.

The output signal in the steady-state will then also be a sinusoid, as long as the system to which the input is applied does not contain any instabilities (the stability of a system is discussed in Chapter 4). So if the steady-state system output, $y(t)$, is obtained in

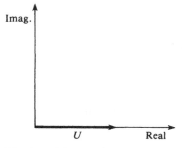

Fig. 3.4 Phasor diagram representation of $u(t) = U \cos \omega t$

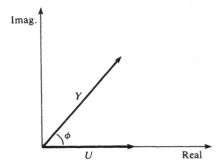

Fig. 3.5   Phasor diagram representation of $y(t) = Y \cos(\omega t + \phi)$, where $y(t)$ is the steady-state output

response to a sinusoidal input acted upon by a system whose transfer function may be described by means of a differential equation as defined in Chapter 2, then the output will be a sinusoid of magnitude Y, with a phase shift denoted by the angle ϕ, as shown on the phasor diagram in Fig. 3.5.

A general polynomial test input was defined in (3.2.8). It is assumed that the system to which the input is applied is defined by a differential equation, then we have that

$$(a_n D^n + a_{n-1} D^{n-1} + \cdots + a_1 D + a_0) y(t)$$
$$= (b_m D^m + b_{m-1} D^{m-1} + \cdots + b_1 D + b_0) u(t)$$

in which the a_i and b_i coefficients determine the characteristics of the system transfer function, as described in Chapter 2. The application of a test input (3.2.8) will result in a steady-state output signal which will also contain enough terms in powers of time t to cause equality in the system differential equation. As an example consider $n = 2$ and $m = 1$ in the above equation with a control input defined by

$$u(t) = d_0 + d_1 t + \frac{d_2 t^2}{2}$$

We have then that:

$$(a_2 D^2 + a_1 D + a_0) y(t) = (b_1 D + b_0)\left[d_0 + d_1 t + \frac{d_2 t^2}{2} \right]$$

$$= b_1 \, d_1 + b_0 \, d_0 + (b_1 \, d_2 + b_0 \, d_1)t + \frac{b_0 \, d_2 t^2}{2}$$

If in general the steady-state output is described by

$$y(t) = c_0 + c_1 t + \frac{c_2 t^2}{2!} + \cdots + \frac{c_k t^k}{k!}$$

then in order to obtain equality in the example equation we must have that

$$y(t) = c_0 + c_1 t + \frac{c_2 t^2}{2!}, \qquad \text{i.e.} \quad k = 2$$

from which it follows that

$$(a_2 D^2 + a_1 D + a_0) y(t) = a_2 c_2 + a_1 c_1 + a_0 c_0 + (a_1 c_2 + a_0 c_1)t + \frac{a_0 c_2 t^2}{2}$$

The steady-state output coefficients c_0, c_1, c_2 can then be found by solving the three equations obtained, i.e.

$$a_0 c_2 = b_0 \, d_2$$

$$a_1 c_2 + a_0 c_1 = b_1 \, d_2 + b_0 \, d_1$$

and

$$a_2 c_2 + a_1 c_1 + a_0 c_0 = b_1 \, d_1 + b_0 \, d_0$$

Notice that the power index k is dependent solely on the complexity of the input signal, i.e. the index p, so

$$k = p \tag{3.2.10}$$

3.2.3   Step, ramp and parabolic inputs

The step input was defined in Section 3.2.1. If such an input is applied to a first-order system transfer function of the form

$$G(s) = \frac{K}{s + a_0}$$

where K is the system gain, then the system output is obtained from $Y(s) = G(s) \cdot U(s)$ to be

$$Y(s) = \frac{K}{s + a_0} \cdot \frac{1}{s}$$

in which $1/s$ is the unit step Laplace transform.

The output is then

$$Y(s) = \frac{K/a_0}{s} - \frac{K/a_0}{s + a_0}$$

or in terms of its inverse transform

$$y(t) = \frac{K}{a_0} - \frac{K}{a_0} \cdot \exp(-a_0 t)$$

the time response of which is shown in Fig. 3.6, where $a_0 > 0$.

It can be considered either that $1/a_0$ seconds is the time taken for the output to reach within $\exp(-1)$ of its final value, i.e. the time to reach 63.2% of the final value, or $1/a_0$ is the time taken to reach the final value if the initial rate of increase is continued.

The final value of the output signal, that is the steady-state output value, can be found

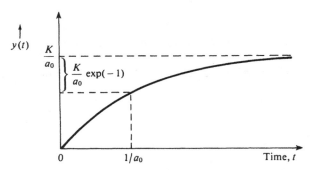

Fig. 3.6   First-order system response to a unit step input

by means of the final value theorem of the Laplace transform, which states that if the limit of a function $f(t)$ as $t \to \infty$ exists, then

$$\lim_{s \to 0} sF(s) = \lim_{t \to \infty} f(t) \tag{3.2.11}$$

So for the first-order example just discussed,

$$\lim_{s \to 0} sY(s) = s \cdot \frac{K}{s + a_0} \cdot \frac{1}{s} \bigg|_{s \to 0} = \frac{K}{a_0}$$

and this agrees with the result found from the inverse Laplace transform as $t \to \infty$; shown in Fig. 3.6.

By applying a unit ramp input to the same system transfer function, the output transform is now given by

$$Y(s) = \frac{K}{s + a_0} \cdot \frac{1}{s^2}$$

and on applying the final value theorem we find that

$$\lim_{s \to 0} sY(s) = s \cdot \frac{K}{s + a_0} \cdot \frac{1}{s^2} \bigg|_{s \to 0} = \infty$$

So, for a first-order system as shown, application of a unit ramp input will result in the system output tending to infinity as $t \to \infty$. This can also be seen by finding the inverse Laplace transform which is

$$y(t) = \frac{K}{a_0} \left[t - \frac{1}{a_0} + \frac{\exp(-a_0 t)}{a_0} \right]$$

such that the output tends to infinity as $t \to \infty$.

A similar result in which the output tends to infinity is witnessed as the first-order system response to a unit parabolic input, given from the final value theorem as

$$\lim_{s \to 0} sY(s) = s \cdot \frac{K}{s + a_0} \cdot \frac{1}{s^3} \bigg|_{s \to 0} = \infty$$

in which $1/s^3$ is the unit parabolic input.

In the following section the response of a second-order system, in particular to a unit step input, is considered at length in order to introduce the possibility of an output response overshooting its final value. For the first-order system considered in this section it is not possible to obtain any overshoot; the output value will always approach its final value from one direction only. The value of K merely part defines the actual final output value for a first-order system whereas the value of a_0, as well as partly defining the actual final output value, characterizes how rapidly that final value is approached and is therefore an important measure of system performance, the figure $1/a_0$ being called the *time constant*.

A general second-order system transfer function can be written as

$$G(s) = \frac{K}{s^2 + a_1 s + a_0} \tag{3.2.12}$$

Note: This is second order because the highest power of $s = 2$.

By analyzing the output of this system in response to a unit step input signal, the final value theorem reveals

$$\lim_{s \to 0} sY(s) = s \cdot \frac{K}{s^2 + a_1 s + a_0} \cdot \frac{1}{s} \bigg|_{s \to 0} = \frac{K}{a_0}$$

which is an identical result to that found for a first-order system. It is worth noting that if $a_0 = 0$ the steady-state output signal tends to infinity, in fact with $a_0 = 0$ the system denominator reduces to $s(s + a_1)$, i.e. it contains a common s factor — it is a type 1 system, see Section 3.4.1, and this factor s, when set to zero, causes the infinite output.

The time response, to a step input, of a first-order system was shown in Fig. 3.6, and this diagram holds for all $a_0 > 0$, different values of a_0 merely affecting the time and magnitude axis scaling. The step input time response of a second-order system is rather more complicated, the type of response obtained being dependent on whether the system denominator roots (the system poles) are real or complex, and this is the subject of the following section.

3.3   Transient response

In the previous section (3.2), the steady-state response of a system was considered to be that part of the total response which remains as time $t \to \infty$. In a similar fashion the transient response of a system can be regarded as that part of the total response which tends to zero as time $t \to \infty$. The total response therefore consists of the steady-state response summed with the transient response. The concept that a system has a transient response which tends to zero implies that the system has certain properties which prevent the output from constantly increasing in magnitude towards infinity. This is the same as requiring that the system is 'stable', a topic discussed in Chapter 4.

All of the analysis in this section is based on the assumptions that (a) the control test input is a unit step function, and (b) the system is represented exactly by a second-order transfer function. Ramp or parabolic inputs could be applied as well as or instead of the step input, however no further important aspects are revealed by such a procedure when compared with a step input response, and for the majority of systems it will be a step response which is most relevant in practice. Restricting the analysis to second-order systems only by no means limits the applicability of the results. Although certain systems are of order greater than two, they can in many cases be modeled adequately by a second-order system, the total response obtained being dominated by that due to a second-order model.

3.3.1 Overdamped response

If the system transfer function has two real denominator roots, then it follows that:

$$G(s) = \frac{K}{s^2 + a_1 s + a_0} = \frac{K}{(s + \alpha)(s + \beta)} \qquad (3.3.1)$$

By applying a unit step input to such a system, the output signal transform is found to be:

$$Y(s) = \frac{K}{(s + \alpha)(s + \beta)} \cdot \frac{1}{s}$$

$$= \frac{K}{\alpha\beta} \left[\frac{1}{s} + \frac{\beta}{(\alpha - \beta)} \cdot \frac{1}{(s + \alpha)} + \frac{\alpha}{(\beta - \alpha)} \cdot \frac{1}{(s + \beta)} \right]$$

or in terms of its time-domain representation, this is

$$y(t) = K\left\{ \frac{1}{\alpha\beta} + \frac{1}{(\alpha - \beta)} \left[\frac{\exp(-\alpha t)}{\alpha} - \frac{\exp(-\beta t)}{\beta} \right] \right\} \qquad (3.3.2)$$

It can be noted from the step response that by setting t to infinity in (3.3.2), and assuming $\alpha > 0$, $\beta > 0$,

$$\lim_{t \to \infty} y(t) = \frac{K}{\alpha\beta} = \frac{K}{a_0}$$

where, from (3.3.1), $a_0 = \alpha\beta$.

This result agrees with that obtained by application of the final value theorem in Section 3.2.3.

A further point to note is that at $t = 0$:

$$y(t) = K\left\{ \frac{1}{\alpha\beta} + \frac{1}{(\alpha - \beta)} \left[\frac{1}{\alpha} - \frac{1}{\beta} \right] \right\} = 0$$

and

$$\frac{dy(t)}{dt} = 0 \bigg|_{t=0}$$

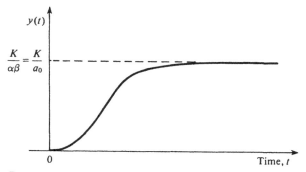

Fig. 3.7 Overdamped response to a step input

The overall time response for the overdamped case is shown in Fig. 3.7, where the initial rate of change of output signal is identically zero.

In common with the response of a first-order system to a step input, the output signal always approaches its final value from one side, as shown here it increases from zero to its final value with no overshoot.

The effect of having a numerator root in the system transfer function, e.g.

$$G(s) = \frac{K(s+\gamma)}{(s+\alpha)(s+\beta)}$$

is to cause the differential $dy(t)/dt$ at time $t = 0$ to be nonzero. It has very little effect on the time response other than that.

3.3.2 Critically damped response

A critically damped response can in a certain sense be regarded as a special or extreme case of the overdamped response. If the two denominator roots of the system transfer function are identical, then

$$G(s) = \frac{K}{(s+\alpha)^2} \tag{3.3.3}$$

When a unit step input is applied to this system the output signal transform is

$$Y(s) = \frac{K}{(s+\alpha)^2} \cdot \frac{1}{s}$$

$$= \frac{K}{\alpha^2} \left[\frac{1}{s} - \frac{1}{(s+\alpha)} - \frac{\alpha}{(s+\alpha)^2} \right]$$

and in terms of its time-domain representation:

$$y(t) = \frac{K}{\alpha^2} \{1 - \exp(-\alpha t) - \alpha t \cdot \exp(-\alpha t)\}$$

or

$$y(t) = \frac{K}{\alpha^2} \{ 1 - \exp(-\alpha t)[1 + \alpha t] \}$$

(3.3.4)

Assuming that $\alpha > 0$, we have that

$$\lim_{t \to \infty} y(t) = \frac{K}{\alpha^2} = \frac{K}{a_0}$$

where

$$a_0 = \alpha^2$$

Initial conditions can be found by setting t to zero in (3.3.4) which gives

$$y(t) = 0$$

and also

$$\left. \frac{\mathrm{d}y(t)}{\mathrm{d}t} = 0 \right|_{t=0}$$

The overall time response for the critically damped case is shown in Fig. 3.8, in which the initial rate of change of output signal is zero.

The critically damped response represents the most rapid type of response which does not result in any overshoot, i.e. the output signal is in Fig. 3.8 never greater than its final value. This property is used to good effect in some measuring instruments and is particularly useful for example in certain machine tool applications or robot welders, in which it would be extremely dangerous to have any overshoot and yet taking a long time to reach the desired position, as would be the case for an overdamped response, would mean a waste of revenue due to less productivity. For many systems however it is desirable for an overshoot to occur before the output settles on its steady-state value, and such a response is described in the next section.

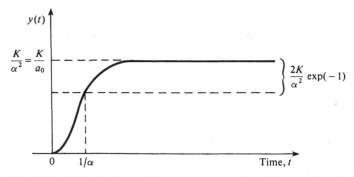

Fig. 3.8 Critically damped response to a step input

3.3.3 Underdamped response

The underdamped response is characterized principally by the output overshooting its final steady-state value following the application of a step input. This is by far the most widely studied type of transient response. More parameters are used to describe the response than in the previous two cases and hence in an academic environment it is a much more fruitful topic as far as exam questions are concerned.

If the two denominator roots of the system transfer function are complex (complex conjugates in fact) then the output signal will exhibit oscillations when a step input signal is applied. Just how quickly these oscillations die down is dependent on the degree of damping, such that if the response is not damped at all the oscillations will in theory carry on for ever. However, as the amount of damping is increased so the oscillations die down more quickly until we reach a critically damped response in which the degree of damping is so great that the oscillations do not even start up!

Assuming that the system transfer function contains complex roots, then we have

$$G(s) = \frac{K}{(s + \alpha + j\beta)(s + \alpha - j\beta)} \tag{3.3.5}$$

or

$$G(s) = \frac{K}{s^2 + 2\alpha s + (\alpha^2 + \beta^2)}$$

which is also

$$G(s) = \frac{K}{(s + \alpha)^2 + \beta^2}$$

such that when compared with the standard polynomial denominator form given in (3.3.1) it follows that in this case

$$a_0 = \alpha^2 + \beta^2 \qquad \text{and} \qquad a_1 = 2\alpha$$

By applying a unit step input to the system (3.3.5) the output signal transform is

$$Y(s) = \frac{K}{(s + \alpha)^2 + \beta^2} \cdot \frac{1}{s}$$

$$= \frac{K}{\alpha^2 + \beta^2} \left[\frac{1}{s} - \frac{(s + \alpha)}{(s + \alpha)^2 + \beta^2} - \frac{\alpha}{(s + \alpha)^2 + \beta^2} \right]$$

In terms of its time-domain representation this is

$$y(t) = \frac{K}{\alpha^2 + \beta^2} \left\{ 1 - \exp(-\alpha t) \cdot \cos \beta t - \frac{\alpha}{\beta} \cdot \exp(-\alpha t)\sin \beta t \right\}$$

or

$$y(t) = \frac{K}{\alpha^2 + \beta^2} - \frac{K \exp(-\alpha t)}{\beta(\alpha^2 + \beta^2)^{1/2}} \left\{ \frac{\beta}{(\alpha^2 + \beta^2)^{1/2}} \cos \beta t + \frac{\alpha}{(\alpha^2 + \beta^2)^{1/2}} \cdot \sin \beta t \right\}$$

and if we define

$$\phi = \tan^{-1} \frac{\beta}{\alpha} \tag{3.3.6}$$

the time response can be written as

$$y(t) = \frac{K}{(\alpha^2 + \beta^2)} \left\{ 1 - \frac{(\alpha^2 + \beta^2)^{1/2}}{\beta} \cdot \exp(-\alpha t) \cdot \sin(\beta t + \phi) \right\} \tag{3.3.7}$$

So the response is a sine wave of frequency β with an exponential damping governed by the factor α and a steady-state value given by $K/(\alpha^2 + \beta^2)$, as shown in Fig. 3.9.

The most usual way to write the system denominator of (3.3.5) is in fact in the form $s^2 + 2\zeta\omega_n s + \omega_n^2$, where

$$s^2 + 2\zeta\omega_n s + \omega_n^2 = s^2 + 2\alpha s + (\alpha^2 + \beta^2) \tag{3.3.8}$$

in which ζ is the damping ratio and ω_n the natural frequency. These definitions arise because:

(a)   We have from (3.3.8) that:

$$\zeta = \alpha/\omega_n \quad \text{and} \quad \omega_n^2 = \alpha^2 + \beta^2$$

such that if $\beta \to 0$, something which causes the roots of the denominator to tend to those of the critically damped system − see (3.3.5) and (3.3.3) − then the damping ratio $\zeta \to 1$. On the other hand if $\alpha \to 0$, so the exponential damping tends to zero and we approach the case of a purely oscillatory response − see (3.3.7), i.e. $\zeta \to 0$. Hence for an underdamped response $0 < \zeta \leqslant 1$, where the damping is heavier as ζ tends to unity. In fact for $\zeta > 1$ this corresponds to the overdamped case discussed in Section 3.3.1 in which the system denominator consists of two nonidentical real roots.

(b)   If the damping ratio is zero then $\zeta = 0 = \alpha$ and $\omega_n = \beta$, such that ω_n is the frequency at which oscillations will occur when there is no damping and in such a case the oscillations would be sustained indefinitely. Note that for a particular response, β is the actual frequency of the oscillations, but this will only be equal to the natural frequency ω_n in the limit as the damping tends to zero.

The response shown in Fig. 3.9 indicates several basic characteristics of a system which are used to describe, in relatively simple terms, the performance of that system for comparison or design purposes. In the initial part of the response it must be remembered however that the phase shift ϕ will have a certain effect such that $\phi \to 90°$ as the damping tends to zero, hence for an undamped system the response will be dictated from time $t = 0$ by a cosine waveform.

The initial part of the response is most often described by means of two terms, firstly the *delay time*, which is the time taken for the output step response to reach 50% of its final value, and secondly the *rise time*, which is the time taken for the output step response to rise from 10 to 90% of its final value. At a later stage in the time response, the *settling time*, shown as T_s in Fig. 3.9, is defined as the time taken for the step

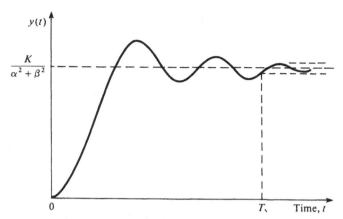

Fig. 3.9 Underdamped response to a step input

response to reach and stay within a specified percentage of its final value. A commonly encountered figure is 5%, and this was employed for the response shown in Fig. 3.9.

Another term used to describe the transient performance of a system is overshoot, or more usually *maximum overshoot*, which is the maximum difference between the transient and steady-state step response values that occurs after the response has first passed its final value. The maximum overshoot is frequently written as a percentage of the steady-state output response value, i.e.

$$\frac{\text{maximum overshoot} \times 100\%}{\text{steady-state value}}$$

The maximum and minimum values of the output response can be found by equating the differential form of (3.3.7) to zero, which results in

$$\frac{\mathrm{d}y(t)}{\mathrm{d}t} = \frac{K}{\beta} \cdot \exp(-\alpha t) \cdot \sin \beta t = 0 \tag{3.3.9}$$

It can be seen that the maximum and minimum values of the differential are dictated by the sine term only, such that they occur when $t = i\pi/\beta$, where $i = 0, 1, 2$, etc. This means that on substitution back into (3.3.7) the magnitude of each maximum and minimum value is given by

$$y(i) = \frac{K}{\alpha^2 + \beta^2} \{1 - \exp(-i\pi\alpha/\beta) \cdot \cos i\pi\} \tag{3.3.10}$$

in which $i = 0, 1, 2$, etc.

At the time instants denoted by the index i, the output response therefore has values of

$$y(i) = \frac{K}{\alpha^2 + \beta^2} \{1(-)^{i+1} \exp(-i\pi\alpha/\beta)\}$$

Relative to its final value the output response values for $i = 1, 2$, etc., are termed overshoots, the initial condition being $y(i) = 0$ for $i = 0$. The maximum overshoot is then the first overshoot, i.e. when the exponential term is at its largest, and this is found from the actual magnitude at $i = 1$ to be

$$\text{maximum overshoot} = \frac{K}{\alpha^2 + \beta^2} \cdot \exp(-\pi\alpha/\beta)$$

where the steady-state value $K/(\alpha^2 + \beta^2)$ has been subtracted from the actual response magnitude in order to obtain the magnitude of overshoot at this time instant. In terms of a percentage this is

$$\text{maximum overshoot (\%)} = \exp(-\pi\alpha/\beta) \times 100\% \qquad (3.3.11)$$

and the time taken to reach this maximum, first, overshoot is $t = \pi/\beta$.

The maximum overshoot value is in fact also the absolute value of the ratio of successive overshoots.

Consider for example the magnitude of the second overshoot divided by the magnitude of the first overshoot, which is

$$\text{successive overshoot ratio} = \frac{\exp(-2\pi\alpha/\beta)}{\exp(-\pi\alpha/\beta)} = \exp(-\pi\alpha/\beta)$$

This is then also an indication of how quickly the oscillations die away in magnitude. It is often written as a logarithmic value, that is

$$\text{logarithmic decrement} = 20 \log_{10} \exp(-\pi\alpha/\beta) \simeq -27.29 \, (\alpha/\beta) \text{ dB/overshoot}$$

A special case of the logarithmic decrement is arrived at when $\alpha = \beta$, in which case we observe a loss of 27.29 dB per overshoot, and this particular relationship between α and β results in what is known as *ideal damping*, with a transfer function

$$G(s) = \frac{K}{(s + \alpha)^2 + \alpha^2} \qquad (3.3.12)$$

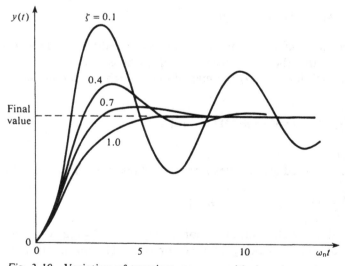

Fig. 3.10   Variation of transient response with damping ratio

However in the design of a system it might, in practice, not be possible to achieve an exact equality between α and β, and this leads to the idea of *acceptable damping*, for which

$$3^{-1/2} < \alpha/\beta < 3^{1/2}$$

To conclude this section, a set of step responses is given in Fig. 3.10 in order to show the difference in signal decay due to the damping factor in each case.

3.4   Errors

The performance of a system is usually assessed in terms of the error between the demanded or required output signal and the actual output signal that is obtained. Any error which does occur can be caused by one or more of a number of reasons, for instance the output measurement could be incorrect because of poor calibration or noise interference. The problem of noise is quite a serious one in many systems, and the objective of some controllers is simply to regulate the variation in output level, caused by disturbances, about a reference value. The effect of noise signals on a system is considered later in this section and the advantages gained by the employment of a feedback control scheme, in terms of disturbance error reduction, are discussed.

Another major source of errors is the employment of an incorrect system model for controller design purposes. It might be that a poor model was calculated in the first instance, although it is quite possible that even though the model was a good one when obtained, the system has aged or has been modified since then and the representation that we have of the system is no longer suitable. A control system design can be investigated in order to find out how sensitive it is to variations in the system description, i.e. if the actual system is not quite the same as we have pictured it to be, by how much is the control design affected? The sensitivity of a control system is also discussed later in this section.

The first classification of error to be considered here is that caused simply by the effect of certain system types on straightforward input signals, namely step, ramp and parabolic inputs. When applying a specific desired reference input to a system it is generally hoped that the system output will settle down to the desired level as steady-state conditions are approached. It is however shown here that this is by no means necessarily the case and is strongly dependent on the *system type*. In the first instance therefore a definition of system type is given.

3.4.1   System type

When discussing a system type it is important to note that reference is always made to the open-loop transfer function of a feedback system, on the assumption that unity feedback is applied as shown in Fig. 3.11.

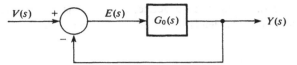

Fig. 3.11   Unity feedback system

If the original block diagram representation of a system contains elements other than unity in its feedback path, the representation must be manipulated into a unity feedback system, by means of the techniques introduced in Chapter 2, in order to obtain $G_0(s)$, the open-loop transfer function — this is shown in detail in the following section.

The transfer function definition given in the previous chapter is now further generalized, by defining $G_0(s)$ as:

$$G_0(s) = \frac{K(s^m + b_{m-1}s^{m-1} + \cdots + b_1 s + b_0)}{s^q(s^n + a_{n-1}s^{n-1} + \cdots + a_1 s + a_0)} \tag{3.4.1}$$

where q, m and n are integers, $a_0, b_0 \neq 0$ and $\{q, m, n \geqslant 0\}$. In many cases it is possible that the denominator and numerator polynomials of order n and m respectively can be factorized, these roots being termed poles and zeros of the transfer function, respectively. K is a gain constant and $q = 0, 1, 2, \ldots$ denotes the type of the system, e.g. a type 1 system is achieved when $q = 1$. Note that the order of the system, usually given by the order of the denominator, is then $n + q$, e.g. if $n = 2$ and $q = 1$, this would signify a third-order system. It is worth noting that it is usual for $n \geqslant m$, and this is the case considered here.

Example 3.4.1

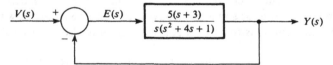

This is a type 1, third-order system.

Example 3.4.2

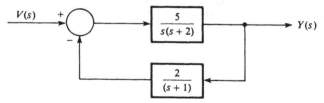

This is a type 0, third-order system.

The second of these examples, i.e. Example 3.4.2, is important in that it categorizes the error considered in this section always to be the difference between the desired and actual outputs $E(s) = V(s) - Y(s)$. In certain other texts, e.g. Kuo (1982), the error is considered to be the difference between the desired output $V(s)$ and the value $H(s) Y(s)$ – it is imperative that the reader is aware of this point. Only when $H(s)$ is unity are the two approaches identical.

Consider the closed-loop equation obtained from the unity feedback system shown in Fig. 3.11; this is given by

$$Y(s) = \frac{G_0(s)}{1 + G_0(s)} V(s) \tag{3.4.2}$$

For a general feedback system, as shown in Fig. 3.12, the closed-loop equation is, however:

$$Y(s) = \frac{G(s)}{1 + G(s)H(s)} V(s) \tag{3.4.3}$$

The equivalent open-loop transfer function $G_0(s)$ can therefore be found from $G(s)$ and $H(s)$ by means of the equation, obtained by connecting (3.4.2) and (3.4.3)

$$G_0(s) = \frac{G(s)}{1 + G(s)[H(s) - 1]} \tag{3.4.4}$$

where it can be noted that if $H(s) = 1$, the unity feedback case, the transfer function $G_0(s) = G(s)$.

3.4.2 Steady-state errors

Consider the unity feedback closed-loop system in Fig. 3.11, in which the error $E(s)$, is found to be

$$E(s) = \frac{1}{1 + G_0(s)} V(s) \tag{3.4.5}$$

such that the error between desired and actual output signals depends on the input signal $V(s)$, i.e. the desired reference output signal, and also on the open-loop transfer function $G_0(s)$ whether this is obtained directly or via equation (3.4.4).

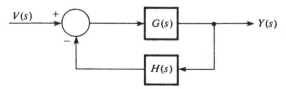

Fig. 3.12 General feedback system

The total error can be regarded as consisting of two parts; firstly the steady-state error can be written as

$$e(t)\,|_{t\to\infty} = sE(s)\,|_{s\to0} \tag{3.4.6}$$

in which $sE(s)$ has no denominator root whose real part is positive, and where (3.4.6) is merely the Laplace transform final value theorem result. The remainder of the total error is then the transient error and this in fact links directly with the transient system response discussed in Section 3.3; it will therefore only be considered briefly here:

$$\text{Transient error} = e(t) - e(t)\,|_{t\to\infty} \tag{3.4.7}$$

Steady-state step error

If a unit step input is applied to a unity feedback system, we have that

$$V(s) = \frac{1}{s}$$

The steady-state error following a unit step input is then

$$sE(s)\,|_{s\to0} = \left[\frac{1}{1 + G_0(s)}\right]\Bigg|_{s\to0} = \frac{1}{1 + [G_0(s)]\,|_{s\to0}} \tag{3.4.8}$$

which is obtained from (3.4.5) and (3.4.6).

If $G_0(s)$ has any common s factors in its denominator the steady-state error to a step input will be zero, i.e.

$$e(t)\,|_{t\to\infty} = \frac{1}{1 + \infty} = 0$$

for $q > 0$, i.e. for a type > 0 system.

For a type 0 transfer function, however, the term

$$[G_0(s)]\,|_{s\to0} = K_s, \quad \text{a constant}$$

Example 3.4.3

Consider the system

$$G_0(s) = \frac{3}{(s + 2)}$$

The steady-state step error is given as

$$sE(s)\,|_{s\to0} = \frac{1}{1 + 1.5} = 0.4$$

where

$$G_0(s)\big|_{s\to0} = \left[\frac{3}{s+2}\right]\bigg|_{s\to0} = K_s = 1.5$$

It is important to note that if the step input is of a magnitude other than unity, the magnitude of the type 0 steady-state step error will be different from that calculated for a unit step input.

For a type 0 system, the steady-state step error can be made smaller by increasing K_s, which can be done by increasing the gain K, e.g. consider the system in Example 3.4.3, but with

$$G_0(s) = \frac{8}{(s+2)}$$

This would make

$$K_s = 4$$

and therefore

$$sE(s)\big|_{s\to0} = \frac{1}{1+4} = 0.2$$

i.e. a smaller steady-state error has been achieved.

Steady-state ramp error

If a unit ramp input is applied to a unity feedback system, we have that

$$V(s) = \frac{1}{s^2}$$

The steady-state error following a unit ramp input is therefore

$$sE(s)\big|_{s\to0} = \left[\frac{1}{1+G_0(s)}\cdot\frac{1}{s}\right]\bigg|_{s\to0} = \frac{1}{[sG_0(s)]}\bigg|_{s\to0}$$

For a type 0 system it is observed that

$$[sG_0(s)]\big|_{s\to0} = 0$$

such that the steady-state ramp error is ∞.
For a type 1 system however, we have

$$[sG_0(s)]\big|_{s\to0} = K_r, \quad \text{a constant}$$

such that the steady-state ramp error is $1/K_r$.

For a type > 1 system though

$$[sG_0(s)]\,|_{s\to0} = \infty$$

such that the steady-state ramp error is 0.

It can be noted that if the ramp input is of a magnitude other than unity, the magnitude of the type 1 steady-state ramp error will be different from that calculated for a unit ramp input.

Steady-state parabolic (acceleration) function

If a unit parabolic input is applied to a unity feedback system, we have that

$$V(s) = \frac{1}{s^3}$$

The steady-state error following a unit parabolic input is therefore

$$sE(s)\,|_{s\to0} = \left[\frac{1}{1 + G_0(s)} \cdot \frac{1}{s^2}\right]\Bigg|_{s\to0} = \frac{1}{[s^2 G_0(s)]}\,\Bigg|_{s\to0}$$

For a type < 2 system it is observed that

$$[sG_0(s)]\,|_{s\to0} = 0$$

such that the steady-state parabolic error is ∞.

For a type 2 system however, we have

$$[sG_0(s)]\,|_{s\to0} = K_a, \quad \text{a constant}$$

such that the steady-state parabolic error is $1/K_a$.

For a type > 2 system though

$$[sG_0(s)]\,|_{s\to0} = \infty$$

such that the steady-state parabolic error is 0.

If the parabolic input is of a magnitude other than unity, the magnitude of the type 2 steady-state parabolic error will be different from that calculated for a unit parabolic input.

The analysis carried out for the three classes of input can in fact be continued further if desired; however, a halt will be called here by summarizing the results obtained in Table 3.1.

In this section we have been concerned primarily with the steady-state error obtained from different system types in response to certain test inputs. In the following section the remaining part of the total error, namely the transient error, is considered.

Table 3.1 Steady-state errors

System type	Unit step input	Unit ramp input	Unit parabolic input
0	$\dfrac{1}{1 + K_s}$	∞	∞
1	0	$\dfrac{1}{K_r}$	∞
2	0	0	$\dfrac{1}{K_a}$
3	0	0	0

3.4.3   Transient errors

The transient response of a system was considered at length in Section 3.3, and as the results in this section are closely related, the analysis is not carried out to such a great depth here.

The transient error is the error which remains when the steady-state error is subtracted from the total error. If the total error between desired and actual outputs is

$$e(t) = v(t) - y(t)$$

then the transient error can be written, in the s-domain, as

$$\text{transient error} = E(s) - [sE(s)\,|_{s\to0}]\,(s) \tag{3.4.9}$$

Transient step error

For a type 0 system, the steady-state error following a unit step input is given by

$$sE(s)\,|_{s\to0} = \frac{1}{1 + K_s} = e(t)\,|_{t\to\infty}$$

with the step input transform $V(s) = 1/s$ applied. The transient error of a type 0 system in response to a step input is therefore

$$\text{transient error} = \frac{1}{1 + G_0(s)} \cdot \frac{1}{s} - \frac{1}{1 + K_s} \cdot \frac{1}{s}$$

in which the transform of the steady-state error has been taken, as shown in (3.4.9). The steady-state error $1/(1 + K_s)$ is simply a scalar value; this means that the denominator characteristics of the transient error transfer function are identical to those of the closed-loop transient response, see Problem 3.4. It follows then that the essential time domain characteristics of the type 0 system transient error following a step input will also be those of the transient response to a step input.

For a type > 0 system, the steady-state error following a unit step input is zero. The transient error is therefore also the total error, i.e.

$$\text{transient error} = \frac{1}{1 + G_0(s)} \cdot \frac{1}{s}$$

and as the closed-loop response to a step input is given by

$$Y(s) = \frac{G_0(s)}{1 + G_0(s)} \cdot \frac{1}{s}$$

it follows that as the denominator of the transient error is also that of the output response, the characteristics of the two responses will be essentially the same.

Transient ramp error

For a type 0 system the steady-state error following a unit ramp input is ∞. The transient error in such a case is therefore only of academic interest and is not really defined well in any useful sense, although it can be regarded as a special case of the following.

For a type 1 system the steady-state error following a unit ramp input is $1/K_r$. The transient error is then, by means of (3.4.9)

$$\text{transient error} = \frac{1}{1 + G_0(s)} \cdot \frac{1}{s^2} - \frac{1}{K_r} \cdot \frac{1}{s}$$

where $K_r \rightarrow 0$ for a type 0 system.

For a type > 1 system the steady-state error following a unit ramp input is zero. The transient error is therefore also the total error, i.e.

$$\text{transient error} = \frac{1}{1 + G_0(s)} \cdot \frac{1}{s^2}$$

A similar analysis can be carried out for a parabolic input.

In conclusion, it must be remembered that the transient error behavior of a system depends on a denominator which is the same as that for a closed-loop transform of the same system, the same input signal being assumed for both cases.

3.4.4   Disturbances

No matter how well the system design and construction phases are carried out and how much protection is introduced, a certain amount of interference caused by disturbances will affect any control system. In some cases it is negligible with respect to the noise-free

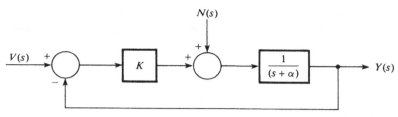

Fig. 3.13   Unity feedback system with disturbance, $N(s)$

(deterministic) part of the system operation and need not be considered further. For a large number of systems, though, it presents a major problem and can result in the total objective of the control design being to reduce the effect of noise on the output signal to an acceptable level.

It is usual for disturbances to be regarded as either (a), a steady nonzero value – these are called load disturbances or offsets – or (b), a random noise signal, which may be corrupted. Disturbances are however considered in a fairly general way in this section and designs particular to either (a) or (b) are not discussed.

Consider a feedback control system, with a selectable gain K, which is affected by the disturbance term $N(s)$, as shown in Fig. 3.13.

The closed-loop transfer function for this system is

$$Y(s) = \frac{K}{(s + \alpha + K)} \, V(s) + \frac{1}{(s + \alpha + K)} \cdot N(s) \tag{3.4.10}$$

which in fact consists of two separate transfer functions, one which relates the signal $V(s)$ to the output $Y(s)$ and the other which relates the noise $N(s)$ to the output $Y(s)$. The total output value is therefore a summation of the desirable signal with the undesirable noise.

Assume the desired input $V(s)$ is a step of magnitude V and the disturbance $N(s)$ is a load disturbance of magnitude N. We would hope that in the steady state the output signal would be equal to the desired value V. By the Laplace transform final value theorem, the steady-state output value is in fact given by:

$$sY(s)|_{s \to 0} = \left[\frac{KV}{(s + \alpha + K)} + \frac{N}{(s + \alpha + K)} \right] \Bigg|_{s \to 0}$$

Hence, if we make K a very large value this will satisfy both the steady-state final value $y \simeq V$ and noise reduction problems. However the $(s + \alpha + K)$ denominator term means that in the time domain the output signal will tend to its final value $\simeq V$ by means of an exponential increase, the time constant of which is $\tau = 1/(\alpha + K)$. In practice such a rapid response may not be possible and the resulting large signal at the output of the gain K may not be achievable.

One of the main reasons for employing a feedback loop in a control system is to reduce the effects of noise on the output signal, and it has been shown here how this is possible merely through the selection of an appropriate gain value, which at the same time reduces the steady-state step error in the example given. Increasing the gain can though cause problems, something which is easily seen if the first-order system example in Fig. 3.13 is replaced by a second-order system, for which an increase in K would result in oscillatory behavior such that for a very high value of K output signal oscillations would be wild and would take some time to settle down.

3.4.5 Sensitivity

When designing a controller for a system it is generally assumed that both the system model itself and the resulting controller description are accurate, exact representations of their real-world physical counterparts. This assumption ignores not only the errors inherent in the modeling and construction stages but also measurement problems, and the fact that either the system itself or its environment will have altered to some extent, since the system was modeled. One of the main reasons behind the employment of adaptive control schemes is the need to vary the control action in accordance with variations in the system characteristics. However, it is by no means always the case that an adaptive controller is necessary or even particularly desirable.

It is important to discover how sensitive system performance is to certain variations or inaccuracies in the parameters used to describe the system. For instance, if it is found that when one particular parameter varies by a few percent from its nominal value the output performance is degraded considerably, then it is sensible to redesign the controller in order to reduce the dependence of output performance on the accuracy of that parameter.

It is shown here how the use of feedback in the control of a system, reduces the sensitivity of that system to variations in the parameters which represent the system's characteristics. When investigating the sensitivity of a system though, the particular system performance that is being measured must be specified as must the parameter(s) which the sensitivity is considered in terms of. For simplicity, the following discussion restricts itself to the sensitivity of the output signal in respect of variations in the system transfer function. This can be summarized by means of the sensitivity function, S_G^Y, defined as

$$S_G^Y = \frac{\partial Y(s)/Y(s)}{\partial G(s)/G(s)} \tag{3.4.11}$$

which is simply the fractional change in output divided by the fractional change in system transfer function.

For an open-loop system, we have that the transfer function relating the input signal transform, $U(s)$, to the output signal transform, $Y(s)$, results in an equation

$$Y(s) = G(s)U(s)$$

from which it follows that for an invariant input signal $U(s)$

$$\frac{\partial Y(s)}{\partial G(s)} = U(s) \tag{3.4.12}$$

such that sensitivity function becomes

$$S_G^Y = \frac{\partial Y(s)/Y(s)}{\partial G(s)/G(s)} = \frac{G(s)U(s)}{G(s)U(s)} = 1 \tag{3.4.13}$$

which means that the output varies in a directly proportional relationship to variations in the open-loop transfer function — or, looked at in another way, if the system transfer function is only accurate to $\pm 10\%$ then so is the system output.

Consider now a closed-loop system, as shown in Fig. 3.12, with the desired reference input transform $V(s)$ and feedback path $H(s)$ considered as invariant. The original closed-loop transfer function is given as

$$Y(s) = \frac{G(s)}{1 + G(s)H(s)} V(s)$$

such that

$$\frac{\partial Y(s)}{\partial G(s)} = \frac{V(s)}{1 + G(s)H(s)} - \frac{H(s)G(s)V(s)}{[1 + G(s)H(s)]^2} = \frac{V(s)}{[1 + G(s)H(s)]^2} \tag{3.4.14}$$

and the sensitivity function becomes

$$S_G^Y = \frac{\partial Y(s)}{\partial G(s)} \cdot \frac{G(s)}{Y(s)} = \frac{V(s)}{[1 + G(s)H(s)]^2} \cdot \frac{[1 + G(s)H(s)]}{V(s)}$$

or

$$S_G^Y = \frac{1}{[1 + G(s)H(s)]} \tag{3.4.15}$$

Hence (3.4.15) shows that the sensitivity of a system, when in closed-loop, is reduced from its open-loop value by a factor $1/(1 + G(s)H(s))$. The selection of feedback $H(s)$ is therefore very important in determining the sensitivity of a closed-loop system to variations in the system transfer function.

If it is the case that the system transfer function is invariant, but the feedback function $H(s)$ varies, we have that

$$\frac{\partial Y(s)}{\partial H(s)} = - \left[\frac{G(s)}{1 + G(s)H(s)} \right]^2 V(s) \tag{3.4.16}$$

with a sensitivity function, dependent on variations in $H(s)$ rather than $G(s)$,

$$S_H^Y = \frac{\partial Y(s)}{\partial H(s)} \cdot \frac{H(s)}{Y(s)} = \frac{-G(s)H(s)}{[1 + G(s)H(s)]} \tag{3.4.17}$$

This means that when $|G(s)H(s)| \gg 1$, the sensitivity function magnitude is equal to 1 and thus the accuracy of the output will be directly dependent on the accuracy of the feedback path, while for the same conditions, from (3.4.15), it will be only weakly linked with the accuracy of the system transfer function. Although this places much more importance on the accuracy of the feedback path, it is usually a desirable feature in that precision and modifications can be much more readily built in to $H(s)$ rather than $G(s)$.

3.5   Servomechanisms

In order to draw together the motor/generator analysis carried out in the previous chapter with the differential system equation properties of damping and steady-state error introduced in this chapter, a position control system, commonly referred to as a servomechanism, is considered here, rather in the form of an extended example.

The reference input signal $V(s)$ and sensed output position $Y(s)$ are provided by rotational potentiometers with gains K_v and K_y respectively. The latter signal is subtracted from the former at a summing junction in order to obtain the error, $E(s)$, between the amplified desired and actual output signals. This error is then itself amplified in order to provide the armature current, $I_a(s)$, for an armature controlled motor, the output of which is the output position $Y(s)$, as shown in Fig. 3.14. The motor is assumed to have a constant field current and also to contain a gearbox placed between the motor and the load. Before considering the overall input-output servomechanism transfer function, the effect of the gearbox, regarded as being part of the motor in Fig. 3.14, will be discussed in detail.

3.5.1   Gears

The block described in Fig. 3.14 as the 'motor' is expanded in Fig. 3.15 to introduce the idea of a gear ratio n, which means that n revolutions of the motor produce 1 revolution of the load, i.e. $\omega_m = n\omega_l$.

The motor impedance is $Z_m(s) = B_m + J_m s$, in which B_m is the resistance of the motor and J_m is the motor inertia. Also the load impedance is $Z_l(s) = B_l + J_l s$, in which B_l is the resistance of the load and J_l is the load inertia.

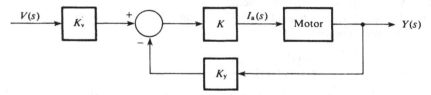

Fig. 3.14   Position control servomechanism

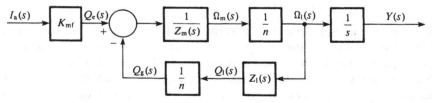

Fig. 3.15   Servomotor in open-loop

$\Omega_m(s)$ is the motor speed input to the gearbox and $\Omega_l(s)$ is the load speed output from the gearbox, also $Q_l(s)$ is the load torque output from the gearbox. Therefore the torque input to the gearbox $Q_g(s)$ must be such that for an ideal, lossless gearbox, power in = power out, that is

$$Q_g(s)\Omega_m(s) = Q_l(s)\Omega_l(s) \tag{3.5.1}$$

or

$$Q_g(s) = \frac{Q_l(s)}{n} \tag{3.5.2}$$

The total torque obtained from the motor, $Q_e(s)$ is in fact split up between that used on the motor and that input to the gearbox, i.e.

$$Q_e(s) = Z_m(s) \cdot \Omega_m(s) + Q_g(s) \tag{3.5.3}$$

or by substitution

$$Q_e(s) = \left[Z_m(s) + \frac{Z_l(s)}{n^2} \right] \Omega_m(s) \tag{3.5.4}$$

The effect of the gearbox is thus to reflect back the load impedance to combine with that of the motor, subject to the square of the gear ratio. Hence, the larger the gear ratio, the smaller the reflections.

This can also be regarded as

$$Q_e(s) = \left[nJ_m + \frac{J_l}{n} \right] \Omega_l(s) + \left[nB_m + \frac{B_l}{n} \right] \Omega_l(s) \tag{3.5.5}$$

or

$$Q_e(s) = Q_a(s) + Q_s(s)$$

in which

$$Q_a(s) = \left[nJ_m + \frac{J_l}{n} \right] s\Omega_l(s) \tag{3.5.6}$$

is the torque required in order to accelerate the load, whereas $Q_s(s)$ is that required in order to keep the speed of the load constant.

When selecting the gear ratio n for a particular servomechanism, a criterion often employed is that of obtaining n such that $Q_a(s)$ is a minimum, and this value of n is then referred to as the ideal gear ratio.

Now the torque required to accelerate the load, $Q_a(s)$, when differentiated gives

$$\frac{dQ_a(s)}{dn} = J_m - \frac{J_l}{n^2}$$

and this must be equal to zero for a minimum of $Q_a(s)$, hence

$$n = \sqrt{\frac{J_l}{J_m}} \tag{3.5.7}$$

It can be noted that:

(a) In the selection of n, to minimize $Q_a(s)$, the torque $Q_s(s)$ is disregarded completely;

(b) It is easy to show, by taking the second differential of $Q_a(s)$ with respect to n, that the value of n given in (3.5.7) results in a minimum value of $Q_a(s)$ rather than a maximum.

3.5.2   Servomechanism transient response

The relationship between torque $Q_e(s)$ and load speed $\Omega_l(s)$ can be viewed as that in the servomotor block diagram, Fig. 3.15, or in terms of the relationships (3.4.4) and (3.5.5) which reduce the servomotor block diagram to that in Fig. 3.16, where the impedance of the load is referred to that of the motor, such that

$$B_r = B_m + B_l/n^2$$

and   (3.5.8)

$$J_r = J_m + J_l/n^2$$

By cascading the blocks shown in Fig. 3.16, we have that

$$Y(s) = \frac{K_{mf}}{ns(B_r + J_r s)} \cdot I_a(s) \qquad (3.5.9)$$

and on substitution for $I_a(s)$, shown in Fig. 3.14 to be

$$I_a(s) = K[K_v V(s) - K_y Y(s)] \qquad (3.5.10)$$

this results in an overall closed-loop equation relating desired output value (reference input) $V(s)$ to actual output $Y(s)$ of

$$Y(s) = \frac{K_{mf}KK_v}{ns(B_r + J_r s) + K_{mf}KK_y} \cdot V(s)$$

or

$$Y(s) = \frac{K_{mf}KK_v/J_r n}{s^2 + \dfrac{B_r}{J_r}s + \dfrac{K_{mf}KK_y}{J_r n}} \cdot V(s) \qquad (3.5.11)$$

By referring to the transient response forms considered in Section 3.3 it is possible to set the transient response of the servomechanism considered, by an appropriate choice of

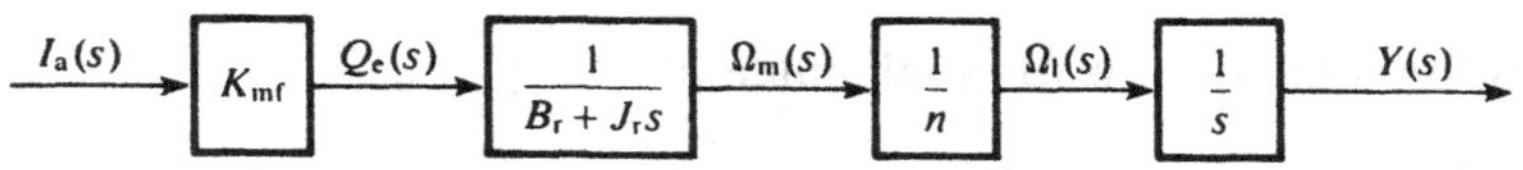

Fig. 3.16   Servomotor with reflected load impedance

the potentiometer/amplifier gains. When $V(s)$ is a step input, i.e. when a desired output position value is asked for, the transient conditions which occur, will be those selected. Consider for example the case of ideal damping discussed in Section 3.3.3. The denominator of (3.3.12) is of the form

$$(s + \alpha)^2 + \alpha^2$$

and if the servomechanism control parameters are to be chosen with this in mind it is required that the denominator of (3.5.11) should also be of this form, i.e.

$$(s + \alpha)^2 + \alpha^2 = s^2 + 2\alpha s + 2\alpha^2$$

such that from (3.5.11)

$$2\alpha = \frac{B_r}{J_r}$$

and

$$2\alpha^2 = \frac{K_{mf} K K_y}{J_r n} = \frac{1}{2} \left[\frac{B_r}{J_r} \right]^2$$

Assuming B_r, J_r, n, K_{mf} and K to have been previously set, we therefore must set the feedback potentiometer gain K_y to

$$K_y = \frac{B_r^2 n}{2 J_r K_{mf} K}$$

Steady-state conditions following a unit step input, $V(s) = 1/s$, can be investigated by means of the final value theorem, whereby from (3.5.11)

$$\lim sY(s)|_{s \to 0} = \frac{K_v}{K_y} = \lim_{t \to \infty} y(t) \qquad (3.5.12)$$

such that the output $y(t)$ will follow its required value, $v(t)$, in the steady-state, as long as the input potentiometer gain, K_v, is equal to that of the feedback potentiometer, K_y.

3.6   Summary

It was shown in this chapter how some fairly simple input signals can be used to test the response of a system and hence to investigate some of the fundamental properties of that system. Following the application of an input signal, the system response can be considered to consist of two separate parts, the steady-state response which remains after any initial response fluctuations have died down, i.e. as $t \to \infty$, and the transient response which is the name given to those initial response fluctuations and although these can persist for quite some time after the input signal is applied, it is considered that they die down to zero as $t \to \infty$.

The analysis of system steady-state and transient responses carried out made use of the fact that once a system transfer function is obtained in terms of the Laplace operator s, on the application of an input signal, also considered in terms of its s-domain representation, it is not necessary to take inverse transforms in order to investigate the time domain properties of the response. Both the steady-state response and transient response can be discussed with reference to s-domain operators only, via the final value theorem and in terms of the degree of damping respectively.

Some of the main reasons for control system error were considered in Section 3.4. Firstly, the importance of system type and the input applied were shown in terms of their effect on the steady-state error between the desired and actual output signals. Secondly, some of the useful properties obtained from the employment of feedback in a control system were highlighted by showing how disturbance corruption and performance sensitivity can be reduced to an acceptable level simply by a suitable selection of feedback terms.

Many of the fundamental performance requirements in the design of a control system have been introduced in this chapter, and these are made use of in the chapters which follow.

Problems

3.1   A unity feedback closed-loop system contains an open-loop transfer function

$$G(s) = \frac{9}{s(s+5)}$$

A factor $D(s) = Ks + 1$ is cascaded with $G(s)$, where K is an adjustable gain, as shown in Fig. 3.17.

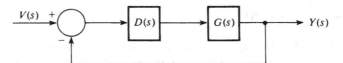

Fig. 3.17

(a)   Find the value of K necessary to provide a critically damped closed-loop system.
(b)   Find the range of values of K which will provide acceptable damping for the closed-loop system.
(c)   What is the magnitude of overshoot which follows a unit step input, when $K = 0.1$?

3.2   A control system with unity feedback has an open-loop transfer function:

$$G(s) = \frac{K(s+12)}{s^2 + 25}$$

where K is a scalar gain.

Calculate the closed-loop denominator and find the range of values for K over which the closed-loop poles are real. With $K = 24$, find the step response of the closed-loop system and evaluate the steady-state error which follows a step response.

3.3   For the control system shown in Fig. 3.18, consider how derivative feedback, with gain K, can be employed to improve the transient performance.

Find the value of K for which the transient response is ideal and the steady-state and transient step errors in this case.

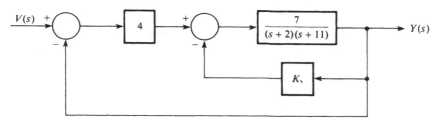

Fig. 3.18

3.4   Show that the denominator of a type 0 system's transient error following a unit step input is also the denominator of the system's closed-loop transient response. By using the open-loop transfer function:

$$G(s) = \frac{K}{(s + 1)(s + 2)}$$

show that (a) the steady-state error to a step input is $K_s = K/2$, and (b) a scalar multiple in the denominator does not affect the answer.

3.5   Given an open-loop, type 1 system:

$$G(s) = \frac{K}{s(s + \alpha)}$$

show that if $K = 100$ and $\alpha = 10$, this will result in a transient step error with damping ratio $\zeta = 0.5$ and natural frequency $\omega_n = 10$. Also, show that the steady-state ramp error which is produced by such a choice is 10%.

3.6   Consider the unity feedback closed-loop system with an open-loop transfer function:

$$G(s) = \frac{K}{s + 3}$$

Find the sensitivity function S_K^Y which relates the sensitivity of the output to variations in the gain K.

Show how this function is modified if a feedback gain of $H(s) = 200$ is employed, rather than $H(s) = 1$ in the unity feedback case.

3.7   For the feedback control system with open-loop transfer function

$$G(s) = \frac{K}{s(s + 20)}$$

and feedback $H(s) = K_1 s + 1$, find the gain value K and amount of derivative feedback K_1 for which the system has ideal transient performance and a ramp error of 5%.

3.8   For a unity feedback system with open-loop transfer function

$$G(s) = \frac{K}{s(\beta s + 1)}$$

show that for the closed-loop system, the damping ratio is

$$\zeta = \frac{1}{2\sqrt{K\beta}}$$

and the natural frequency

$$\omega_n = \sqrt{\frac{K}{\beta}}$$

3.9   A unity feedback closed-loop system contains an open-loop transfer function:

$$G(s) = \frac{K}{s(s + 25)}$$

Find the value of K which will provide a steady-state ramp error of 5%, i.e. $K_r = 20$. For this value of K obtain (a) the steady-state error to a unit step input, and the transient errors produced by (b) a unit step input, and (c) a unit ramp input.

3.10   Find the steady-state parabolic error for the unity feedback closed-loop system whose open-loop transfer function is

$$G(s) = \frac{2.5}{s^2(2s + 1)}$$

3.11   Find the steady-state step and ramp errors for the unity feedback closed-loop system whose open-loop transfer function is

$$G(s) = \frac{s + 4}{s(s + 5)}$$

3.12   Consider the system shown in Fig. 3.19, where $N(s)$ is an unwanted noise signal and

$$G(s) = \frac{1}{s(s + 1)} \qquad D(s) = \frac{K(s + 1)}{(s + \alpha)}$$

Determine values for K and α which will achieve a critically damped output

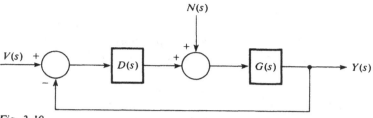

Fig. 3.19

response to a unit step input $V(s)$, and a disturbance effect on the output signal equal to 1/10 that of a step signal $N(s)$.

3.13   A unity feedback closed-loop system contains an open-loop transfer function:

$$G(s) = \frac{0.03K}{s^2 + 0.35s + 0.03}$$

What value of K will result in:

(a)   Steady-state step error $= 0.02$?;
(b)   Ideally damped transient step error?;

For $K = 20$, what are the steady-state and transient errors if a unit step input signal is applied?

3.14   Consider the system diagram shown in Problem 3.12, in which $G(s) = 1/Js$ and $N(s)$ is a ramp signal of magnitude 0.5 units. It is required that the steady-state signal value is equal to 25 units $\pm 2\%$.

If

$$D(s) = \frac{K(s + \alpha)}{s}$$

find values for K, α and input signal, $V(s)$, step magnitude, in order to provide a damping ratio $\zeta = 0.5$.

3.15   For a particular armature current controlled servomechanism, the error between the reference position and sensed output position is amplified by a gain of K Nm/rad to achieve the total torque obtained from the motor. By reflecting back the load impedance to combine with that of the motor, the total inertia is J kgm^2 and the total resistance R Nm/rad/s with a gearbox ratio $n:1$.

Find the range of values for K over which acceptable damping is obtained.

3.16   Consider a unity feedback closed-loop system in which the open-loop transfer function is

$$G(s) = \frac{4(s + 3)K}{s(s^2 + 4s - 4)}$$

If $K = 10$, find the steady-state error to a ramp input of magnitude 0.3. If $K = 0.5$, find a compensator $D(s)$ which when connected in cascade with $G(s)$ will produce a damping ratio $\zeta = 1.75$ and a natural frequency $\omega = 1.25$ rad/s.

3.17   An armature current controlled servomechanism is subject to a total inertia of 4 kgm^2 and a total resistance of 3 Nm/rad/s with a gearbox ratio 1:1. The error between desired position and sensed output position is amplified by a gain K Nm/rad, where $0 < K < 10$. Find a value for K which results in a suitable damping.

3.18   A field current controlled servomechanism is subject to a total inertia of 15 kgm^2 and a total resistance of 2 Nm/rad/s (referred to the motor) with a gearbox ratio 10:1. Find a value for the amplifier gain K Nm/rad which results in ideal damping.

3.19   A field voltage controlled servomechanism is subject to a field resistance of 2 Ω and a field inductance of 4 H, with a motor resistance of 1 Nm/rad/s and a motor inertia of 2 kgm^2. The constant relating torque to field current is 25 Nm/A, and the load has a resistance of 10 Nm/rad/s with an inertia 15 kgm^2. Find the output speed following a 10 V step input.

3.20   An armature voltage controlled servomechanism is subject to a motor resistance of 0.2 Nm/rad/s and a motor inertia of 0.4 kgm^2, with a load resistance of 10 Nm/rad/s and a load inertia of 120 kgm^2. The motor armature has a resistance equal to 3 Ω and an inductance equal to 6 H, the field current is a constant 2.5 A and the electromechanical coupling constant is 25 Nm/A^2.

Find the transfer function which relates output load speed to input armature voltage, when the ratio of the gearbox is ideal.

Further reading

Bower, J. L. and Schultheiss, P. M., *Introduction to the design of servomechanisms*, Chapman and Hall, 1958.

Brown, G. S. and Campbell, D. P., *Principles of servomechanisms*, Chapman and Hall, 1950.

Chestnut, H. and Mayer, R. W., *Servomechanisms and regulating system design*, Wiley (Vols 1 and 2), 1951.

D'Azzo, J. J. and Houpis, C. H., *Linear control system analysis and design*, McGraw-Hill, 1983.

DiStefano, J. J., Stubberud, A. R. and Williams, I. J., *Feedback and control systems*, McGraw-Hill (Schaum), 1976.

Doebelin, E. O., *Control system principles and design*, John Wiley & Sons, 1985.

Dorf, R. C., *Modern control systems*, 6th ed., Addison-Wesley, 1992.

Elloy, J.-P. and Piasco, J.-M., *Classical and modern control*, Pergamon Press, 1981.

Frank, P. M., *Introduction to system sensitivity theory*, Academic Press, 1978.

Healey, M., *Principles of automatic control*, Hodder and Stoughton, 1981.

Kuo, B. C., *Automatic control systems*, 3rd ed., Prentice-Hall, 1990.

Ogata, K., *System dynamics*, Prentice-Hall, 1978.

Oppenheim, A. V. and Willsky, A. S., *Signals and systems*, Prentice-Hall, 1983.

Power, H. M. and Simpson, R., *Introduction to dynamics and control*, McGraw-Hill, 1978.

Raven, F. H., *Automatic control engineering*, 2nd ed., McGraw-Hill, 1990.

Richards, R. J., *An introduction to dynamics and control*, Longman, 1979.

Sinha, N. K., *Control systems*, Wiley Eastern, 1994.

4

System stability and performance

4.1   Introduction

The problem of control system design is one of specifying a number of performance objectives which it is required that the controlled system must meet. The vast majority of objectives are plant dependent, e.g. a certain automobile is required to travel at 150 km/h; they can usually be posed, however, in a way which is more generally applicable – in this case the system output (speed) must be able to reach a specified level (150 km/h). Many different objectives exist, even in the general sense, those commonly encountered being dealt with in this book.

One of the most important requirements for a control system is that it be stable. This means that if a bounded input or disturbance (i.e. no bigger than some finite value) is applied to a linear, time-invariant, system, if it is a stable system then its output will also be bounded. System stability can be looked at in another way by considering the application of an impulse function at the system input. If the response of a system to an impulse input approaches zero as time $t \rightarrow \infty$ then the system is stable, conversely if its response tends to infinity magnitude as $t \rightarrow \infty$ then it is unstable. There is, however, a third possibility in that a system response to an impulse might tend to some finite but nonzero value as $t \rightarrow \infty$, in this case the system is said to be critically or marginally stable, the same definition applying were the system output to oscillate between two finite values – the oscillations neither increasing nor decreasing with respect to time. Certain bounded inputs when applied to a critically stable system can produce an unbounded output signal and therefore critically stable systems should not be regarded in the same way as stable systems, although it must be remembered that they should also not be regarded as unstable – merely as a special case.

Consider a system transfer function in terms of numerator roots (zeros) and denominator roots (poles), such that

$$G(s) = \frac{K(s + z_1) \ldots (s + z_m)}{(s + p_1) \ldots (s + p_n)}$$

where K is a scalar gain and the poles are either real or occur in complex-conjugate pairs.

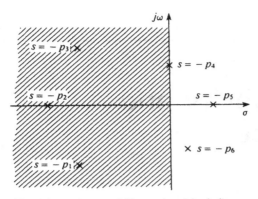

Fig. 4.1 s-plane stability region (shaded)

If the system output $Y(s)$ is related to the system input $U(s)$ by means of

$$Y(s) = G(s) \cdot U(s)$$

then when an impulse is applied at the input, $U(s) = 1$, we have that:

$$y(t) = \sum_{i=1}^{n} K_i e^{-p_i t}$$

further terms being necessary where multiple poles exist. It follows that if any one pole real part is negative, the output will tend to infinity irrespective of the remaining poles. Further, if all pole real parts are positive then the output will tend to zero as $t \to \infty$. A positive real part for a particular pole p_i means that the denominator root $s = -p_i$ will lie in the left half of the s-plane, as shown in Fig. 4.1. In Fig. 4.1 $s = -p_1$, $s = -p_2$ and $s = -p_3$ are all within the stability region; however, if a pole existed at either $s = -p_5$ or $s = -p_6$ then the system which contained such a pole would be unstable.

It does not matter how many poles of a system lie in the left half of the s-plane, if one or more lie in the right half then that system is unstable.

Notes:

1. If the poles of a system transfer function all lie in the left half of the s-plane then that system is stable.
2. A system is critically stable if one or more poles lie on the imaginary axis of the s-plane, subject to no poles lying in the right half of that plane. (The pole at $s = -p_4$ in Fig. 4.1 is an example.)
3. It is only the poles of a system transfer function which are important as far as stability is concerned, system zeros are irrelevant (as long as no zeros are exactly equal to and therefore cancel with any right-half plane poles).
4. The poles of a system are the polynomial roots obtained when the system denomi-

nator is equated with zero. The system denominator is known as the *characteristic polynomial* and when it is equated with zero it is known as the *characteristic equation*.

5. The concept of stability is applicable to both open- and closed-loop systems. For a particular system, if it is open-loop stable, this does not necessarily mean that it is also closed-loop stable, furthermore if it is open-loop unstable this does not necessarily mean that it is also closed-loop unstable.

If a system is to be tested to find out whether or not it is stable, this can be done by finding the roots of the system's characteristic equation, although this will, in most cases, be difficult for other than low-order systems. It is shown in Section 4.2 that it is in fact not necessary to find the characteristic equation roots when testing for stability and that a technique exists for testing a polynomial to ascertain whether or not there are any roots which do not lie in the left half of the *s*-plane. The Routh–Hurwitz testing criterion merely requires the formation of a table from the characteristic polynomial coefficients, and this procedure can also be used when a system is stable to find out just how stable it is by testing to see whether or not any roots which lie in the left half of the *s*-plane are to the right of a specified real value $s = -\alpha$; this concept is known as *relative stability*.

The location of system poles in the *s*-plane gives an indication not only of the stability and relative stability of that system but also of its transient performance. If it is intended to control the system in a closed-loop form then it is useful to see how the closed-loop characteristic equation roots move around in the *s*-plane as certain controller parameters are varied. The movement of each root is called the locus of that root and gives a graphical representation when plotted in the *s*-plane. In this chapter it is shown how the loci of characteristic equation roots can be plotted simply by means of a set of readily applicable rules to follow. It is assumed that the gain K is the coefficient which can be varied and hence it is the variation of K which causes the loci. Further, it is assumed that the system $G(s)$ is employed in a unity feedback system; this means that if the feedback $H(s)$ is other than unity then the equivalent forward transfer function within a unity feedback system must be obtained by algebraic means or block manipulations before proceeding. However, as it is only the closed-loop denominator which is of importance, this means that it is only necessary to cascade $H(s)$ with $G(s)$, i.e. obtain $G(s)H(s)$, and to proceed as though this were the original open-loop system.

4.2   The Routh–Hurwitz criterion

The stability of a system, whether it be open-loop or closed-loop, can be investigated by inspection of the transfer function denominator roots (poles). However, the roots may well not be directly available, the denominator being represented simply in terms of an *n*th order polynomial. If this is so and *n* is a fairly low value then the poles can be found without too much computational effort, however for values of *n* greater than five no direct factoring methods exist and obtaining even approximate roots can be an extremely time-consuming process.

The Routh–Hurwitz criterion presents a method by which the stability of a linear, time-invariant system may be investigated without the need to calculate the transfer function poles. Further, the method consists entirely of an algebraic technique and hence no inexact graphical stages are required. The procedure is in fact applicable to any general polynomial which contains real coefficients and it merely tests for roots which lie in the right half of the s-plane; in terms of a transfer function denominator this means that the procedure tests to see if the system is unstable.

In order to find the poles of a transfer function, the denominator is equated to zero, hence realizing the characteristic equation. Thus we have that

$$s^n + a_{n-1}s^{n-1} + a_{n-2}s^{n-2} + \cdots + a_2s^2 + a_1s + a_0 = 0 \qquad (4.2.1)$$

in which $\{a_i: i = 0, \ldots, n-1\}$ are constant real numbers. An assumption has been made in (4.2.1) in that if initially there is a constant real term a_n associated with s^n, then both sides of the characteristic equation must be divided through by this factor in order to obtain the normalized form given in (4.2.1): note that the coefficient a_n could be positive or negative, but by definition must be nonzero. Another point worth making is that the transfer function for which the polynomial in (4.2.1) is the denominator is assumed to be proper, i.e. there are assumed to be no exact common factors between the numerator and denominator. The point is only important if there are any transfer functions zeros which lie in the right half of the s-plane or on the imaginary axis.

An initial test by inspection can be carried out on the polynomial of (4.2.1), and this is that all coefficients of the polynomial must be positive definite, i.e.

$$\{a_i > 0: i = 0, \ldots, n-1\}$$

If this test fails and either at least one coefficient is non-existent, i.e. it is zero, and/or at least one coefficient is negative, this means that roots of the polynomial exist either in the right half of the s-plane or on the imaginary axis and the system is then considered not to be stable.

If the polynomial of (4.2.1) is found to pass the first test then this does not mean that there are no roots which lie in the right-half plane, it merely means that we must also carry out the full Routh–Hurwitz stability test as described in the next section. Where a polynomial has been found to fail the initial test, however, it has been shown that the system is not stable (although this could mean it is critically stable), and there is therefore no need to apply the full test.

4.2.1   Routh–Hurwitz stability test

The coefficients of the characteristic equation shown in (4.2.1) are rewritten to form the first two rows of the ROUTH ARRAY, which is defined in full as follows:

$$
\begin{array}{c|cccc}
s^n & 1 & a_{n-2} & a_{n-4} & \cdots \\
s^{n-1} & a_{n-1} & a_{n-3} & a_{n-5} & \cdots \\
s^{n-2} & b_1 & b_2 & b_3 & \cdots \\
s^{n-3} & c_1 & c_2 & c_3 & \cdots \\
\vdots & \vdots & \vdots & \vdots & \cdots \\
s^1 & \cdot & \cdot & \cdot & \cdots \\
s^0 & \cdot & \cdot & \cdot & \cdots \\
\end{array}
$$

It is of no consequence whether the a_0 term actually appears on the top or next to top row.

The remaining rows of the array are constructed by

$$b_1 = a_{n-2} - \frac{1}{a_{n-1}} \cdot a_{n-3}, \qquad b_2 = a_{n-4} - \frac{1}{a_{n-1}} \cdot a_{n-5}, \text{ etc.}$$

$$c_1 = a_{n-3} - \frac{a_{n-1}}{b_1} \cdot b_2, \qquad c_2 = a_{n-5} - \frac{a_{n-1}}{b_1} \cdot b_3, \text{ etc.}$$

So, the coefficients of each row depend only on the coefficients in the two rows immediately above. Elements in the array are calculated until only zeros appear; in the vertical direction the array should then contain $(n + 1)$ rows in total.

The Routh–Hurwitz criterion can be stated as follows:

(a)   The roots of the characteristic equation all lie in the left-hand side of the s-plane if all the elements in the first column of the Routh array are positive definite.

(b)   If there are any negative elements in the first column, then if we start at the top of the array and work to the bottom by comparing successive elements in the first column, the number of sign changes indicates the number of roots which lie in the right-hand side of the s-plane.

Note: The most important aspect of the Routh–Hurwitz criterion is that the system is stable if there are no sign changes in the first column of the Routh array.

Example 4.2.1

Consider the polynomial equation

$$s^4 + 6s^3 + 11s^2 + 6s + 15 = 0$$

The Routh array is

$$
\begin{array}{c|ccc}
s^4 & 1 & 11 & 15 \\
s^3 & 6 & 6 & \\
s^2 & 10 & 15 & \\
s^1 & -3 & & \\
s^0 & 15 & & \\
\end{array}
$$

A system with such a denominator polynomial is therefore an unstable system because of the negative sign in the first column of the Routh array. Further, as there are two sign changes, i.e. $+10 \rightarrow -3 \rightarrow +15$, there are two roots which lie in the right-hand side of the s-plane.

Example 4.2.2

Consider the polynomial equation

$$s^3 + 6s^2 + 11s + 6 = 0$$

The Routh array is

s^3	1	11
s^2	6	6
s^1	10	
s^0	6	

A system with such a denominator polynomial is therefore a stable system because there are no sign changes in the first column of the Routh array.

4.2.2 Special cases

In the analysis carried out in the previous section the problem of a zero term appearing in the first column was completely avoided. If a zero does appear it is then not possible to continue forming elements of the array because all elements of the next row are necessarily infinity by means of the definitions given. However, when a zero appears in the first column there are two possibilities, either the remaining terms in that particular row will all be zero or at least one of them will be nonzero. The latter case will be dealt with first.

Consider a polynomial for which a zero appears in the first column of the Routh array, although the remainder of the row in which the zero appears does not consist entirely of zeros. The original characteristic polynomial should then be multiplied by a factor $(s + \alpha)$ in which the real constant $\alpha > 0$, however it must be checked that for any particular selection of α the original polynomial does not contain a factor $(s - \alpha)$. The new polynomial, with a factor $(s + \alpha)$ should then be tested for stability via the Routh array. Normally a selection such as $\alpha = 1$ is perfectly suitable, although one must be certain that the original polynomial does not then contain a factor $(s - 1)$, something which is very easy to test.

Example 4.2.3

Consider the polynomial equation:

$$s^4 + 3s^3 + 2s^2 + 6s + 2 = 0$$

The first three rows of the Routh array are

s^4	1	2	2
s^3	3	6	
s^2	0	2	

and so a zero has appeared in the first column. Let us choose the multiplying factor $s + \alpha = s + 1$, after first checking in the equation above that $s = 1$ is not a factor of the original polynomial.

It follows that

$$(s^4 + 3s^3 + 2s^2 + 6s + 2)(s + 1) = s^5 + 4s^4 + 5s^3 + 8s^2 + 8s + 2 = 0$$

The full Routh array, formed from the new polynomial, is then:

s^5	1	5	8
s^4	4	8	2
s^3	3	7.5	
s^2	-2	2	
s^1	10.5		
s^0	2		

A system with such a denominator polynomial is therefore an unstable system because of the negative sign in the first column of the Routh array. Further, as there are two sign changes, i.e. $+3 \rightarrow -2 \rightarrow +10.5$, there are two roots which lie in the right-hand side of the s-plane. The multiplying factor merely consisted of a single root at $s = -1$, i.e. in the left-hand side of the s-plane, both of the roots in the right-hand side of the s-plane are therefore roots of the original polynomial.

A second special case exists when all the elements in a row of the Routh array are zero. This is caused by pairs of complex-conjugate roots, in which the real part is not necessarily nonzero, and/or by pairs of real roots with opposite signs.

When a row of zeros occurs, an auxiliary polynomial must be formed with the coefficients in the row of the array immediately above the row of zeros. The coefficient in the first column is associated with the power of s which denotes that particular row; this will always be even. The second coefficient is then associated with the power of s two less than that denoting the row, and so on until the final coefficient in the row is associated with s^0. The auxiliary polynomial is in fact a factor of the original polynomial and its order indicates the number of root pairs.

The procedure to be carried out when a row of zeros occurs is (a) form the auxiliary polynomial, (b) divide the original polynomial by the auxiliary polynomial and (c) test the remaining polynomial by means of the Routh array.

Example 4.2.4

Consider the polynomial equation

$$s^6 + 2s^5 + 4s^4 + 2s^3 + 7s^2 + 8s + 12 = 0$$

The first four rows of the Routh array are

s^6	1	4	7	12
s^5	2	2	8	
s^4	3	3	12	← Auxiliary polynomial coefficients
s^3	0	0		← Row of zeros

and so a row of zeros has appeared.

The auxiliary polynomial is formed, from the coefficients of the row directly above the row of zeros, as:

$$\text{Auxiliary polynomial} = 3s^4 + 3s^2 + 12 = 3(s^4 + s^2 + 4)$$
$$= 3(s^2 + \sqrt{3}s + 2)(s^2 - \sqrt{3}s + 2)$$

So $s^4 + s^2 + 4$ is a factor of the original polynomial. If this auxiliary polynomial is divided into the original we obtain a new equation

$$s^2 + 2s + 3 = 0$$

i.e.

$$(s^2 + 2s + 3)(s^4 + s^2 + 4) = s^6 + 2s^5 + 4s^4 + 2s^3 + 7s^2 + 8s + 12$$

The 'new' remainder polynomial, $s^2 + 2s + 3$, can now be tested in the Routh array:

s^2	1	3
s^1	2	
s^0	3	

which shows the remainder polynomial to be stable with roots which all lie in the left-hand side of the s-plane.

Note: When forming the remainder polynomial, any common multiplying constant in the auxiliary polynomial — in example 4.2.4 there was a factor 3 — can be neglected.

4.2.3 Relative stability

Employment of the Routh–Hurwitz criterion as considered thus far merely tells us whether or not a polynomial has any roots which do not lie in the left-hand side of the s-plane, i.e. whether the system, for which the polynomial is its transfer function denominator, is stable or not. It is often useful, however, with a stable system, to find out how close the poles are to the imaginary axis, i.e. to find out how far away the system is from being unstable, without actually solving for the roots.

Introducing the constant real value β, the concept of relative stability can be viewed as one in which the characteristic equation is tested to find out whether or not any roots lie to the right of an imaginary axis which passes through the point $s = -\beta$, where $\beta \geqslant 0$, as shown in Fig. 4.2.

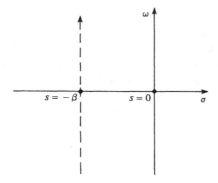

Fig. 4.2 Imaginary axis shifting

Testing of the characteristic equation by means of the Routh array can in fact be carried out if the substitution $s = r - \beta$ is made, such that the equation becomes one in factors of r rather than s. The point $s = -\beta$ then occurs when $r = 0$. The method of testing for relative stability by employing the Routh array is shown by the following illustrative example.

Example 4.2.5

Consider the polynomial equation:

$$s^3 + 5.5s^2 + 8.5s + 3 = 0$$

The Routh array formed from this polynomial is

s^3	1	8.5
s^2	5.5	3
s^1	175/22	
s^0	3	

A system with such a denominator polynomial is therefore a stable system because there are no changes of sign in the first column of the Routh array.

The imaginary axis is now shifted from $s = 0$ to $s = -1$ by making the substitution $s = r - 1$, the new polynomial equation is

$$r^3 + 2.5r^2 + 0.5r - 1 = 0$$

and although it is now apparent by the negative sign in the polynomial that at least one root of this polynomial lies to the right of the point $r = 0$, the Routh array will be formed for completeness.

r^3	1	0.5
r^2	2.5	-1
r^1	0.9	
r^0	-1	

From the Routh array it is apparent that as only one change of sign occurs in the first column, so only one root of the polynomial lies to the right of the $r = 0$ $(s = -1)$ point. Also, from the previous Routh array it was shown that no roots exist to the right of the $s = 0$ point, thus the root in question must lie between the $s = -1$ and $s = 0$ points.

4.3   Root loci construction

The Routh–Hurwitz criterion is a method for testing the stability of a system by investigating merely the system transfer function denominator. Further, the system transient response analysis carried out in Chapter 3 also concentrated specifically on the transfer function denominator. The importance of this polynomial is highlighted again in the remaining sections of this chapter, although in a different vein from the previous section; it is the position in the s-plane of the individual roots that is central to the discussion.

It was shown in the previous chapter how the transient performance of a system is dependent on the poles of the system's transfer function. When all system and controller parameters are fixed, a particular response pattern will always be obtained. If, however, as is often the case, at least one parameter such as an amplifier gain can be varied, it is of great interest to see how the poles of the transfer function vary as the parameter varies, and hence how the performance of the system is affected by certain values of that parameter. Looked at in a slightly different way one can investigate the effect of parameter variations, such as those due to ageing or temperature drift, by considering the resulting movement of poles. The movement of poles, or more generally roots of a polynomial, about the s-plane can be shown in a graphical sense by means of a locus drawn for each root.

The root locus method involves plotting the s-plane loci of polynomial roots caused by variations in a particular parameter. The technique gives important information with respect to system stability and performance without the need to continually solve the characteristic equation to find the root positions for a large number of values of the varying parameter.

Although the root locus procedure is appropriate for the investigation of root positions when one parameter of any general polynomial is varying, in this chapter we will restrict ourselves to the case of the denominator roots belonging to a unity feedback closed-loop system. Consider then such a feedback system with an open-loop transfer function:

$$G(s) = K \frac{B(s)}{A(s)} = \frac{K(s^m + b_{m-1}s^{m-1} + \cdots + b_1 s + b_0)}{s^n + a_{n-1}s^{n-1} + \cdots + a_1 s + a_0} \tag{4.3.1}$$

where K is an adjustable gain value.

The closed-loop transfer function obtained when $G(s)$ is placed within a unity feedback system is given by

$$\frac{G(s)}{1 + G(s)} = \frac{KB(s)}{A(s) + KB(s)} \tag{4.3.2}$$

Hence the closed-loop denominator is $A(s) + KB(s)$, such that the characteristic equation of this feedback system can be written as

(a) $A(s) + KB(s) = 0$

or $\tag{4.3.3}$

(b) $1 + G(s) = 0$

The position in the s-plane of the roots obtained as a solution to (4.3.3) will therefore vary as the gain K is varied. A plot of the locus of each root as K varies then forms the basis of the root locus analysis method.

It is readily observed from (4.3.3(a)) that the roots of the characteristic polynomial will tend to those of the open-loop denominator, $A(s)$, as $K \to 0$, whereas they will tend to those of the open-loop numerator, $B(s)$, as $K \to \infty$. This initial result in general forms a starting point for all root loci plots, in that for a particular closed-loop system the root loci will commence at the open-loop poles and end at the open-loop zeros, as K varies from zero to infinity.

It is also apparent from (4.3.3) that as

$$G(s) = K \frac{B(s)}{A(s)} = -1$$

it follows that

$$|G(s)| = 1 \tag{4.3.4}$$

and for all $K > 0$

$$\underline{/G(s)} = (2k + 1)\pi \tag{4.3.5}$$

in which

$$k = 0, \pm 1, \pm 2, \ldots$$

whereas if $K < 0$

$$\underline{/G(s)} = 2k\pi \tag{4.3.6}$$

In terms of the poles and zeros of the transfer function, the definitions (4.3.4), (4.3.5) and (4.3.6) can be reconsidered by letting

$$G(s) = \frac{K(s + z_1) \ldots (s + z_m)}{(s + p_1) \ldots (s + p_n)} \tag{4.3.7}$$

in which $\{z_i: i = 1, \ldots, m\}$ are the open-loop zeros and $\{p_i: i = 1, \ldots, n\}$ are the open-loop poles, all of which are either real or occur in complex-conjugate pairs.

The magnitude condition (4.3.4) can now be written as

$$|G(s)| = \frac{K \prod\limits_{i=1}^{m} |s + z_i|}{\prod\limits_{i=1}^{n} |s + p_i|} = 1 \tag{4.3.8}$$

i.e. the magnitude of the transfer function, which must equal unity, is proportional to the product of the individual zero term magnitudes divided by the product of the individual pole term magnitudes for any value s.

So any point in the s-plane which is on a root locus for a particular system must satisfy (4.3.8) for that system. Also, if there exists a point in the s-plane which lies on a root locus for a particular system, then the value of K which results in a root at the point can be found from (4.3.8). This is known as the *magnitude criterion*.

By means of the definition (4.3.7) we have that

$$\underline{/G(s)} = \sum_{i=1}^{m} \underline{/(s + z_i)} - \sum_{j=1}^{n} \underline{/(s + p_j)} \Big\} \begin{aligned} &= (2k + 1)\pi \quad \text{for} \quad K > 0 \\ &= 2k\pi \quad \text{for} \quad K < 0 \end{aligned} \tag{4.3.9}$$

Hence, for any point in the s-plane to lie on a root locus of a system, the sum of the angles from the zeros to that point minus the sum of the angles from the poles to that point must satisfy (4.3.9). This is known as the *angle criterion*.

Armed with the magnitude and angle criteria, along with loci start and finish points, given a system transfer function, it is possible to construct the required root loci. In general though it is not necessary to carry out the laborious exercise of plotting each locus with great accuracy by searching for all points which satisfy the angle and magnitude criteria. Although it might be appropriate to perform careful plotting and multiple calculations around certain points of interest, root loci are usually constructed by making use of cardinal points and asymptotic properties in forming a sketch of the loci, and it is root locus construction by this means that is considered here.

4.3.1   Root loci: some basic properties

Property 1

Because the open-loop poles and zeros are all real or occur in complex pairs, the polynomial $A(s) + KB(s)$ can only have roots which are either real or occur in complex pairs. The root loci will therefore be symmetrical about the real axis.

Property 2

The total number of loci for a particular system is equal to the number of poles n of the open-loop transfer function, assuming that $n \geqslant m$ by definition.

Property 3

If $K > 0$, then for a point on the real axis to lie on a root locus there must exist an odd number of finite poles and zeros to the right of the point.

If $K < 0$, then for a point on the real axis to lie on a root locus there must exist an even number of finite poles and zeros to the right of the point.

This property follows from the angle criterion, and can be regarded slightly differently in that if there are no points which exist on the real axis to the left of an odd number of finite poles and zeros, then if $K > 0$, no part of a root locus exists on the real axis; a similar result being true for $K < 0$ if even is substituted for odd.

Property 4

The number of loci which tend to infinity is given by the number of open-loop poles n minus the number of open-loop zeros m, assuming that $n \geqslant m$ by definition.

Example 4.3.1

In order to illustrate some of the properties discussed thus far, consider the following example:

$$G(s) = \frac{K(s+2)}{(s+1)(s+3)}$$

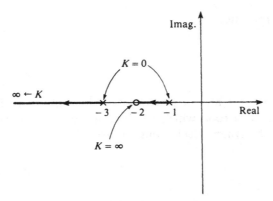

Fig. 4.3 Root loci for Example 4.3.1

The characteristic equation for this open-loop system is given by:

$$(s+1)(s+3) + K(s+2) = 0$$

For $K = 0$ the two loci will therefore start at the open-loop poles, i.e. $s = -1$ and -3.

If we are just considering $K > 0$, then the two loci will end at $s = -2$ and $s = -\infty$. The latter end point can be confirmed by making reference to Property 3.

The direction of increasing K is shown by the arrows on the root loci of Fig. 4.3.

Notice that, as directed by Property 3, the loci only exist on the real axis, to the left of an odd number of poles and zeros.

Consider the point

$$s = -1.4$$

The angle criterion is satisfied because the angle to that point from both the zero at $s = -2$ and the pole at $s = -3$ is $0°$, whereas the angle from the pole at $s = -1$ is π. We then have from (4.3.9) that

$$\underline{/G(s)} = 0° - \pi - 0° = -\pi = (2k+1)\pi; \quad k = -1$$

and from the magnitude criterion

$$|G(s)| = \frac{K(0.6)}{(0.4)(1.6)} = 1 \tag{4.3.10}$$

in which 0.6 is the magnitude to the point at $s = -1.4$ from the zero at $s = -2$, and 0.4 is the magnitude from the pole at $s = -1$ whereas 1.6 is the magnitude from the pole at $s = -3$.

It follows from (4.3.10) that a value of $K = 32/3$ will result in a root on the locus at $s = -1.4$.

4.3.2   Branches and asymptotes

Each separate root locus is also referred to as a branch, and Property 4, given in the previous section, states that $n - m$ of these branches will tend to infinity magnitude. In Example 4.3.1 only one branch tends to infinity, as $K \to \infty$, and this does so along the negative real axis, i.e. at an angle π. In fact for $n > m$ each of the branches involved will tend to infinity magnitude by means of a straight line asymptote, the angle of the asymptote being given, when $K > 0$, by

$$\phi_{k+1} = \frac{(2k + 1)\pi}{n - m} \tag{4.3.11}$$

in which $k = 0, 1, 2, \ldots, (n - m - 1)$. Whereas if $K < 0$, we have

$$\phi_{k+1} = \frac{2k\pi}{n - m} \tag{4.3.12}$$

The total number of asymptotes is therefore equal to $n - m$, for either $K > 0$ or $K < 0$.

The asymptotes do not emanate from the s-plane origin, but rather are centered on a point on the real axis given by

$$s = \sigma = \frac{b_{m-1} - a_{n-1}}{n - m} \tag{4.3.13}$$

where b_{m-1} and a_{n-1} defined in (4.3.1) are also equal to

$$b_{m-1} = \sum_{i=1}^{m} z_i \quad \text{(sum of zeros)}$$

and

$$a_{n-1} = \sum_{i=1}^{n} p_i \quad \text{(sum of poles)}$$

respectively, in which z_i and p_i are defined in (4.3.7). For complex poles/zeros only the real part is required.

Example 4.3.2

Consider the system transfer function

$$G(s) = \frac{K(s + 2)}{(s + 1)(s + 3)(s + 4)}, \quad K > 0$$

such that $n = 3$ and $m = 1$, therefore $n - m = 2$ so two branches (loci) will tend to infinity magnitude following asymptotes with angles

$$\phi_1 = \frac{\pi}{2} \quad \text{and} \quad \phi_2 = \frac{3\pi}{2}$$

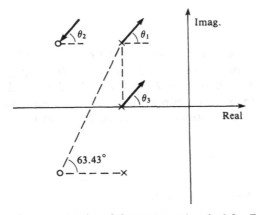

Fig. 4.4 Angles of departure and arrival for Example 4.3.3

and these asymptotes are centered on a point at

$$s = \sigma = \frac{2 - (1 + 3 + 4)}{2} = -3$$

A complete root loci diagram, indicating the calculated asymptotes, is shown in Fig. 4.6.

4.3.3   Angles of departure and arrival

Previously the angle criterion was defined in (4.3.9). This can be interpreted as, for $K > 0$, the sum of the angles from each zero to a particular point on a root locus minus the sum of the angles from each pole to that same point, is equal to $(2k + 1)\pi$.

Simply by rearranging (4.3.9) a condition is found which determines the angle from any one specific pole or zero to the locus as it departs from or arrives at that pole or zero respectively.

The angle of departure of a locus from a pole at $s = -p_1$ is given by

$$\underline{/(s + p_1)} = \sum_{i=1}^{m} \underline{/(s + z_i)} - \sum_{j=2}^{n} \underline{/(s + p_j)} - (2k + 1)\pi \tag{4.3.14}$$

and, in the same fashion, the angle of arrival of a locus to a zero at $s = -z_1$ is given by

$$\underline{/(s + z_1)} = (2k + 1)\pi + \sum_{j=1}^{n} \underline{/(s + p_j)} - \sum_{i=2}^{m} \underline{/(s + z_i)} \tag{4.3.15}$$

The index $k = 0, \pm 1, \ldots$, in these two equations is not really of any great significance here, other than for completeness of the equations. It is probably easier in practice to assume $k = 0$ initially and then to take away the required multiples of 2π when an answer is obtained.

Example 4.3.3

Consider the system transfer function:

$$G(s) = \frac{K(s + 2 - j)(s + 2 + j)}{(s + 1)(s + 1 - j)(s + 1 + j)}, \qquad K > 0$$

To find the angles of departure and arrival, due to symmetry of the loci around the real axis, only one complex pole and one complex zero need to be considered.

     Referring to Fig. 4.4, θ_1 is that angle of departure of the locus from the pole at $s = -1 + j$, and this is found from (4.3.14) to be:

$$\theta_1 = (0° + 63.43°) - (90° + 90°) - 180°$$
$$= 63.43° - 360°$$

or

$$\theta_1 = \underline{63.43°}$$

Similarily, θ_2 is the angle of arrival of the locus at the zero at $s = -2 + j$, and this is found from (4.3.15) to be:

$$\theta_2 = 180° + (180° + 135° + 116.57°) - 90°$$
$$= 360° + 161.57°$$

or

$$\theta_2 = \underline{161.57°}$$

Finally θ_3 is the angle of departure of the locus from the pole at $s = -1$, and this is found from (4.3.14) to be:

$$\theta_3 = (-45° + 45°) - (-90° + 90°) - 180°$$

so

$$\theta_3 = \underline{-180°}$$

i.e. along the negative real axis.

4.3.4   Breakaway points

Breakaway points are points on root loci at which multiple order roots occur; they can therefore be considered as points at which two or more branches either join or split. It is usually the case that breakaway points occur on the real axis as shown in Fig. 4.5,

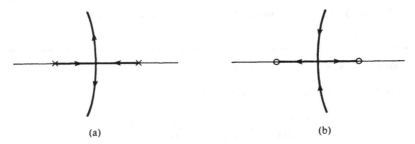

Fig. 4.5 Breakaway points on the real axis

although this is not necessarily true, it being possible for breakaway points to exist in complex conjugate pairs.

By reference to (4.3.3), all breakaway points must satisfy

$$\frac{\mathrm{d}K}{\mathrm{d}s} = \frac{\mathrm{d}}{\mathrm{d}s}\left[-\frac{A(s)}{B(s)}\right] = \frac{\mathrm{d}}{\mathrm{d}s}\left[\frac{B(s)}{A(s)}\right] = 0 \qquad (4.3.16)$$

as this gives the maximum and minimum values of K for which the roots are real, noting that in order to be a breakaway point a solution to (4.3.16) must also comply with (4.3.3). It is necessary to remember this latter point because not all solutions to (4.3.16) satisfy (4.3.3) and therefore not all solutions to (4.3.16) are breakaway points.

The angle at which branches approach or leave a breakaway point is dependent on the root multiplicity, such that if r is the number of roots which exist at a breakaway point, the branches approach with an angle between them of $360°/r$ and leave with an angle between them of $360°/r$, remembering that symmetry about the real axis must be preserved.

Example 4.3.4

Consider the system transfer function:

$$G(s) = \frac{K(s+2)}{(s+1)(s+3)(s+4)}$$

where

$$B(s) = (s+2)$$

and

$$A(s) = (s+1)(s+3)(s+4)$$

Breakaway points for the root loci obtained from this system can be found by means of the solution to

$$\frac{\mathrm{d}}{\mathrm{d}s}\left[-\frac{(s+1)(s+3)(s+4)}{(s+2)}\right] = 0$$

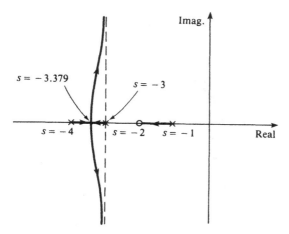

Fig. 4.6 Complete root loci for Examples 4.3.2/4

such that

$$2(s^3 + 8s^2 + 19s + 12) - (s + 2)(3s^2 + 16s + 19) = 0$$

or

$$s^3 + 6s^2 + 13s + 14 = 0$$

and it follows that:

$$(s + 3.379)(s + 1.311 + j1.558)(s + 1.311 - j1.558) = 0$$

On substitution of the solution $s = -3.379$ back into (4.3.3) it is found that $K = 0.406$. However if the complex solution is substituted back into (4.3.3) a solution cannot be found, and therefore only the point $s = -3.379$ is a breakaway point

Having calculated the breakaway point for this example, by reference to Example 4.3.2 the complete root loci for the system considered are shown in Fig. 4.6.

The fairly simple example considered here highlights the problem of finding a breakaway point by a purely analytical method, in that it is not an easy problem to find the solution to the third-order equation which results. For a higher-order system therefore, the problem becomes much more difficult and it may well not be possible to obtain a solution. In practice it is most likely that an approximate solution can be obtained by graphical means from the remainder of the loci; an exact answer can then be found if necessary by employment of for example a Newton–Raphson iteration.

4.3.5 Worked example

In order to explain more fully how the root loci are plotted for a particular

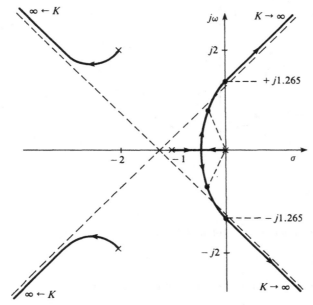

Fig. 4.7 Root loci for the worked Example 4.3.5
(poles at $s = 0$, $s = -1$ and $s = -2 \pm j2$)

system transfer function, an extended example is considered in detail here. The open-loop transfer function is given by:

$$G(s) = \frac{K}{s(s+1)(s^2 + 4s + 8)}; \quad K > 0 \qquad (4.3.17)$$

and this means that:

1. The total number of loci is equal to $n = 4$, and these will start from the four open-loop poles, which are shown plotted in Fig. 4.7.
2. The number of loci which tend to infinity is given by $n - m$, and as the number of zeros $m = 0$, this means that all four loci tend to infinity as $K \to \infty$.
3. Only the real axis between the poles at $s = 0$ and $s = -1$ has an odd number of poles to the right, and therefore this is the only part of the real axis on which loci exist.
4. Branches will tend to infinity by means of asymptotes, the angle of each one being given by:

$$\phi_{k+1} = \frac{(2k+1)\pi}{4} \quad \text{for} \quad k = 0, 1, 2, 3$$

i.e.

$$\phi = \frac{\pi}{4}, \frac{3\pi}{4}, \frac{5\pi}{4} \quad \text{and} \quad \frac{7\pi}{4}$$

Also, the asymptotes are centered on a point on the real axis, found to be

$$s = \sigma = \frac{b_{m-1} - a_{n-1}}{n - m} = \frac{-5}{4}$$

in which $b_{m-1} = 0$ and $a_{n-1} = 5$, where the latter is found from the open-loop denominator $s^4 + 5s^3 + 12s^2 + 8s$.

The asymptotes are shown by broken lines in Fig. 4.7.

5. Since there are no finite zeros, there are no angles of arrival to calculate. However, angles of departure of the loci from the complex poles at $s = -2 \pm j2$ need to be calculated.

   The angle of departure of the locus from the pole at $s = -2 + j2$ is given by

$$\underline{/(s + 2 - j2)} = -\underline{/(s + 0)} - \underline{/(s + 1)} - \underline{/(s + 2 + j2)} - (2k + 1)\pi$$
$$= -135° - 116.57° - 90° - (2k + 1)\pi$$
$$= -341.57° - 180° = -161.57°$$

Similarly, the angle of departure of the locus from the pole at $s = -2 - j2$ is given by

$$\underline{/(s + 2 + j2)} = -\underline{/(s + 0)} - \underline{/(s + 1)} - \underline{/(s + 2 - j2)} - (2k + 1)\pi$$
$$= -225° - 243.43° - 270° - (2k + 1)\pi$$
$$= -738.43° - 180° = -198.43° = +161.57°$$

6. Breakaway points for the root loci are obtained as the solution to:

$$\frac{d}{ds}\{-s(s + 1)(s^2 + 4s + 8)\} = 0$$

that is

$$\frac{d}{ds}(s^4 + 5s^3 + 12s^2 + 8s) = 0$$

such that

$$4s^3 + 15s^2 + 24s + 8 = 0$$

and the result of importance, which is expected from the loci already plotted, to lie on the real axis between $s = 0$ and $s = -1$, is calculated to be

$$s = -0.4402$$

The angle at which the branches approach or leave this breakaway point is given by $360°/2$, in which 2 is the number of roots which exist at $s = -0.4402$. This means that the loci will arrive at the breakaway point with an angle π between them – and as they arrive along the real axis this property is certainly true. Because the loci also leave with an angle π between them, this means that they must leave at angles of $\pi/2$ and $-\pi/2$ to the real axis – in order to retain symmetry of the loci about the real axis.

7. Imaginary axis crossing: the points at which the loci cross the imaginary axis and the value(s) of K at those points can be found from the Routh–Hurwitz criterion as follows: the characteristic polynomial equation (closed-loop denominator) for the transfer function (4.3.17) is found to be:

$$s^4 + 5s^3 + 12s^2 + 8s + K = 0$$

For which the Routh array is:

$$
\begin{array}{c|ccc}
s^4 & 1 & 12 & K \\
s^3 & 5 & 8 & \\
s^2 & 10.4 & K & \\
s^1 & \lambda & & \\
s^0 & K & &
\end{array}
$$

where

$$\lambda = 8 - 5K/10.4$$

A system with such a denominator is therefore stable as long as $K > 0$ and

$$8 - \frac{5K}{10.4} > 0$$

i.e. $16.64 > K$

Hence when $K = 16.64$ roots will exist on the imaginary axis, and these are the points we are looking for. Substituting $K = 16.64$ into the characteristic polynomial equation, results in

$$s^4 + 5s^3 + 12s^2 + 8s + 16.64 = 0$$

and this has roots on the imaginary axis of

$$s = \pm j\sqrt{8/5} = \pm j\, 1.265$$

8. As an extension to the previous calculation, the value of gain K can in fact be calculated at any point on the root loci, on condition that, having obtained the critical points/angles in (1)–(7), the remainder of the loci are drawn with good accuracy, by following the angle criterion. The magnitude criterion then means that for any point on a root locus, see (4.3.8),

$$K = \frac{\displaystyle\prod_{i=1}^{n} |s + p_i|}{\displaystyle\prod_{i=1}^{m} |s + z_i|} \qquad (4.3.18)$$

i.e. the gain K is equal to the product of the magnitude from each pole to that point, divided by the product of the magnitude from each zero to that point.

Consider then the case when the two right-hand complex roots are looked at in terms of natural frequency and damping, i.e. $s^2 + 2\zeta\omega_n s + \omega_n^2$. For example, let the damping ratio $\zeta = 0.5$, then

$$s^2 + 2\zeta\omega_n s + \omega_n^2 = s^2 + \omega_n s + \omega_n^2$$

which gives roots at

$$s = -\frac{\omega_n}{2} \pm j\sqrt{3}\,\frac{\omega_n}{2}$$

and these lie at angles of $\pm 60°$ (i.e. $\pm\cos^{-1}\zeta$) to the negative real axis, as shown in Fig. 4.7.

A calculation of (4.3.18) results in:

$$K = 4.114$$

with $\omega_n = 0.695$.

Also the two other complex root positions are found, from this value of K, to be

$$s = -2.153 \pm j\,1.973.$$

These are fairly simply obtained, in this case, from the closed-loop denominator — only a second-order equation needing to be solved

The real part of each of the roots denotes, in the time domain, how quickly their effect dies away, and for this example the left-hand root pair real part values are much larger than those for the right-hand roots, which means that these latter roots will tend to represent the system response as a whole.

The example looked at in this section highlights the main points to be considered when plotting the root loci for a system transfer function and provides a set of rules (1) to (8) which need to be investigated for root loci in general. In the following section the application of root loci plots to specific problem areas is discussed.

4.4   Application of root loci

Although a strict set of rules was outlined in the previous section for the construction of root loci, it is in fact more likely that an experienced control system designer (certainly one who has read this book) will be able very quickly to plot approximate loci for most lower order systems, without the need to follow all of the rules or to make accurate measurements. Because of this, the technique of root locus plotting is a very useful tool with which to investigate system transfer functions without the requirement of detailed algebraic calculations. Hence, the effect of modifying or adding system poles and zeros can be witnessed by considering the effect the change will have on the root loci. In this section the use of root loci in finding an approximate solution to the roots of any polynomial is discussed, the advantage of the technique being to greatly simplify the procedure necessary.

Later in the section, the occurrence of a system pure time delay, i.e. a transport delay, is investigated in terms of its effect on the root loci of a transfer function. Finally, it is shown how the root locus method is of great use in the study of system parameter variations and their effect on system performance in the sense of control system sensitivity.

4.4.1   Roots of a polynomial

The technique of root loci plotting is considered here merely as a method by which approximate solutions can be found to the roots of a polynomial. Certainly it is fairly simple, as is shown here, to obtain a very rough idea as to the complex/real nature of the roots and their approximate position. The root locus procedure itself becomes quite difficult to carry out on high order transfer functions and this means that it is not really suitable in the analysis of high order polynomials. The explanation is given by means of an example polynomial, which although it has been selected because of its exact integer root solution, it has certainly not been chosen to show the root locus method at its best.

Consider the polynomial equation

$$s^3 + 7s^2 + 20s + 24 = 0 \tag{4.4.1}$$

This can also be written as

$$\frac{20(s + 1.2)}{s^2(s + 7)} = -1 \tag{4.4.2}$$

which is in the same form as (4.3.3) with the gain $K = 20$. If, therefore, the figure 20 in the numerator of (4.4.2) is replaced by a general $K > 0$, a root loci plot can be made for which the required polynomial solution exists when $K = 20$.

Let the transfer function under consideration be:

$$G(s) = \frac{K(s + 1.2)}{s^2(s + 7)} = -1 \tag{4.4.3}$$

and this has a total of $n = 3$ loci which start at the poles $s = 0$, $s = 0$ and $s = 7$, ending at the zero $s = -1.2$, with the remaining two loci tending to ∞ as $K \to \infty$. By reference to Section 4.3.5, rule 3 shows that the only part of the real axis on which a locus exists is that between the pole at $s = -7$ and the zero at $s = -1.2$. Also the branches which tend to infinity will do so at an angle of asymptotes given by $\pm \pi/2$, and a center point $s = (1.2 - 7)/(3 - 1) = -2.9$. The root loci plot for the transfer function (4.4.3) is shown in Fig. 4.8, for which angles of departure − rule 5 − have also been calculated ($\pm \pi/2$).

The root loci plot shows that one of the three roots is real and lies between $s = -1.2$ and $s = -7$. Further, the remaining two roots can be seen to form a complex pair with a real part lying between $s = 0$ and $s = -2.9$. The roots for $K = 20$ are in fact $s = -3$ and

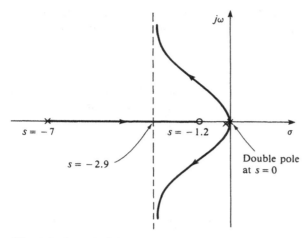

Fig. 4.8 Root loci for the transfer function (4.4.3)

$s = -2 \pm j2$ and it would be expected that with the approximate root values obtained from the root loci in Fig. 4.8, the exact solution can be found without too much difficulty.

From the starting point of the original polynomial (4.4.1) the gain K was arbitrarily chosen to be equal to 20, the parameter associated with the s-term. It would have been equally correct to set K to be any one of the polynomial parameters, and to obtain a different root loci from the alternative starting point, e.g. consider (4.4.1) rewritten as:

$$\frac{24}{s(s^2 + 7s + 20)} = -1 \tag{4.4.4}$$

such that if the numerator 24 is substituted by a gain K, then a plot can be made of the resulting root loci. The three new loci then start at the poles $s = 0$ and $s = -3.5 \pm j\sqrt{31}/2$, each one of the loci in this case tending to infinity magnitude.

So, to conclude this section, the point is made that when obtaining approximate solutions to polynomials by means of root loci, it is worthwhile selecting a K which simplifies the procedure as much as possible. In the example shown, the use of (4.4.4) rather than (4.4.2) would have necessitated the calculation and plotting of complex poles as starting points for two of the loci; the equation (4.4.2) was therefore preferred.

4.4.2 Time delay systems

A large number of systems exist for which there is a time delay between an input signal variation and the witnessing of an effect on the system output signal in response to that input variation, this time delay is also known as dead time or transport lag. If T seconds is the transport delay, the lag can be represented in the s domain by means of the exponential function, exp{ – *Ts*}. The overall open-loop transfer function of a system with a transport delay can therefore be written, in a modified form of (4.3.1), as

$$\bar{G}(s) = G(s)\exp\{-Ts\} = K\frac{B(s)}{A(s)}\exp\{-Ts\} \tag{4.4.5}$$

where it must be remembered that it is assumed that unity feedback is to be applied and *K* is an adjustable gain. The closed-loop denominator, i.e. the characteristic polynomial, resulting from this system then produces the characteristic equation:

$$A(s) + KB(s)\exp\{-Ts\} = 0$$

or

$$K\frac{B(s)}{A(s)}\exp\{-Ts\} = -1 \tag{4.4.6}$$

Because *s* consists of a real and imaginary part, i.e. $s = \sigma + j\omega$, it follows that

$$K\frac{B(s)}{A(s)}\exp\{-T\sigma\}\exp\{-j\omega T\} = -1 \tag{4.4.7}$$

Since $|\exp(-j\omega T)| = 1$, the magnitude condition for this system becomes:

$$\left| K \cdot \frac{B(s)}{A(s)} \right| \exp\{-T\sigma\} = 1 \tag{4.4.8}$$

or

$$|G(s)| = \exp\{+T\sigma\}$$

which compares with (4.3.4).

Also the phase component of the term exp{ – *T*σ} is zero, therefore the angle condition for this system becomes:

$$\angle\left[K\frac{B(s)}{A(s)}\exp\{-j\omega T\} \right] = (2k+1)\pi \tag{4.4.9}$$

for *K* > 0 and in which *k* = 0, ± 1, ± 2, ...

Hence, the final angle condition can now be written as

$$\angle G(s) = (2k+1)\pi + \omega T \tag{4.4.10}$$

Armed with the versions of angle and magnitude criteria applicable for systems with a time delay, consider the problem of plotting the root loci for the open-loop system characterized by the transfer function:

$$Y(s) = \frac{K\exp\{-3s\}}{s(s+2)} U(s); \quad \text{where} \quad K > 0, \ T = 3 \text{ s} \tag{4.4.11}$$

The characteristic equation for this system is:

$$s(s+2) + K\exp\{-3s\} = 0 \tag{4.4.12}$$

1. The term exp{ – *Ts*} can be written in terms of an infinite series in powers of *s*, hence the characteristic equation has an infinite number of roots – resulting in an infinite

number of loci. When $K = 0$, the roots of (4.4.12) give $s = 0$ and $s = -2$, the open-loop poles, as loci starting points, but $\sigma = -\infty$ also gives $K = 0$ and therefore further starting points when $\omega = \pm \pi/T, \pm 3\pi/T, \ldots$ − from (4.4.10) with $(2k + 1)\pi = \omega T$.

2. With exp$\{-Ts\}$ written as an infinite series, there are an infinite number of open-loop zeros, $m > n$, and therefore an infinite number of branches tend to infinity magnitude. From (4.4.8) $K \to \infty$ as $\sigma \to \infty$: from (4.4.10) with $(2k + 1)\pi = \omega T$, this is when $\omega = \pm \pi/T, \pm 3\pi/T, \pm 5\pi/T, \ldots$.

3. Existence of loci on the real axis is dependent on identical conditions to those given previously ($\omega = 0$ on the real axis), so for this problem loci exist between the poles at $s = 0$ and $s = -2$, as this is the only part of the real axis for which an odd number of poles lie to the right.

4. Branches will tend to infinity by means of asymptotes, the angle of each one being π (when $\sigma = +\infty$), the asymptotes are not in this case centered at points on the real axis → due to the asymptote angles being π, as shown in Fig. 4.9.

5. Angles of departure and arrival do not need to be calculated here − the only applicable poles being those on the real axis.

6. Breakaway points for the root loci are obtained as the solution to

$$\frac{dK}{ds} = \frac{d}{ds}\{s(s + 2)\exp\{+Ts\}\} = 0 \qquad (4.4.13)$$

as can be seen with the help of (4.4.6).

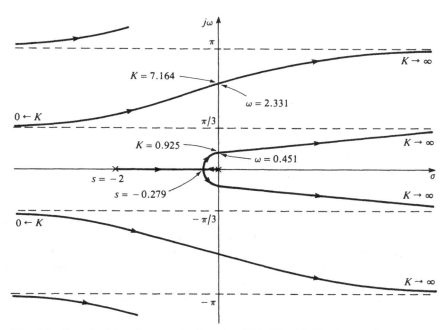

Fig. 4.9 Root loci for the transfer function (4.4.11) with $T = 3$ seconds

We have from (4.4.13) that

$$T \exp\{+Ts\}\left[s^2 + 2s + \frac{2s}{T} + \frac{2}{T}\right] = 0 \tag{4.4.14}$$

and we know from Fig. 4.9 that the desired solution is expected to lie between the limits $0 > s > -2$. With $T = 3$ seconds the solution to the quadratic part (4.4.14) is

$$s = \frac{-8 \pm \sqrt{40}}{6}$$

and this gives the required solution of a breakaway point at:

$$\underline{s = -0.279}$$

The angle at which the branches approach or leave this breakaway point is given by $360°/2$. The loci therefore leave with an angle π between them which means they must leave at angles $\pi/2$ and $-\pi/2$ to the real axis — in order to retain symmetry.

7., 8. Points at which the loci cross the imaginary axis cannot be calculated by means of the Routh–Hurwitz criterion, since the characteristic equation (4.4.12) is not algebraic in terms of powers of s. For this example however it is fairly straightforward to obtain both ω and K values where the loci cross the imaginary axis, because at these points the characteristic equation must be:

$$s^2 + \omega^2 = s^2 + 2s + K \exp\{-3s\} = 0 \tag{4.4.15}$$

and by substituting $s = +jw$ or $s = -jw$ we obtain

(a) $2\omega = K \sin 3\omega$

and   (4.4.16)

(b) $\omega^2 = K \cos 3\omega$

from which the following solution pairs occur:

$$\omega = \pm 0.451, \quad K = 0.925$$
$$\omega = \pm 2.331, \quad K = 7.164$$
$$\omega = \pm 4.333, \quad K = 20.680$$
$$\omega = \pm 6.384, \quad K = 42.693, \text{ etc.}$$

It is worth noting from these results:

(a)   For any one value of K there are an infinite number of possible solutions. For the closed-loop system to be stable the characteristic equation roots must lie in the left half of the s-plane. This can only be assured when $0 < K < 0.925$ — stability properties are therefore given by the value of K corresponding to the imaginary axis crossing nearest the origin.

(b)   Solutions for $K < 0$ can also be obtained from the equation set (4.4.16).

(c)   It might be expected that the root loci would cross the imaginary axis at values of ω equal to the half-way point between associated asymptote values. This is not the case however, the crossings occurring at $\omega = \pm 2.331$, ± 4.333, etc. and not $\omega = \pm 2\pi/3$ [2.094], $\pm 4\pi/3$ [4.189], etc.

(d) The value of K can be calculated at any point on the loci by graphical means, as was done previously, although K at the imaginary axis crossing points can be calculated algebraically, and hence more accurately, without too much difficulty, as was shown in the example discussed.

4.4.3 Sensitivity to parameter variations

In Section 3.4.5 the sensitivity of a system output $Y(s)$ to variations in the system's transfer function $G(s)$ was defined as:

$$S_G^Y = \frac{\partial Y(s)/Y(s)}{\partial G(s)/G(s)} \qquad (4.4.17)$$

The root loci plotting procedure examines the variation in closed-loop roots (poles) as one specified parameter, usually a gain K, is varied. Combining the ideas of root loci plotting and sensitivity analysis, it is convenient to investigate the sensitivity of a particular root location (position) to variations in one system parameter or other, not necessarily the gain K. If the root position under investigation is that at $s = -\alpha$, i.e. $s + \alpha$ is a root of the closed-loop denominator, then let us consider in the first instance its sensitivity to variations in gain K, where the unity feedback system open-loop transfer function is

$$G(s) = \frac{K(s + z_1) \dots (s + z_m)}{(s + p_1) \dots (s + p_n)} \qquad (4.4.18)$$

The sensitivity of the root at $s = -\alpha$ to the gain K is then

$$S_K^\alpha = \frac{\partial \alpha}{\partial K/K} \qquad (4.4.19)$$

Note that here we are not interested in the fractional change in closed-loop pole position, but merely the change in its position.

Further, the characteristic equation of the system (4.4.18) can be written as

$$(s + p_1) \dots (s + p_n) + K[(s + z_1) \dots (s + z_m)] = (s + \alpha)X(s) = 0 \qquad (4.4.20)$$

in which $X(s)$ is the remainder of the closed-loop denominator, once the factor $(s + \alpha)$ has been removed.

It follows then that

$$K = \frac{(s + \alpha)X(s) - [(s + p_1) \dots (s + p_n)]}{(s + z_1) \dots (s + z_m)}$$

from which

$$\frac{\partial K}{\partial \alpha} = \frac{X(s)}{(s + z_1) \dots (s + z_m)}$$

or

$$\frac{\partial \alpha}{\partial K/K} = \frac{K[(s + z_1) \dots (s + z_m)]}{X(s)} = \frac{G(s)}{1 + G(s)} \cdot (s + \alpha)$$

where $(s + \alpha)$ is also a denominator factor.

It is necessary however to make a slight modification to the root sensitivity function before continuing, in that the differentiation procedure assumes that only a small change occurs in the root position ($\partial \alpha$), hence the root sensitivity function calculated is only valid in the neighborhood of $s = -\alpha$.

The root sensitivity function is thus redefined as:

$$S_K^\alpha = \frac{\partial \alpha}{\partial K/K} \bigg|_{s = -\alpha} \tag{4.4.21}$$

which results in

$$S_K^\alpha = \frac{(s + \alpha)G(s)}{1 + G(s)} \bigg|_{s = -\alpha} \tag{4.4.22}$$

which is a complex measure whose magnitude and direction gives an indication of how that root position will change for a small change in gain K. Note that (4.4.22) can also be written as

$$S_K^\alpha = \frac{K[(s + z_1) \dots (s + z_m)]}{X(s)} \bigg|_{s = -\alpha}$$

Hence the sensitivity function can be calculated directly from the root loci plot by finding the distance from each zero to the root located at $s = -\alpha$, and also the distance from each of the other closed-loop pole positions (with open-loop gain K) to the root at $s = -\alpha$. The root sensitivity is then found as the gain K multiplied by the product of distances from each zero to $s = -\alpha$ and divided by the product of distances from each of the other closed-loop poles to $s = -\alpha$.

As an example, consider the open-loop system transfer function:

$$G(s) = \frac{K}{s(s + 2 + j\sqrt{2})(s + 2 - j\sqrt{2})} \tag{4.4.23}$$

in which K is nominally equal to 4.

It follows then that the closed-loop expression is

$$\frac{G(s)}{1 + G(s)} = \frac{4}{(s + 2)(s + 1 + j)(s + 1 - j)}$$

If we wish to investigate the effect of a small change in gain K from its value 4, on the closed-loop pole located at $s = -\alpha = -1 - j$, then we have

$$s_{K=4}^{(1+j)} = \frac{4}{(s+2)(s+1-j)} \bigg|_{s=-1-j}$$

or

$$s_{K=4}^{(1+j)} = \frac{2j}{1-j} = j - 1$$

From this we can conclude that if K increases slightly then the parameter $\alpha = 1 + j$ will vary slightly in the direction of $\underline{/135°}$. The corresponding closed-loop root, which is originally located at $s = -\alpha$, will then vary slightly in the direction of $-(j-1)$, i.e. in the direction of $\underline{/-45°}$. This sensitivity function therefore gives an indication in terms of magnitude and phase of the increase in parameter value α caused by an increase in gain K. Because the closed-loop pole is given by $s = -\alpha$, the sensitivity function shows the decrease in root magnitude and an angle equal to π + the phase change in root position caused by an increase in gain K. The effect on the pole position $s = -1 - j$ in the example is shown in Fig. 4.10 in terms of an increase in gain K.

For a small decrease in gain K, the magnitude of the sensitivity function is the same as that for a small increase in gain, however the angle becomes π + that for a small increase in K, i.e. the sensitivity function is merely negated when ∂K is in the negative direction.

The discussion thus far has restricted itself to changes in α due to changes in the gain K. It might be, however, that the variation in α is caused by the movement of a closed-loop zero, noting that the closed-loop zeros are also open-loop zeros, assuming no pole-zero cancellations take place. The sensitivity of the parameter α, where $s + \alpha$ is a closed-loop denominator factor, to changes in a zero value z_i is given by

$$S_{z_i}^\alpha = \frac{\partial \alpha}{\partial z_i} \bigg|_{s=-\alpha} \tag{4.4.24}$$

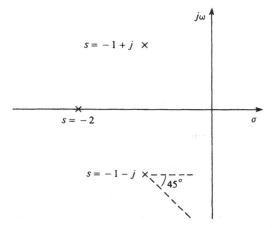

Fig. 4.10 Use of sensitivity function on pole locations

and this is

$$\frac{\partial \alpha}{\partial z_i}\bigg|_{s=-\alpha} = \frac{K(s+z_1)\dots(s+z_m)}{(s+z_i)X(s)}\bigg|_{s=-\alpha}$$

which can also be written as

$$\frac{\partial \alpha}{\partial z_i}\bigg|_{s=-\alpha} = \frac{G(s)}{1+G(s)}\frac{(s+\alpha)}{(s+z_i)}\bigg|_{s=-\alpha} \qquad (4.4.25)$$

or

$$S_{z_i}^{\alpha} = \frac{S_k^{\alpha}}{z_i - \alpha} \qquad (4.4.25)$$

It must be noted that the sensitivity function $S_{z_i}^{\alpha}$ denotes variations in the parameter α in response to variations in the parameter z_i. Both parameters must be negated in order to measure the variations in closed-loop pole $s=-\alpha$ in response to variations in the open-loop $s=-z_i$.

Finally in this section, the sensitivity of α to changes in the open-loop pole value p_i is considered by means of the sensitivity function

$$S_{p_i}^{\alpha} = \frac{\partial \alpha}{\partial p_i}\bigg|_{s=-\alpha} \qquad (4.4.26)$$

which is

$$\frac{\partial \alpha}{\partial p_i}\bigg|_{s=-\alpha} = \frac{(s+p_1)\dots(s+p_n)}{(s+p_i)X(s)}\bigg|_{s=-\alpha}$$

and this is also

$$\frac{\partial \alpha}{\partial p_i}\bigg|_{s=-\alpha} = \frac{G(s)}{1+G(s)}\frac{(s+\alpha)}{(s+p_i)}\frac{1}{G(s)}\bigg|_{s=-\alpha}$$

or

$$S_{p_i}^{\alpha} = \frac{S_k^{\alpha}}{p_i - \alpha} \qquad (4.4.27)$$

noting that $G(s)^{-1}\big|_{s=-\alpha} = 1$.

The sensitivity function $S_{p_i}^{\alpha}$ denotes variations in the parameter α in response to variations in the parameter p_i. Both parameters must be negated in order to measure the variations in closed-loop pole $s=-\alpha$ in response to variations in the open-loop pole $s=-p_i$, i.e. if the parameter p_i increases from $p_i=2$ to $p_i=2.1$ then this means that an open-loop pole moves from $s=-2$ to $s=-2.1$.

Root sensitivity analysis is a very useful tool with which to investigate the effects on a closed-loop root of variations in various system parameters. It has particular application when from a root loci plot, one root is shown to lie in the left half of the s-plane, but to be near the imaginary axis. It is then possible to see how slight variations in certain of the system parameters could cause that root to migrate towards or even over the imaginary axis, thereby affecting the stability properties of the system.

4.5   Compensation based on a root locus

In this chapter it has been shown how the root loci of a system transfer function can be plotted in the s-plane by means of following a simple procedure in the form of a set of rules. It is then a fairly straightforward task to employ a root loci plot in the desig of a desired control system in order to meet a performance specification. It is shown directly here how it is sometimes possible to achieve the required design objective merely by selecting, from the closed-loop denominator root loci, an appropriate open-loop gain K. However, in many cases the performance specification cannot be met by only a choice of gain K, and the result of this is that a controller block must be connected to the original open-loop system such that the overall closed-loop transfer function can be tailored by means of choosing appropriate controller parameters. Later in this section it is shown how controller parameters can be selected by employing information obtained from a root loci plot, such that overall closed-loop denominator root loci are shifted to ensure that they follow pre-defined paths at specific points.

A controller design is carried out here to show initially that it is sometimes possible to meet certain closed-loop system performance objectives simply by a selection of system open-loop gain K. Consider the open-loop transfer function:

$$G(s) = \frac{K}{s(s+5)} \tag{4.5.1}$$

It is required that the steady-state error to a unit ramp input is no more than (a) 33% or (b) 5%. Further there must be no more than a 10% overshoot of its final value in the unity feedback closed-loop system's response to a step input.

It can be noted that if the closed-loop function is written as

$$\frac{Y(s)}{V(s)} = \frac{K}{s^2 + 2\zeta\omega_n s + \omega_n^2} \tag{4.5.2}$$

then by using Fig. 3.10 as a guide, it follows that $\zeta \geqslant 0.6$ in order to achieve a 10% overshoot at most, also we have that $2\zeta\omega_n = 5$ and $\omega_n^2 = K$. The root loci for the system (4.5.1), (4.5.2) are drawn in Fig. 4.11 by following the rules presented in Section 4.3.

When the two roots are complex, $s = -2.5 \pm j\beta$, where $\beta^2 = -2.5^2 + \omega_n^2$ and $\omega_n \cos\theta = 2.5$. Because $\zeta = 2.5/\omega_n$, it follows that $\zeta = \cos\theta$ or $\theta = \cos^{-1}\zeta$, such that when $\zeta = 0.6$, $\theta = 53.13°$, $\omega_n = 4.167$, $\beta = 3.33$ and $K = 17.36$. For any $\zeta > 0.6$ then $\omega_n < 4.167$ and $K < 17.36$: K must therefore lie in the range $0 < K < 17.36$ in order to satisfy the desired maximum overshoot value.

The steady-state error following a unit ramp input is $1/K_r$ where

$$K_r = \lim_{s \to 0} sG(s) = K/5$$

such that for condition (a)

$$K_r = K/5 > 3$$

or

$$K > 15$$

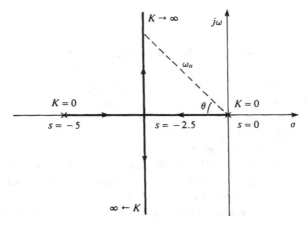

Fig. 4.11   Root loci plot for $G(s)$ of (4.5.1)

Hence a K can be selected which meets both the maximum overshoot requirements and provides a maximum ramp error of less than 33%, as long as $15 < K < 17.36$.

For the steady-state ramp error requirements made in condition (b) it is necessary that $K_r = K/5 > 20$, which means that $K > 100$. It is therefore not possible for both the maximum overshoot requirements and the maximum ramp error requirement (b) to be met simply by a selection of open-loop gain K, and therefore in this case further controller flexibility, in the form of other selectable parameters must be used.

4.5.1   First-order compensation

Attempting to meet two or more control system performance criteria with only one controller parameter, the gain K, will in many cases result in an unsatisfactory final selection, as far as at least one of the criteria is concerned. The problem is made worse if further performance objectives are stipulated and/or the open-loop system is of higher order. Fortunately, it is often the case that when a system is of a higher order its response can be approximated with reasonable accuracy by means of a low (first, second or third) order model, and hence controller complexity is not necessarily increased by any great extent.

As the majority of systems to be controlled can be adequately represented by a model of order no higher than three, it is a consequence that in addition to a selection of forward gain K, a cascade controller consisting of one pole and one zero is suitable for many control purposes, as shown in Fig. 4.12.

In theory it is then possible to simply cancel an undesired open-loop pole of $G(s)$ with the zero at $s = -b$, and to replace it with a pole at $s = -a$, it being assumed in general that a and b are positive definite real scalar values. There are however two drawbacks to this seemingly sensible idea, the first being that there is not always an obvious

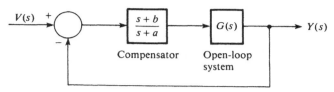

Fig. 4.12   Lag/lead cascade compensator

relationship between the open-loop system/controller pole-zeros and the specified performance objectives made in terms of the closed-loop system. The second drawback is a practical one in that it is not often the case that an 'exact' representation of the open-loop system is available, in which case any cancellation of an open-loop pole with the controller zero will itself not be an exact cancellation. Although this may not be serious when the open-loop pole in question lies in the left half of the s-plane, if the pole lies in the right half of the s-plane a controller zero will be introduced which also lies in the right half of the s-plane and this would result in an unstable mode being present in the closed-loop system.

4.5.2   Phase and magnitude compensation

The compensator introduced in the previous section can be employed effectively to alter the system phase and/or magnitude characteristics by means of a root loci plot, the effect that the compensator has being dependent on the magnitude of the coefficients a and b. This can best be seen by making reference to the angle and magnitude root locus criteria considered in (4.3.9) and (4.3.8) respectively. The cascading of a compensator of the form $(s + b)/(s + a)$ can therefore be viewed in terms of how the compensator affects the original, uncompensated, loci in terms of magnitude and phase.

Regarding Fig. 4.13, points s' and s'' are considered to be two points on an originally uncompensated root loci plot, and the compensator has been designed such that $a > b \geqslant 0$.

In terms of the angle criterion the phase of the point s' is affected by the compensator pole-zero pair by means of the relationship $\underline{/s'}$ = original $\underline{/s'}$ + $\underline{/(s + b)}$ − $\underline{/(s + a)}$. Hence for the compensator with $a > b$ it follows that for all s' with $\omega > 0$ we have $\underline{/(s + b)} > \underline{/(s + a)}$, and thus the phase of s' is shifted in the positive phase angle direction as shown by the arrow in Fig. 4.13. The compensator is known as a *lead* compensator because of this phase shifting property. Note that for all s'' with $\omega < 0$ we have $\underline{/(s + a)} > \underline{/(s + b)}$ and thus the phase of s'' is shifted in the negative phase angle direction − a result which can be deduced from the symmetrical nature of the loci. An example of the effect of a lead compensator is shown in terms of the open-loop system transfer function

$$G(s) = \frac{K}{(s + 3)(s + 5)} \tag{4.5.3}$$

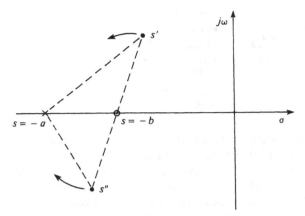

Fig. 4.13 Effect of a lead compensator

with a lead compensator whose transfer function is

$$\frac{s+b}{s+a} = \frac{s+1}{s+2}$$

The uncompensated closed-loop system root loci are shown in Fig. 4.14(a) whereas those with the compensator applied are shown in Fig. 4.14(b).

Regarding Fig. 4.15, points s' and s'' are two points on an originally uncompensated root loci plot, and the compensator has been designed such that $b > a \geqslant 0$. The angle criterion then means that the phase of the point s', when $\omega > 0$, is shifted in the negative phase angle direction as shown by the arrow in Fig. 4.15. This is due to the fact that when $b > a$, we have $\underline{/(s+a)} > \underline{/(s+b)}$ and $\underline{/s'} = $ original $\underline{/s'} + \underline{/(s+b)} - \underline{/(s+a)}$. The compensator is known as a *lag* compensator because of this phase shifting property.

Note that for all s'' with $\omega < 0$ we have $\underline{/(s+b)} > \underline{/(s+a)}$ and thus the phase of s'' is shifted in the positive phase angle direction. An example of the effect of a lag

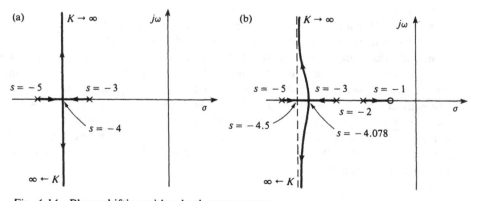

Fig. 4.14 Phase shifting with a lead compensator

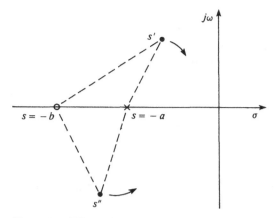

Fig. 4.15 Effect of a lag compensator

compensator is shown in terms of the open-loop transfer function (4.5.3) with a lag compensator whose transfer function is

$$\frac{s+b}{s+a} = \frac{s+2}{s+1}$$

The uncompensated closed-loop system root loci are shown in Fig. 4.16(a) (also in Fig. 4.14(a)) whereas those with the compensator applied are shown in Fig. 4.16(b).

Magnitude compensation by means of root loci plots is generally not a complicated procedure and is usually a case of selecting a pole-zero pair such that steady-state error conditions are achieved without seriously affecting the system's closed-loop transient response. Hence both the compensator pole and zero must lie to the left of the s-plane imaginary axis for stability, but well to the right of any real axis breakpoints. The effect of a compensator selected for steady-state error magnitude conditions can be seen by reference to the example initially considered at the start of Section 4.5, i.e. that given in

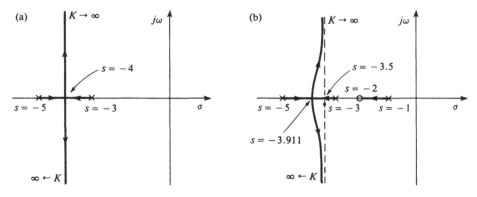

Fig. 4.16 Phase shifting with a lag compensator

(4.5.1). A requirement is that the steady-state error to a unit ramp input is no more than 5% and it was shown that use of a simple gain K could not satisfactorily achieve this objective whilst also trying to meet maximum overshoot conditions.

The steady-state error following a unit ramp input is $1/K_r$ where

$$K_r = \lim_{s \to 0} \left\{ sG(s) \left[\frac{s+b}{s+a} \right] \right\}$$

or

$$K_r = \lim_{s \to 0} \left\{ \frac{K(s+b)}{(s+5)(s+a)} \right\} = \frac{Kb}{5a}$$

If it is remembered that we needed $K < 17.36$ to satisfy the overshoot conditions (less than 10%), then selecting $K = 10$, we have that

$$20 < K_r = \frac{2b}{a} \quad \text{or} \quad b > 10a$$

Further, the original breakaway point was at $s = -2.5$ (see Fig. 4.11), so let us select b to be 0.2 and $a = 0.01$ such that the pole at $s = -0.01$ and zero at $s = -0.2$ lie well to the right of $s = -2.5$.

The new root loci plot with a lag compensator

$$\frac{s+b}{s+a} = \frac{s+0.2}{s+0.01}$$

is shown in Fig. 4.17.

The effect of the compensator pole-zero pair is magnified in Fig. 4.17 in order to show the loci movement, but nevertheless it can readily be seen that the complex loci are very little affected and hence the transient response is not significantly altered. In particular

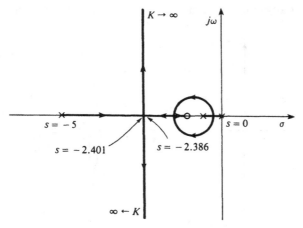

Fig. 4.17 Compensated root loci plot for $G(s) = K/s(s+5)$

when $K = 10$ a root lies at $s = -0.233$ and hence

$$s^2 + 2\zeta\omega_n s + \omega_n^2 = s^2 + 4.787s + 8.984$$

Thus $\omega_n = 2.997$ and $\zeta = 0.799$ which is certainly greater than the 0.6 value required for a 10% overshoot maximum. This means that with very little effort the steady-state ramp error requirement has been satisfied and this has been possible while also satisfying the maximum transient overshoot condition in response to a step input.

4.6   Summary

The topics included in this chapter have been concerned primarily with the study of closed-loop denominator characteristics. It must be said however that the Routh–Hurwitz criterion, which is used to test for system stability, is merely testing the roots of a polynomial and is therefore applicable to both open- and closed-loop system transfer functions. The complexity of the Routh–Hurwitz stability testing method described in Section 4.2 does not increase as the order of the polynomial increases, it merely means that more calculations are necessary. This is a major selling point for the method because higher-order systems, whose characteristic equation factors (roots) are not readily obtainable, can be tested in a relatively straightforward manner to show their stability and even relative stability properties.

A graphical display of relative stability properties is given by plotting the root loci of a closed-loop system transfer function denominator, as the forward gain K is varied. By employing a computer it is not too difficult to obtain a complete root loci plot for a given transfer function simply by plotting the root positions for a large number of K values, although problems can exist for higher-order systems. Often it is not in fact necessary to resort to a full root loci plot, as it is usually the case that a sketch of the loci can be rapidly drawn, with only one or two points being required to any great accuracy. In this chapter it has been shown how the root locus technique can readily incorporate systems which exhibit a transport delay and how the roots of a system are sensitive to parameter variations. Also the possibility of designing a controller by making use of root loci plots was introduced in Section 4.5, and although for many systems the simple lead/lag type of compensator is of sufficient complexity, it may well be necessary, in order to tailor the final root loci to meet a particular performance objective, to employ a multi-pole/multi-zero compensator.

Problems

4.1   For an open-loop system transfer function

$$G(s) = \frac{1}{s(s^3 + 6s^2 + 11s + 6)}$$

show, by means of the Routh–Hurwitz criterion, that if the system is connected

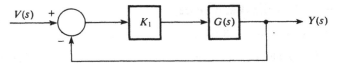

Fig. 4.18

up with an amplifier, of gain K_1, in the unity feedback loop (Fig. 4.18) then the closed-loop system will be stable for $10 > K_1 > 0$.

4.2   Determine whether or not the open-loop system described by the transfer function $G(s)$ is stable, where

$$G(s) = \frac{1}{(s+3)(s-1)(s+2)}$$

A negative feedback loop is applied with transfer function $H(s) = K_1$, in which K_1 is a scalar gain value. Find the closed-loop transfer function and find the ranges of values for K_1 over which the closed-loop system is stable.

4.3   A unity feedback control system has an open-loop transfer function:

$$G(s) = \frac{K(s+2)}{s(s+1)(s^2+2s+2)}$$

Find the range of values for K over which the closed-loop system is stable.

4.4   Determine the conditions for which the feedback control system, with the block diagram shown in Fig. 4.19, is stable.

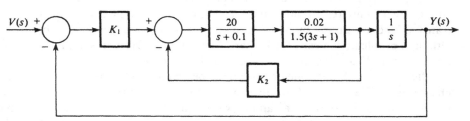

Fig. 4.19

4.5   Determine whether or not the systems with the transfer functions given are stable.

(a) $\dfrac{s+1}{s^5 + 3s^4 + 4s^3 + 4s^2 + 3s + 1}$

(b) $\dfrac{1}{s^5 + 2s^4 + 6s^3 + 10s^2 + 8s + 12}$

(c) $\dfrac{1}{s^5 + 2s^4 + 3s^3 + 6s^2 + 10s + 15}$

4.6   Find the number of roots of the polynomial equation

$$s^3 + 2s^2 + 11s + 20 = 0$$

which lie between $s = 0$ and $s = -1$.

4.7   Find the conditions for which the feedback control system, with the block diagram as shown in Fig. 4.20, is stable.

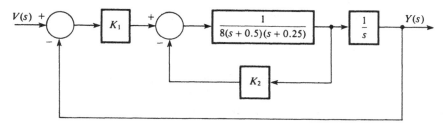

Fig. 4.20

4.8   Find the range of values of the gain K for which the unity feedback closed-loop system is stable when the open-loop transfer function is:

(a) $G(s) = \dfrac{2(s + 1)K}{2s^3 + s^2 + 8s + 1}$

(b) $G(s) = \dfrac{2(1 - s)K}{2s^3 + s^2 + 8s + 1}$

4.9   The open-loop transfer function of a unity feedback system is:

$$G(s) = \frac{K}{s(0.5s + 1)(2s + 1)}$$

Determine the range of values for K over which the closed-loop system will be stable.

Show that if the feedback is made equal to $(s + 3)$ rather than unity, then the stability range will be increased.

4.10   The open-loop transfer function of a system is given by:

$$G(s) = \frac{1}{s(s + 1)(s + 3)}$$

If a variable gain controller whose transfer function is

$$D(s) = K \frac{(s + 1.5)}{(s + 2)}$$

is cascaded with $G(s)$ and a unity negative feedback loop applied, for what values of K will the closed-loop system be stable?

4.11   Draw the root locus diagram for a unity feedback system whose open-loop transfer function is

$$G(s) = \frac{K(s+9)^2}{(s+1)^3}$$

If the open-loop zeros lie at $s = -9.1$ rather than $s = -9$ what effect does this have on the diagram?

Similarly, what happens when the zeros lie at $s = -8.9$?

4.12   A unity feedback system has an open-loop transfer function

$$G(s) = \frac{2.5K(s+2)}{(s-1)(s+1)}$$

Draw the root locus diagram for this system and find the range of values for K over which the closed-loop system is stable.

4.13   Draw the root locus diagram for the system

$$G(s) = \frac{K(s+2)}{s(s+1)}$$

and show the exact breakaway point values.

4.14   Draw the root locus diagram for the system

$$G(s) = \frac{K}{s(s+1)(s+2)}$$

and find the value of K which gives the complex closed-loop pole pair a damping ratio of $\zeta = 0.5$.

4.15   Draw the root locus diagram for the system

$$G(s) = \frac{K}{s(s+4)}$$

How is the diagram affected, when a cascade compensator of the form

$$D(s) = \frac{s+2}{s+6}$$

is employed within a unity feedback loop?

4.16   Consider the open-loop system transfer function

$$G(s) = \frac{10K}{s(s+1)(s+4)}$$

If a cascade compensator of the form

$$D(s) = \frac{s+b}{s+4}$$

is employed, by means of a root locus diagram, find values for K and b which produce the dominant closed-loop pole pair

$$s = -1 \pm \sqrt{3}j$$

4.17   Draw a root locus diagram for the open-loop system

$$G(s) = \frac{1}{s(s+1)}$$

when employed within a unity feedback loop. What is the effect on the root loci when a cascade compensator of the following form is applied

$$D(s) = \frac{s+b}{s+a} \qquad a, b > 0?$$

Consider four cases: (i) $1 < a < b$, (ii) $1 < b < a$, (iii) $b < a < 1$, (iv) $a < b < 1$, with particular attention paid to stability properties.

Consider also the cases: (v) $a < 1 < b$ and (vi) $b < 1 < a$.

4.18   Draw the root locus diagram for the open-loop system

$$G(s) = \frac{K}{s(s^2 + 4s + 8)}$$

when employed within a unity feedback loop.

What is the effect on the loci when a cascade compensator of the form

$$D(s) = \frac{s+10}{s+1}$$

is applied?

Conversely, what happens if the compensator is altered to

$$D(s) = \frac{s+1}{s+10}?$$

4.19   By gain compensation only, within a unity feedback loop, it is desired to control the position of a radio telescope whose open-loop transfer function is

$$G(s) = \frac{K}{(s+7)(s+100)}, \qquad K > 0$$

Draw the root locus diagram for this system and show how, by a suitable choice of cascade compensation, a faster system step response can be achieved. What effect does the employment of such a compensator have on the system's stability properties?

4.20   Consider the open-loop system transfer function

$$G(s) = \frac{K}{s(s+10)}, \qquad K > 0$$

Plot the root locus for this system when the system is placed within a unity feedback loop. Find a suitable cascade compensator which will ensure that both (a) the closed-loop step response has a maximum overshoot of no more than 20% of the final output value and (b) the steady-state error to a unit ramp input is less than 5%.

4.21   Find the sensitivity function for the dominant poles in the closed-loop system of Problem 4.14 with regard to an increase in gain K (when the damping ratio is $\zeta = 0.5$). How does the result compare with the sensitivity function obtained with regard to a decrease in gain K?

Further reading

Brogan, W. L., *Modern control theory*, Prentice-Hall, 1991.

Dickenson, B. W., *Systems analysis and design*, Prentice-Hall, 1991.

Dorf, R. C., *Modern control systems*, 6th ed., Addison-Wesley, 1992.

Evans, W. R., 'Graphical analysis of control systems', *Trans. AIEE*, **67**, 1948, pp. 547–51.

Evans, W. R., 'Control systems synthesis by the root locus method', *Trans. AIEE*, **69**, 1950, p. 66.

Evans, W. R., *Control system dynamics*, New York, McGraw-Hill, 1954.

Hurwitz, A., 'On conditions under which an equation has only roots with negative real parts', in *Selected papers on mathematical trends in control theory*, Dover, 1964.

Jacobs, O. L. R., *Introduction to control theory*, Oxford University Press, 1974.

Johnson, C. D., *Process control instrumentation technology*, Wiley, 1977.

Kuo, B. C., *Automatic control systems*, 3rd ed., Prentice-Hall, 1990.

Mitchell, J. R. and McDaniel Jr., W. L., 'A generalized root locus following technique', *IEEE Trans. on Automatic Control*, **AC-15,** 1970, pp. 483–5.

Mitra, S. K., *Analysis and synthesis of linear active networks*, Wiley, 1969.

Nelson, R. C., *Flight stability and control*, McGraw-Hill, 1989.

Phillips, C. and Harbor R., *Feedback control systems*, Prentice-Hall, 1991.

Reid, J. G., *Linear system fundamentals*, McGraw-Hill, 1983.

Remec, M. J., 'Saddle points of a complete root locus and an algorithm for their easy location in the complex frequency plane', *Proc. Natl. Electronics Conf.*, **21**, 1965, pp. 605–8.

Richards, R. J., *An introduction to dynamics and control*, Longman, 1979.

Routh, E. J., *Dynamics of a system of rigid bodies*, Macmillan, 1892.

Westcott, J. H. and Boothroyd, A. R., 'Application of the electrolytic tank to servomechanisms design', *Automatic and Manual Control*, Butterworth, 1957.

Wojcik, C., 'Analytical representation of root locus', *Trans. ASME J. Basic Engineering*, Ser. D., **86**, 1964.

5

Frequency response analysis

5.1   Introduction

In Chapter 4, control system analysis and design is carried out by investigating the movement of closed-loop roots about the s-plane, in response to a coefficient variation, for the most part a variation in the open-loop gain K. A fundamental assumption is made that the system open-loop transfer function $G(s)$ must be of unity magnitude and must have a phase shift of $-180°$. An alternative approach is to set the gain K as a constant value and to consider the magnitude and phase of $G(s)$ as they vary in response to variations in the roots of $G(s)$. In terms of stability analysis though, the point at which the system has unity magnitude and a phase shift of $-180°$ is still critical as far as stability is concerned as is thoroughly discussed in this chapter.

Remembering that each location in the s-plane consists of a real and imaginary part, i.e. $s = \sigma + j\omega$, an important case arises when $\sigma = 0$, i.e. $s = j\omega$. In Section 3.2.2 it was shown that for a sinusoidal input signal $u(t) = U\cos\omega t$, which is of magnitude U, then for a stable system the steady-state output signal is also a sinusoid, of magnitude Y, i.e. $y(t) = Y\cos(\omega t + \phi)$ where ϕ denotes a phase shift. In fact, by setting $s = j\omega$, it is found that $Y = |G(j\omega)|U$ and $\phi = \underline{/G(j\omega)}$, where $|G(j\omega)|$ is the magnitude of $G(s)$ when $s = j\omega$; see Problem 5.7.

So, if a sinusoidal input signal of frequency ω rad/s is applied at the input of a stable system, the steady-state output signal will also be a sinusoid of the same frequency ω. The output magnitude Y will be the result of the input magnitude U multiplied by the gain $|G(j\omega)|$ of the transfer function at the frequency ω, and the phase shift ϕ will be the phase $\underline{/G(j\omega)}$ of the transfer function at the frequency ω. This is termed the *frequency response* of the system $G(j\omega)$ when values for $|G(j\omega)|$ and $\underline{/G(j\omega)}$ are found for all frequencies ω. A short example is given at the end of this section: see Example 5.1.1.

Frequency response data values for system gain and phase, at many frequency points, may be obtained either by direct analytical means when an exact transfer function is known, or by taking experimental results. Once the data is obtained however, there are several available techniques which can be employed for system analysis and subsequently for compensator design. Each of these, essentially graphical, techniques makes use of the same data values, but presents the results in a different way. In a certain sense, therefore, the three design techniques described in this chapter, namely Bode, Nyquist and Nichols, merely represent different ways of looking at the same system character-istics. However, each of them has properties which make them useful under certain circumstances, and therefore it is worthwhile becoming familiar with the different approaches, even though for a particular case study one of the approaches will probably be much more appropriate than the others.

All three frequency response methods discussed in this chapter have a common element, in that the frequency response of the system open-loop transfer function $G(j\omega)$ is plotted. Stability results obtained from each technique then apply to the unity feedback closed-loop system which contains $G(j\omega)$ in its forward path. Where a compensator $D(j\omega)$ is also employed, it is considered here that this is connected in cascade with the system open-loop transfer function, such that the forward path becomes $D(j\omega)G(j\omega)$ for which a frequency response plot can be obtained. Stability results then apply to the unity feedback closed-loop system with $D(j\omega)G(j\omega)$ in its forward path.

Example 5.1.1

Consider the transfer function:

$$G(s) = \frac{3}{s+2}$$

The frequency response of this system is obtained by setting $s = j\omega$, such that

$$G(j\omega) = \frac{3}{j\omega + 2}$$

then

$$|G(j\omega)| = \frac{3}{(\omega^2 + 4)^{1/2}} \quad \text{and} \quad \underline{/G(j\omega)} = -\tan^{-1}\left[\frac{\omega}{2}\right]$$

which produces the data in Table 5.1.

Table 5.1 Frequency response data for Example 5.1.1

ω	0	2	10	∞
$\|G(j\omega)\|$	1.5	1.061	0.294	0
$\underline{/G(j\omega)}$	0	$-45°$	$-78.7°$	$-90°$

5.2   The Bode diagram

When an input signal, whose magnitude varies sinusoidally with respect to time, is applied to a linear system which is stable, the output of the system will itself be a sinusoid in the steady-state, although its magnitude will be some multiple of the input signal and there may be a phase shift. If the magnitude of the input signal is held constant and a different frequency of signal is chosen, then another output magnitude and corresponding phase shift will be found. For a constant magnitude input signal the resultant output magnitude or gain can therefore be found for a wide range of frequency values and a plot made of the different values, a similar procedure being carried out to form a separate plot of the corresponding phase shifts. A system transfer function can therefore be characterized in terms of its frequency response considered in the form of two plots, one of gain-v-frequency and the other of phase-v-frequency, noting the employment of gain, which is not dependent on input signal magnitude, rather than output magnitude which is.

If a complete gain-v-frequency plot has been obtained over a wide range of frequencies for a particular system, whether by a series of measurements or by calculations from a transfer function model, modification of the system by changing the steady-state gain K or adding a further pole, will mean that all of the values plotted must be multiplied by a relevant factor in order to avoid having to take a further series of measurements. Even so, if the exact nature of the modification is not known, the latter course of action may well be necessary. A much better approach is to use logarithmic values of gain such that any modification can be regarded as an addition, a much simpler process to carry out graphically when compared to multiplication, as can be seen in this section. Indeed, for the design of feedback amplifiers much work was carried out by H. W. Bode employing the logarithmic value of gain plotted vertically on a linear scale against a horizontal nonlinear logarithmic frequency scale. Such frequency response plots, when combined with corresponding plots of phase shift on a linear axis against logarithmic frequency, are therefore referred to as *Bode plots*.

5.2.1   Logarithmic plots

For both Bode gain and phase-v-frequency plots, the vertical gain or phase axis is linearly scaled whereas the horizontal axis is drawn to a logarithmic frequency scale. Hence the frequency response can be presented over a much wider range of frequencies than would be possible over a linear scale.

The system transfer function, in response to a sinusoidal input signal of frequency ω rad/s can be written as

$$G(j\omega) = |G(j\omega)| \underline{/G(j\omega)} \tag{5.2.1}$$

such that for any frequency ω, the gain can also be expressed in logarithmic form as:

$$\text{gain (dB)} = 20 \log_{10} | G(j\omega) | \qquad (5.2.2)$$

where the gain is written in terms of its standard unit the decibel (dB).

As a simple example, consider $| G(j\omega) | = 2$ which results in a gain $= 6.02$ dB, or if $| G(j\omega) | = 1/\sqrt{2}$ then the gain $= -3.01$ dB, the negative sign indicating a gain of less than unity, which can also be regarded as a loss or attenuation.

Example 5.2.1

Consider the system transfer function:

$$G(j\omega) = K \qquad (5.2.3)$$

where K is a scalar gain value, $K > 0$.

The gain of this system is thus $20 \log_{10} K$ dB over all frequencies, with a phase angle of $0°$. The Bode plots for this system are therefore shown in Fig. 5.1.

It is worth noting that for $K < 0$ the Bode gain plot will be identical to that in Fig. 5.1, as $| K | = | - K |$, however the phase plot, although still merely a horizontal line, will be at $180°$ rather than the $0°$ line shown in Fig. 5.1.

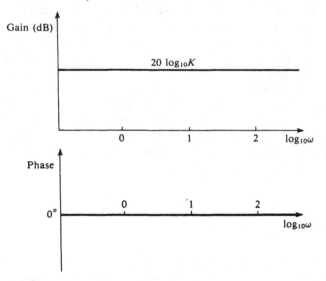

Fig. 5.1   Bode plots for the system (5.2.3)

Example 5.2.2

Consider the system transfer function:

$$G(j\omega) = Kj\omega \qquad (5.2.4)$$

The phase of which will be $+90°$ over all frequencies, when $K > 0$, and whose logarithmic gain is:

$$\text{gain (dB)} = 20 \log_{10} | K\omega | = 20 \log_{10} K + 20 \log_{10} \omega \qquad (5.2.5)$$

such that when

$$\omega = 0.1, \qquad \text{gain (dB)} = 20 \log_{10} K - 20$$
$$\omega = 1 \quad , \qquad \text{gain (dB)} = 20 \log_{10} K + 0$$
$$\omega = 10 \ , \qquad \text{gain (dB)} = 20 \log_{10} K + 20$$

The Bode plots for the system (5.2.4) are shown in Fig. 5.2, where the slope of the gain plot is given as $+20$ dB/decade, that is the gain increases by 20 dB every time the frequency ω is increased by a multiplying factor of 10.

For a system transfer function of the form $G(j\omega) = K[j\omega]^2$, the phase shift is $+180°$ over all frequencies, for $K > 0$, with

$$\text{gain (dB)} = 20 \log_{10}K + 40 \log_{10} \omega$$

Hence the gain plot will still pass through the point $20 \log_{10} K$ when $\omega = 1$ rad/s, but the straight line will have a slope equal to $+40$ dB/decade.

Similarly, for a general system transfer function of the form

$$G(j\omega) = K[j\omega]^p \qquad (5.2.6)$$

it follows that the phase will be $[+90°] \times p$ over all frequencies and the gain:

$$\text{gain (dB)} = 20 \log_{10} K + 20p \log_{10} \omega \qquad (5.2.7)$$

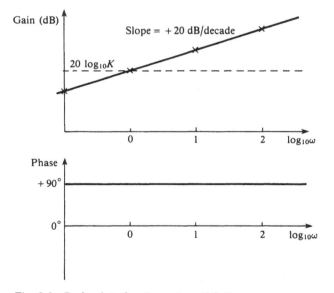

Fig. 5.2 Bode plots for the system (5.2.4)

such that it passes through the point $20 \log_{10} K$ when $\omega = 1$ rad/s and the straight-line plot has a slope of $+20p$ dB/decade.

The phase and gain results for the system (5.2.6) in fact hold for all integers $\{p = 0, \pm1, \pm2, \ldots\}$, which means that for p negative this will result in a negative phase shift and a straight-line plot with negative slope, as is shown in the following example.

Example 5.2.3

Consider the system transfer function:

$$G(j\omega) = K[j\omega]^{-1} = K/j\omega \tag{5.2.8}$$

The phase of which will be $-90°$ over all frequencies, when $K > 0$, and whose logarithmic gain is:

$$\text{gain (dB)} = 20 \log_{10} K - 20 \log_{10} \omega \tag{5.2.9}$$

The Bode plots for the system (5.2.8) are shown in Fig. 5.3, where the slope of the gain plot is -20 dB/decade.

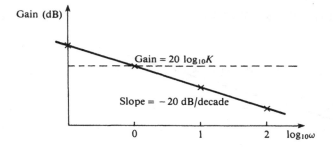

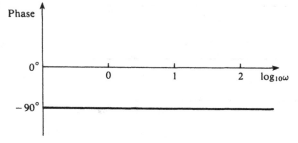

Fig. 5.3   Bode plots for the system (5.2.8)

5.2.2 Asymptotic approximations

Thus far we have merely considered Bode plots for systems consisting of a gain K and either poles or zeros which lie at the origin of the s-plane. In order to generalize the results it is necessary also to consider firstly systems with poles and/or zeros which lie along the s-plane real axis and secondly systems with complex roots. An important feature to be aware of is that because logarithmic gain plots are being obtained, so the effect of each individual pole or zero can be found separately, the effects simply being added in order to obtain the total system frequency response.

Consider the system transfer function:

$$G(j\omega) = \frac{K}{(j\omega + a)} \tag{5.2.10}$$

To simplify the analysis of this system, its response will be considered at high and low frequencies in the first instance. For very low frequencies, when $\omega \ll a$, the phase shift exhibited by the system is $0°$ and the gain is:

$$\text{gain (dB)} = 20 \log_{10} [K/a]: \quad \omega \ll a \tag{5.2.11}$$

However, for very high frequencies, when $\omega \gg a$, the phase shift exhibited by the system is $-90°$ and the gain is:

$$\text{gain (dB)} = 20 \log_{10} [K/a] - 20 \log_{10} [\omega/a]: \quad \omega \gg a \tag{5.2.12}$$

where the a term has been used as a denominator of both K and ω in order to link (5.2.12) with (5.2.11).

The two gain values are in fact both equal to $20 \log_{10}(K/a)$, when $\omega = a$ in (5.2.12), such that an asymptotic approximation of the complete frequency response for the system (5.2.10), is shown in Fig. 5.4. The true response is in fact very close to the asymptotic approximation, the worst error occurring when $\omega = a$, at which point

$$G(j\omega) = \frac{K/a}{j + 1}: \quad \omega = a$$

The phase shift exhibited by this transfer function when $\omega = a$ is $-45°$, with an associated gain of:

$$\begin{aligned}
\text{gain (dB)} &= 20 \log_{10} [K/a] - 20 \log_{10}[\sqrt{2}] \\
&= 20 \log_{10} [K/a] - 3.01 \text{ dB}
\end{aligned}$$

Hence the true response is approximately 3 dB below the asymptotic sketch at the point of largest error, when $\omega = a$. This point, known as the 3 dB point, can be employed such that in order to quickly sketch the actual frequency response gain plot for a system such as (5.2.10), having obtained the high- and low-frequency asymptotes it is then only necessary to ensure that the actual plot passes through the 3 dB point. The frequency at which this occurs, in this case $\omega = a$, is known as the *break frequency*, although it is occasionally also referred to as the corner frequency.

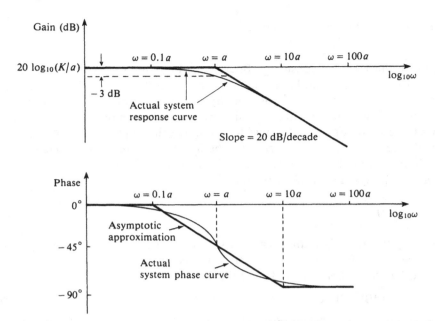

Fig. 5.4 Bode plots with asymptotes for the system (5.2.10)

The phase-v-frequency plot, also shown in Fig. 5.4 for the system (5.2.10), is known for both low and high frequencies and also when $\omega = a$. A simple approximation to the actual plot can be obtained by joining together the plot at $\omega = a/10$ with that at $\omega = 10a$, the line therefore also passes through the point $\omega = a$, i.e. the break frequency. This straight line approximation is never more than $6°$ away from the actual phase plot, and is sufficient for many purposes. It will be seen later though that it is sometimes necessary to find a more accurate plot of the phase characteristics and in certain cases exact calculations employing several values of ω are required.

The results given for the system (5.2.10) can be generalized by considering a system of the form:

$$G(j\omega) = \frac{K}{[j\omega + a]^p} \qquad (5.2.13)$$

where $\{p = 0, \pm 1, \pm 2, \ldots\}$.

For very low frequencies, when $\omega \ll a$, the phase shift will be $0°$, with a gain of:

$$\text{gain (dB)} = 20 \log_{10}\left[\frac{K}{a^p}\right]: \qquad \omega \ll a \qquad (5.2.14)$$

However, for very high frequencies, when $\omega \gg a$, the phase shift is $-90° \times p$, noting that when p is negative this will result in a positive phase shift. The gain is then:

$$\text{gain (dB)} = 20 \log_{10}\left[\frac{K}{a^p}\right] - 20p \log_{10}\left[\frac{\omega}{a}\right]: \qquad \omega \gg a \qquad (5.2.15)$$

The break frequency, $\omega = a$, is therefore unaltered by the integer p, although the slope of the gain-v-frequency plot will be $-20p$ dB/decade for all $\omega > a$. Hence for $p = -1$ the Bode gain and phase plots will be simply a mirror image about the horizontal axis, of those given in Fig. 5.4.

Example 5.2.4

Consider the system transfer function:

$$G(j\omega) = \frac{5(j\omega + 2)}{(j\omega + 3)} \qquad (5.2.16)$$

which can also be written as

$$G(j\omega) = \frac{10}{3} \cdot \frac{(1 + j\omega/2)}{(1 + j\omega/3)}$$

For very low frequencies the phase shift exhibited by the system is $0°$ and the gain is:

$$\text{gain (dB)} = 20 \log_{10}(10/3): \qquad \omega \ll 2, 3$$

However for very high frequencies, the phase shift exhibited by the system is $0°$,

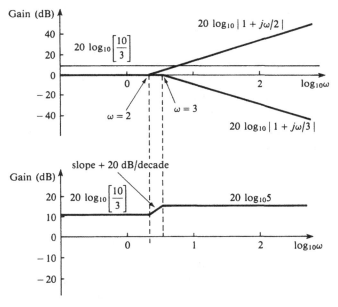

Fig. 5.5 Individual term and overall Bode gain plots for system (5.2.16) – asymptotes only shown: note change of scale

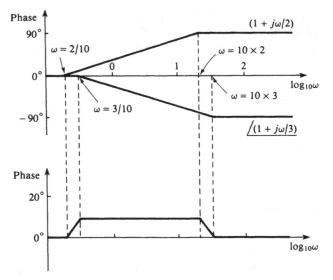

Fig. 5.6   Individual term and overall Bode phase plots for system (5.2.16) – asymptotes only shown: note change of scale

while the gain is:

$$\text{gain (dB)} = 20 \log_{10}(10/3) + 20 \log_{10}(\omega/2) - 20 \log_{10}(\omega/3)$$

$$= 20 \log_{10}\left\{\frac{10}{3}\frac{\omega}{2}\cdot\frac{3}{\omega}\right\} = 20 \log_{10} 5$$

Armed with these checks on the low and high frequency values, the individual gain and phase frequency responses can be obtained separately and simply added to give the overall response as shown in Figs 5.5 and 5.6. This procedure of summing the individual pole and zero plots can be carried out for a system with any number of poles and zeros, although the gain must be calculated at high and low frequencies, as was done in the example. A special case must be treated in a slightly different way, and that is the effect of complex roots.

5.2.3 Complex roots

If a system exists in which two roots are real, the effect of these roots can be treated separately and merely added to find the overall Bode plot. Consider for example a system transfer function with unity numerator and a denominator equal to $(s + \beta_1)(s + \beta_2)$, in which $\beta_2 > \beta_1$. The Bode gain plot for such a system is shown in Fig. 5.7.

The slope of the Bode plot is -20 dB/decade over the frequency range between the

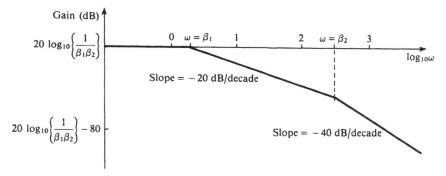

Fig. 5.7 Bode gain plot for the system

$$G(j\omega) = \frac{1}{(j\omega + \beta_1)(j\omega + \beta_2)}$$

– asymptotes shown for $\beta_2 > \beta_1$

two break frequencies, $\omega = \beta_1$ and $\omega = \beta_2$, after which it becomes -40 dB/decade. Hence for a plot in which $\beta_1 = \beta_2$ the slope will become -40 dB/decade directly $\omega = \beta_1 = \beta_2$.

Not all roots are simply real values, they can occur in complex pairs as can be seen from the polynomial equation:

$$s^2 + 2\zeta\omega_n s + \omega_n^2 = 0 \qquad (5.2.17)$$

which produces complex roots when $0 < \zeta < 1$. In particular consider a system transfer function of the form

$$G(s) = \frac{K}{s^2 + 2\zeta\omega_n s + \omega_n^2} \qquad (5.2.18)$$

Asymptotic approximations of the frequency response for this system cannot be produced with any great deal of accuracy for all $0 < \zeta < 0.5$, a simple but rough approximation being to consider the plot to be that pertaining to the case when two real and equal roots exist, i.e. $\zeta = 1$, such that the denominator becomes $(s + \omega_n)^2$. As can be seen from the Bode plots in Fig. 5.8, which show the gain and phase-v-frequency plots for various values of damping factor ζ, the error incurred, by employing the suggested simple approximation, gets larger as $\zeta \to 0$. From (5.2.18) we have that:

$$G(j\omega) = \frac{K/\omega_n^2}{\left(1 - \dfrac{\omega^2}{\omega_n^2}\right) + j2\zeta\dfrac{\omega}{\omega_n}} \qquad (5.2.19)$$

and hence:

$$\text{gain (dB)} = 20\log_{10}\left[\frac{K}{\omega_n^2}\right] - 20\log_{10}\left[\left(1 - \frac{\omega^2}{\omega_n^2}\right)^2 + \left(2\zeta\frac{\omega}{\omega_n}\right)^2\right]^{1/2} \qquad (5.2.20)$$

The first part of this expression, i.e. $20\log_{10}(K/\omega_n^2)$, is merely a gain value providing a figure which is constant for all frequencies, ω, and which has $0°$ phase shift, and

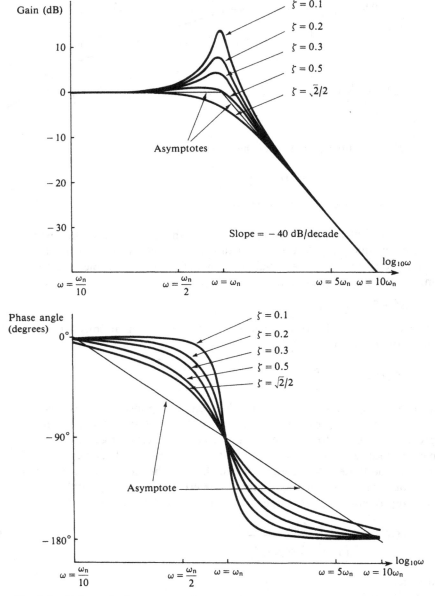

Fig. 5.8 Gain and phase versus frequency plots for a quadratic denominator with varying damping factor, ζ

therefore which can be simply added to the remaining plot obtained from the second part of the expression in (5.2.20). Asymptotes for the remainder of (5.2.20) can be obtained by considering firstly low frequencies for which $\omega \ll \omega_n$ whereby

$$- 20 \log_{10}\left[\left(1 - \frac{\omega^2}{\omega_n^2}\right)^2 + \left(2\zeta\frac{\omega}{\omega_n}\right)^2\right]^{1/2} \simeq 0 \text{ dB } \underline{/0^\circ} \qquad (5.2.21)$$

and secondly high frequencies for which $\omega \gg \omega_n$ whereby

$$- 20 \log_{10}\left[\left(1 - \frac{\omega^2}{\omega_n^2}\right)^2 + \left(2\zeta\frac{\omega}{\omega_n}\right)^2\right]^{1/2} \simeq - 40 \log_{10}\frac{\omega}{\omega_n} \text{ dB } \underline{/180^\circ} \qquad (5.2.22)$$

If $\omega = \omega_n$ then the gain of (5.2.22) is 0 dB and this is therefore the frequency at which the two asymptotes cross. It can be seen that if ω is increased through a decade, i.e. $\omega = 10\omega_n$, the magnitude of (5.2.22) becomes $- 40$ dB and hence the high frequency asymptote has a slope equal to $- 40$ dB/decade.

It must be remembered that the plot shown in Fig. 5.8 is merely that for the second part of (5.2.20), and the gain-v-frequency plot must be modified by adding the constant gain value caused by K/ω_n^2, at all frequencies. Similarly, if the transfer function of (5.2.19) is cascaded with further denominator and/or numerator terms, then the final Bode plots can be obtained, as was demonstrated in Example 5.2.4, by adding the plots obtained separately for each of the terms.

5.3   Compensation using the Bode plot

In the root locus technique described principally in Section 4.3, the loci of closed-loop denominator roots are plotted in the s-plane, as the gain K varies in the system open-loop transfer function $G(s)$, where

$$G(s) = K\frac{B(s)}{A(s)} \qquad (5.3.1)$$

in which $B(s)$ and $A(s)$ are the open-loop numerator and denominator polynomials respectively. When $G(s)$ is configured within a unity feedback system, we obtain the characteristic equation

$$1 + G(s) = 0 \qquad (5.3.2)$$

by setting the closed-loop denominator to zero, and this means that for any point in the s-plane to exist on a root locus, it must satisfy both

$$| G(s)| = 1 \qquad (5.3.3)$$

and, for all $K > 0$

$$\underline{/G(s)} = (2k + 1)\pi \qquad (5.3.4)$$

in which $k = 0, \pm 1, \pm 2, \dots.$

If the system frequency response is of concern, as is the case when obtaining a Bode plot, the only part of the s-plane considered is the imaginary axis, and due to the symmetrical properties exhibited by a system transfer function to positive and negative frequencies, only positive frequencies need be taken into account. By setting $s = j\omega$ in (5.3.2) we are imposing a condition such that when a system satisfies both

$$| G(j\omega) | = 1 \qquad\qquad (5.3.5)$$

and, for all $K > 0$

$$\underline{/G(j\omega)} = (2k + 1)\pi \qquad\qquad (5.3.6)$$

then that system is critically stable. This condition of critical stability only applies for the particular gain value K and frequency ω which result in a solution to both (5.3.5) and (5.3.6), a point which is hopefully made a little clearer in Section 5.3.1.

With reference to the unity feedback system shown in Fig. 5.9, $E(j\omega)$ denotes the error between the desired, $V(j\omega)$, and actual, $Y(j\omega)$, output signal values and $G(j\omega) = G_0(j\omega)$ is the open-loop transfer function.

A basic assumption is therefore made, as with the root locus technique, that a unity feedback loop is applied. If in fact the feedback loop is other than unity, the equivalent open-loop transfer function within a unity feedback loop must first be obtained, as shown in Section 3.4.1 [*Note:* In fact for the analysis carried out in Sections 5.3.1, 5.3.2 and 5.3.3 for closed-loop stability purposes, it is sufficient to consider the frequency response of $G(j\omega)H(j\omega)$, where $H(j\omega)$ is the feedback transfer function, rather than calculating the equivalent unity feedback open-loop function, because similar stability results hold: see Appendix 4]. We can write down, from Fig. 5.9, the error equation:

$$E(j\omega) = - G(j\omega)E(j\omega) + V(j\omega) \qquad\qquad (5.3.7)$$

and if $G(j\omega)$ is such that (5.3.6) is satisfied, it follows that:

$$E(j\omega) = | G(j\omega) | E(j\omega) + V(j\omega) \qquad\qquad (5.3.8)$$

Neglecting, due to its lack of relevance to this particular topic, the input signal $V(j\omega)$, any existing error will tend to decrease, i.e. the closed-loop system is stable, as long as $| G(j\omega) | < 1$; conversely if $| G(j\omega) | > 1$ the closed-loop system is unstable, resulting in an increase in the error signal. For the case $| G(j\omega) | = 1$ the closed-loop system is critically stable and (5.3.5) is also satisfied.

Assume a particular open-loop system to be of the form

$$G(j\omega) = K\frac{B(j\omega)}{A(j\omega)} \qquad\qquad (5.3.9)$$

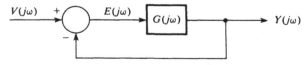

Fig. 5.9 Unity feedback loop

This system is closed-loop critically stable when employed within a unity feedback loop, subject to the fact that both (5.3.5) and (5.3.6) hold for $K = K' > 0$ and $\omega = \omega'$. If then, the system (5.3.9) is configured in the same way, with $\omega = \omega'$, it follows that if $K = \bar{K}$, where $K' > \bar{K} > 0$, the system is now closed-loop stable because

$$| G(j\omega') | = \left| \bar{K}\, \frac{B(j\omega')}{A(j\omega')} \right| < 1$$

on condition that $\underline{/G(j\omega')} = 180°$.

5.3.1   Gain and phase margins

An open-loop transfer function, when connected within a unity feedback loop, will be closed-loop critically stable if it satisfies both (5.3.5) and (5.3.6). If however the transfer function satisfies (5.3.6) and exhibits a gain K such that it is closed-loop stable, i.e. $| G(j\omega) | < 1$, then the difference between $| G(j\omega) |$ and unity, at a phase angle of 180°, gives an indication of how stable the system is, i.e. the further $| G(j\omega) |$ is from unity, the more K can be increased before the closed-loop system becomes unstable. Because $20 \log_{10} | G(j\omega) | = 0$ dB when $| G(j\omega) | = 1$, so the *gain margin* is defined as the amount (in dB) by which the magnitude of $G(j\omega)$ must be increased in order to be equal to 0 dB, when $\underline{/G(j\omega)} = \pm 180°$. It can therefore simply be taken as the inverse Bode magnitude of $G(j\omega)$ at the frequency for which the phase angle of $G(j\omega)$ is 180°.

At a particular frequency, a transfer function could satisfy the magnitude condition (5.3.5) but the resultant phase angle at that frequency may well be something other than 180°. As can be seen in the following example, for a stable system the phase angle will be less negative than $-180°$ (this is the same as more positive than $+180°$), and because $20 \log_{10} | G(j\omega) | = 0$ dB when $| G(j\omega) | = 1$, so the *phase margin* is defined as the additional phase lag required to make $\underline{/G(j\omega)} = \pm 180°$ at the frequency for which the magnitude of $G(j\omega)$ is equal to 0 dB, i.e. it is the extra phase lag which would be necessary to make the system unstable. The phase margin can simply be taken as the difference between the Bode phase of $G(j\omega)$ and $\pm 180°$ at the frequency for which $| G(j\omega) | = 0$ dB.

Although strictly intended to describe the relative stability of a unity feedback closed-loop system, both the gain and phase margins can be used to describe the relative instability of a system by considering for the gain margin the amount by which the magnitude of $G(j\omega)$ must be decreased in order to be equal to 0 dB; and for the phase margin the additional phase lead (negative phase lag) required to make $\underline{/G(j\omega)} = \pm 180°$ at the frequency for which the magnitude of $G(j\omega)$ is 0 dB. For an unstable closed-loop system both the gain and phase margins are therefore given as negative values, whereas for a stable closed-loop system they are both positive.

An example is now given to show how the gain and phase margins are measured, remembering that for good accuracy the asymptotic approximation to both Bode gain and phase plots must be corrected as much as possible before measurements are taken.

Example 5.3.1

Consider the open-loop transfer function:

$$G(j\omega) = \frac{20}{j\omega(j\omega + 2)(j\omega + 5)} \qquad (5.3.10)$$

The Bode (a) gain and (b) phase versus frequency plots for this system are shown in Fig. 5.10 (a) and (b) respectively.

From Fig. 5.10(b) it can be seen that the phase angle of the response is equal to $-180°$ at a frequency of $\omega_1 = 3.2$ rad/s, and by reference to Fig. 5.10(a) it can be seen that the gain of the response at this frequency is -11 dB; the gain margin is therefore 11 dB.

From Fig. 5.10(a) it can be seen that the gain of the response is equal to 0 dB at a frequency of $\omega_2 = 1.5$ rad/s, and by reference to Fig. 5.10(b) it can be seen that the phase angle of the response at this frequency is $-144°$; the phase margin is therefore $180° - 144° = 36°$.

If the numerator gain of $G(j\omega)$ in (5.3.10) is made greater than 20, this will have the effect of shifting the whole gain-v-frequency plot up by the same increase in gain for all frequencies, hence the 0 dB crossover point will occur at a higher frequency. The phase-v-frequency plot is however unaffected by a change of gain, hence ω_2 will approach ω_1 as the open-loop gain is increased, resulting in an unstable closed-loop system when the gain is increased to such an extent that $\omega_2 > \omega_1$, ($\omega_2 = \omega_1$ giving critical stability).

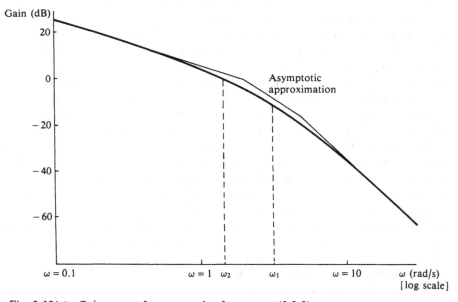

Fig. 5.10(a) Gain versus frequency plot for system (5.3.9)

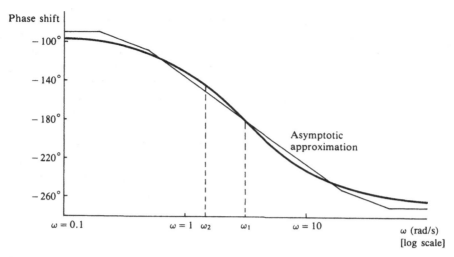

Fig. 5.10(b)   Phase angle versus frequency plot for system (5.3.10)

5.3.2   Gain compensation

It is usually the case that certain requirements exist concerning the performance of a closed-loop control system, often these arise due to specifications made in terms of stability properties such as the gain and phase margin. In so far as the Bode gain and phase versus frequency plots are concerned the result of this is that a controller selection is made in order to reshape the original Bode plot, obtained simply from the open-loop transfer function, such that the appropriate specifications can be satisfied.

Where performance specifications are relatively simple it is in many cases possible to meet the requirements merely by adjusting the system open-loop gain K, within a unity feedback loop. In practice this might actually mean that an amplifier is placed in cascade with the open-loop system, resulting in an amplification of the error signal between the desired and actual output signals.

So, for a system of the type $G(j\omega) = KB(j\omega)/A(j\omega)$, an increase in gain K will result in the gain-v-frequency plot being shifted up by an appropriate amount over all frequencies, whereas a decrease in gain K will result in the gain-v-frequency plot being shifted down by an appropriate amount over all frequencies. A variation in the gain K however, has no effect whatsoever on the phase-v-frequency plot.

Consider again the system described in (5.3.10) and employed in Example 5.3.1. Initially the system has a phase margin of $36°$, the specification is made though that a phase margin of $46°$ is required. It is therefore necessary that at the frequency for which the phase shift is $180° + 46° = -134°$, the gain must be $|G(j\omega)| = 0$ dB.

With reference to Fig. 5.10(b), the phase shift is $-134°$ at a frequency of

$\omega = 1.1$ rad/s, and on Fig. 5.10(a) the gain at this frequency is found to be approximately $+3.5$ dB. It follows therefore that if the gain is reduced by 3.5 dB over all frequencies, then at a frequency of $\omega = 1.1$ rad/s the magnitude will be 0 dB, thus achieving the desired objective of a phase margin $= 46°$. Now 3.5 dB corresponds, by taking antilogs, to a gain of 1.496, and hence the system open-loop gain of 20 must be divided by 1.496 to result in a new open-loop gain of 13.37. So whereas in the first instance $20 \log_{10} 20 = 26.02$ dB, now $20 \log_{10} 13.37 = 22.52$ dB, i.e. 3.5 dB down on its previous value.

To summarize these results: in the first instance the open-loop system defined in (5.3.10), with $K = 20$, when connected in a unity feedback loop, results in a phase margin $= 36°$; when the gain of the open-loop system is reduced to $K = 13.37$, the phase margin is then $46°$, once a unity feedback loop is employed.

5.3.3 Lag/lead compensation

Although it is sometimes possible to meet one particular specification simply by adjustment of the system open-loop gain, in practice such a solution might in fact not be possible due, for instance, to saturation effects when a large amplification factor is necessary. Further, if more than one specification is made it will almost always be the case that with only one control parameter to select, i.e. the open-loop gain, then either only one specification can be met or a compromise must be made between the specifications. Such an eventuality is neither a desirable nor a necessary feature, as with very little effort a slightly more powerful controller can be employed in cascade with the open-loop system as shown in Fig. 5.11.

Because it is connected in cascade with the open-loop system, the Bode frequency response plots for the compensator can be simply added to those of the open-loop system in order to obtain the response of the total plant. The response of the compensator itself is in fact fairly straightforward to analyze if it is noted that at low frequencies ($\omega \to 0$) the compensator gain, whether lag or lead, is b/a, whereas at high frequencies ($\omega \to \infty$) the gain is unity. It can also be seen from the compensator in Fig. 5.11 that the phase shift caused by the compensator itself will be $0°$ at both high and low frequencies, the lag effect – a negative phase contribution, or the lead effect – a positive phase contribution, only being apparent in the mid-frequency range.

For a *lag* compensator, the phase angle will be negative in the mid-frequency range, which means that the effect of the denominator $j\omega + a$, in the compensator, is witnessed

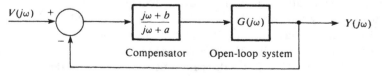

Fig. 5.11 Lag/lead cascade compensator

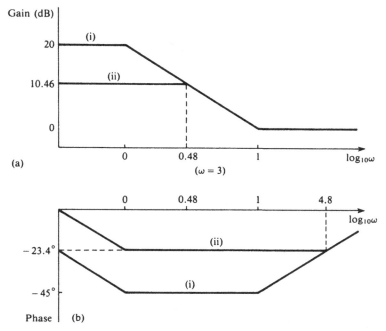

Fig. 5.12 Gain and phase versus frequency asymptote plots for a lag compensator with
(i) $a = 1$, $b = 10$ and (ii) $a = 3$, $b = 10$

(a) Gain versus frequency
(b) Phase versus frequency

at a lower frequency than the numerator $j\omega + b$, hence $a < b$ for a lag compensator. The Bode gain and phase-v-frequency asymptote plots for a lag compensator are shown in Fig. 5.12(a) and (b) respectively.

It is often the case that as well as the essential components of a lag compensator shown by Fig. 5.12, the system open-loop gain K can be adjusted or the compensator has itself an adjustable gain, which has the same resultant effect. This means that the phase versus frequency plot, and the extent to which a negative phase shift is introduced, is selected by the ratio b/a, as indeed is the relative increase in gain between low and high frequencies. The actual gain value at any frequency is though still something which can be selected by means of K, once an appropriate ratio of $b : a$ has been chosen. The effect of a lag compensator is therefore firstly to introduce a negative phase component in the mid-frequency range and secondly to increase the importance of the lower frequency response when compared to the higher frequency response; it can therefore be thought of as a low-pass filter.

The result of employing a lag compensator in terms of gain and phase margin values is dependent on the choice of break frequency values, $\omega = a$ and $\omega = b$: in particular their relationship with the frequency at which the system gain is 0 dB and that at which the system phase is $-180°$. In general though the lag compensator tends to slow down the

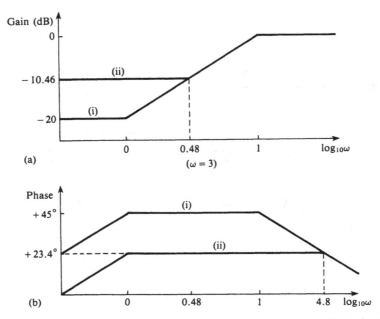

Fig. 5.13 Gain and phase versus frequency asymptote plots for a lead compensator with (i) $a = 10$, $b = 1$ and (ii) $a = 10$, $b = 3$

(a) Gain versus frequency
(b) Phase versus frequency

response of a system due to an emphasis being placed on low frequency (slower) modes at the expense of high frequency (faster) modes and this tends to improve relative stability properties. It is therefore quite possible to either increase or decrease both gain and phase margin values, by employing a lag compensator, the end result being dependent on the values a, b and K, as can be seen in Example 5.3.2.

For a *lead* compensator, the phase angle will be positive in the mid-frequency range which means that the effect of the numerator $j\omega + b$, in the compensator, is witnessed at a lower frequency than the denominator $j\omega + a$, hence $a > b$ for a lead compensator. The Bode gain and phase-v-frequency asymptote plots for a lead compensator are shown in Fig. 5.13(a) and (b) respectively.

The phase-v-frequency plot, and the extent to which a positive phase shift is introduced by a lead compensator, is selected by the ratio b/a, remembering that now $a > b$. Indeed this ratio also determines the relative increase in gain between high and low frequencies, the actual gain value being adjustable by means of the open-loop system gain, K. The effect of a lead compensator is therefore firstly to introduce a positive phase component into the mid-frequency range (the phase shift being zero as both $\omega \rightarrow 0$ and $\omega \rightarrow \infty$) and secondly to increase the higher frequency response when compared to the lower frequency response, it can therefore be thought of as a high-pass filter.

The result of employing a lead compensator in terms of gain and phase margin values is dependent on the choice of break frequency values $\omega = a$ and $\omega = b$, and their relationship with both the frequency at which the system gain is 0 dB and that at which the system phase is $-180°$. In general though the lead compensator tends to speed up the response of a system due to an emphasis being placed on high frequency (faster) modes at the expense of low frequency (slower) modes. It is possible to either increase or decrease both gain and phase margin values, by employing a lead compensator, the end result being dependent on the values a, b and K.

Example 5.3.2

The objective in the use of lag/lead cascade compensation is to reshape the Bode gain and phase-v-frequency plots of an open-loop system in order that the unity feedback closed-loop system meets desired performance requirements in terms of gain and phase margins and low frequency gain.

Consider the open-loop transfer function:

$$G(j\omega) = \frac{10K}{j\omega(j\omega + 1)(j\omega + 2)} \qquad (5.3.11)$$

where K is an adjustable gain.

It is required that a unity feedback system be designed to produce a minimum gain margin value of 10 dB and a minimum phase margin of $30°$.

The uncompensated system's Bode gain and phase-v-frequency responses with $K = 1$, are shown in Fig. 5.14(a) and (b) respectively, where it can be seen that within an uncompensated unity feedback system the gain margin is approximately -8 dB and the phase margin approximately $-18°$, hence the

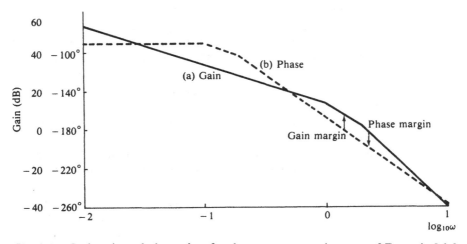

Fig. 5.14 Bode gain and phase plots for the uncompensated system of Example 5.3.2

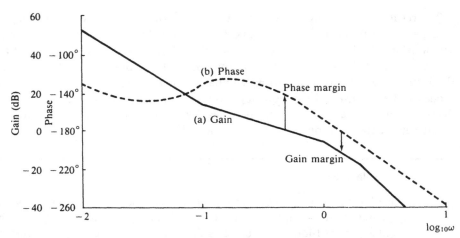

Fig. 5.15 Bode gain and phase plots for the phase lag compensated system of Example 5.3.2

system is unstable. Note that the results are found here simply in terms of asymptotic approximations to both the gain and phase versus frequency plots.

In order to meet the performance specifications, a phase *lag* compensator can be used such that (a) the overall gain is reduced at higher frequencies ($\omega > 0.1$), and (b) the negative phase shift introduced by the compensator occurs at a lower frequency.

Initially the gain margin is measured at approximately $\omega = 1.4$ rad/s when the phase is $-180°$. If the gain is lowered at this frequency by 20 dB with no resultant change in phase, a gain margin of $+12$ dB would be provided. Such a decrease in gain is achieved by selecting $K = 0.1$ and if it is ensured that $b/a = 10$, there will be no change in gain at low frequencies.

Such a choice of gain K will also mean that the total gain will now be 0 dB at a frequency of $\omega = 0.48$ rad/s (the gain was previously $+20$ dB at this frequency). At this frequency the phase of the uncompensated system is $42°$ more positive than $180°$, i.e. a phase margin of $42°$ (well over the required minimum of $30°$) would be obtained if no alteration was made to the phase characteristics. A choice of $10a = b < 0.048$ would ensure that almost no phase change occurs at $\omega = 0.48$ rad/s, although as we are well over the required $30°$ a fairly small negative phase shift at $\omega = 0.48$ rad/s is acceptable and this is provided by the selection $10a = b = 0.1$. The resultant effect obtained by employing a lag compensator of the form $(j\omega + 0.1)/(j\omega + 0.01)$ along with $K = 0.1$ is shown in Fig. 5.15(a) and (b), and it can be seen that the frequency at which the gain is 0 dB is now 0.48 rad/s (as opposed to $\omega = 2.2$ rad/s for the uncompensated system). Further, a gain margin of $+12$ dB and a phase margin of $38°$ are obtained, thus meeting the performance requirements. So employment of a lag compensator has resulted in a lower gain crossover frequency, that is the frequency at which the gain is 0 dB.

The *bandwidth* of a control system is the range of frequencies for which the gain of that system does not fall below -3 dB. For many control systems the gain does not fall below -3 dB at all in the low frequency end of the spectrum, and hence the bandwidth is simply given by the high frequency cut-off (-3 dB) point. In Example 5.3.2 the bandwidth of the uncompensated system is $\omega = 2.42$ rad/s, i.e. that is the frequency at which the gain is equal to -3 dB, whereas once the phase lag compensator is applied, the bandwidth is reduced to $\omega = 0.6$ rad/s. The introduction of a phase lag compensator has therefore resulted in a reduction in bandwidth.

The bandwidth of a system gives an indication of that system's speed of response, such that the larger the bandwidth is, so that system's response is faster. For the phase lag compensator in Example 5.3.2 the bandwidth was reduced, which means that the response speed of the system was slowed down. Slowing the response down had, in the example, the effect of improving the relative stability properties by increasing both the gain and phase margins.

It is however possible to increase the gain and phase margins as well as increasing the bandwidth, by means of applying a phase lead (rather than lag) compensator to the uncompensated system in Example 5.3.2. This is in fact considered in Problem 5.2, in which the lead compensator can be effected by ensuring that the peak of the phase lead characteristics occurs in the frequency neighborhood of the uncompensated system's gain crossover frequency. The gain crossover frequency, once the system has been phase lead compensated, will then increase from its original value, resulting in an increased bandwidth.

It is worth noting that, where necessary, several lag or lead compensators can be placed in cascade, such that the logarithmic gain effects and phase modifications of the individual compensators can be added to achieve much larger values of gain and phase shift in the important crossover region. It is also possible to employ a lag-lead compensator, by cascading lag and lead compensators. The effect of this is to retain the gain and phase characteristics of the uncompensated system at both high and low frequencies, while for a band of mid-range frequencies the system's gain is either increased (band pass filter) or decreased (band limiting filter). The Bode characteristics of the lag-lead compensator can be found straightforwardly by adding the separate lag and lead Bode plots.

5.4   The Nyquist stability criterion

In Section 5.2 it was explained how the steady-state response of a stable, linear system to a sinusoidal input of constant magnitude can be found, in terms of system gain and phase characteristics, over a wide range of frequency values. The gain and phase characteristics can then be plotted separately against frequency, hence forming Bode plots, once logarithmic values of gain and frequency are employed.

Once gain and phase versus frequency values have been obtained for a system response, an alternative form in which the results could be presented is by means of plotting gain versus phase over the range of frequencies involved. By ensuring that the response of the system is considered over all frequencies from $0 \leqslant \omega \leqslant \infty$, then a plot of system gain versus phase is known as the *polar* or *Nyquist plot*, the response being termed the *frequency response*.

If the transfer function, $G(s)$, of a system is known, then its frequency response can be found analytically simply by solving for $G(j\omega)$, in terms of gain and phase, over a large number of frequency values. This can, however, result in a long drawn-out plotting process, even when carried out by means of a computer. In fact for the vast majority of system transfer functions it is not necessary to solve for a large number of frequency values; a few important, simple to evaluate, points will produce an adequate representation. A procedure applicable, in order to simplify Nyquist plots, is given in the following section.

5.4.1   Frequency response plots

There are four important points to be considered when plotting the frequency response of a system. Once the system transfer function has been evaluated in terms of gain and phase at these points, the complete plot can subsequently be sketched by joining the points. The points are:

1. Start of plot when $\omega = 0$
2. End of plot when $\omega = \infty$
3. Crossing of the real axis
4. Crossing of the imaginary axis

So for each of these points, and no more, the gain and phase of the system transfer function is found.

The transfer function, $G(j\omega)$, can be regarded in two ways. Firstly in terms of polar coordinates, i.e. magnitude and phase, and secondly in terms of cartesian coordinates, i.e. real and imaginary parts. The cartesian coordinate technique is particularly useful as far as finding real and imaginary axis crossing points is concerned, by setting the imaginary and real part to zero respectively. However, by viewing $G(j\omega)$ in terms of polar coordinates, individual factors can be dealt with separately both in terms of gain and phase, as can be seen in the following example.

Example 5.4.1

Consider the system transfer function:

$$G(s) = \frac{20(s + 0.5)}{s(s + 3)(s + 1)}$$

In order to investigate the system's frequency response, set $s = j\omega$, such that:

$$G(j\omega) = \frac{20(j\omega + 5)}{j\omega(j\omega + 3)(j\omega + 1)}$$

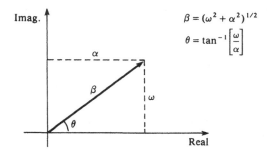

Fig. 5.16   Magnitude and phase for an individual factor

then

$$|G(j\omega)| = \frac{20(\omega^2 + 25)^{1/2}}{\omega(\omega^2 + 9)^{1/2}(\omega^2 + 1)^{1/2}}$$

and

$$\angle G(j\omega) = \tan^{-1}\left[\frac{\omega}{5}\right] - 90° - \tan^{-1}\left[\frac{\omega}{3}\right] - \tan^{-1}\left[\frac{\omega}{1}\right]$$

in which the individual factors have been dealt with separately by applying Pythagoras's theorem for the gain and by taking the tangent of the angle, i.e. as shown in Fig. 5.16. Note that when a factor of $G(j\omega)$ is a zero, its magnitude appears as a numerator term and its phase contribution is positive, whereas when a factor is a pole, its magnitude appears as a denominator term and its phase contribution is negative.

Alternatively the transfer function $G(j\omega)$ can be considered in terms of real and imaginary parts, which are found by multiplying both the numerator and denominator of $G(j\omega)$ with complex conjugate denominator roots. Thus

$$G(j\omega) = \frac{-20j(j\omega + 5)(3 - j\omega)(1 - j\omega)}{\omega(\omega^2 + 9)(\omega^2 + 1)}$$

then

$$G(j\omega) = \frac{-20[\omega(\omega^2 + 17) + j(-\omega^2 + 15)]}{\omega(\omega^2 + 9)(\omega^2 + 1)}$$

so in terms of real and imaginary parts, this is:

$$\text{real } [G(j\omega)] = \frac{-20(\omega^2 + 17)}{(\omega^2 + 9)(\omega^2 + 1)}$$

and

$$\text{imaginary } [G(j\omega)] = \frac{20(\omega^2 - 15)}{\omega(\omega^2 + 9)(\omega^2 + 1)}$$

The frequency plot can now be sketched by means of the four important points:

1. $\omega = 0$

$$|G(j\omega)| = \infty; \qquad \underline{/G(j\omega)} = -90°$$

$$\text{real } [G(j\omega)] = -\frac{340}{9}; \qquad \text{imaginary } [G(j\omega)] = -\infty$$

The plot therefore starts at infinity magnitude, with a phase angle of $-90°$, but also from an asymptote $-340/9$ along the real axis.

2. $\omega = \infty$

$$|G(j\omega)| = 0; \qquad \underline{/G(j\omega)} = -180°$$

$$\text{real } [G(j\omega)] = 0; \qquad \text{imaginary } [G(j\omega)] = 0$$

which gives information showing how the plot approaches the origin.

3. Real axis crossing

$$\underline{/G(j\omega)} = 0° \text{ or } \pm 180°; \qquad \text{imaginary } [G(j\omega)] = 0$$

A solution is found when $\omega^2 = 15$, i.e. $\omega = +\sqrt{15}$ as $0 \leqslant \omega \leqslant \infty$. Then

$$\text{real } [G(j\omega)] = \frac{-20(15 + 17)}{(15 + 9)(15 + 1)} = -\frac{5}{3}$$

Note: The solution is found most easily here by setting the imaginary part of $G(j\omega)$ to zero.

4. Imaginary axis crossing

$$\underline{/G(j\omega)} = \pm 90°; \qquad \text{real } [G(j\omega)] = 0$$

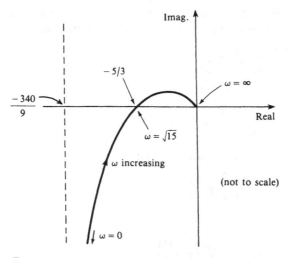

Fig. 5.17 Nyquist plot for Example 5.4.1

There are no real solutions for ω, and hence the plot does not cross the imaginary axis.

Armed with the four points of interest, the complete Nyquist plot can be sketched as in Fig. 5.17.

5.4.2   The Nyquist criterion

Before discussing the Nyquist criterion in detail, it must first be pointed out that, in common with other techniques (e.g. root locus, Bode), a basic assumption is made that a unity feedback loop is to be applied. By plotting the frequency response for an open-loop system transfer function, the Nyquist criterion provides a test for closed-loop stability when that open-loop system is connected within a unity feedback system.

If the actual closed-loop system contains a feedback term $H(j\omega)$ which is other than unity, then the equivalent open-loop transfer function $G_0(j\omega)$ within a unity feedback loop should first be obtained, as shown in Section 3.4.1. It is sufficient though, for the purposes of the Nyquist criterion, to plot the frequency response of $G(j\omega)H(j\omega)$, because similar stability results hold, see Appendix 4. Throughout the remainder of this section, reference to the transfer function $G(j\omega)$ implies either the open-loop function $G_0(j\omega)$ or the function $G(j\omega)H(j\omega)$, whichever method is preferred.

The Nyquist plot considered thus far (for $0 \leqslant \omega \leqslant \infty$) must be generalized by also considering negative frequencies, such that a continuous response diagram is obtained.

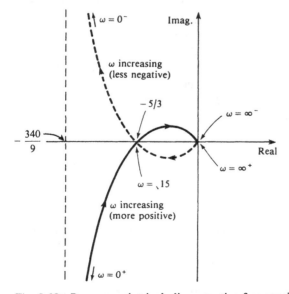

Fig. 5.18   Response plot including negative frequencies

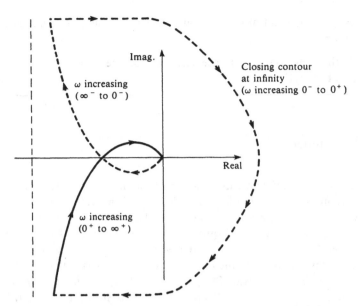

Fig. 5.19   Response plot including closing contour at infinity

For all real functions $G(j\omega)$, we have that $G(-j\omega) = G^*(j\omega)$, (the complex conjugate), see Problem 5.3. So, if for a certain frequency ω', $G(j\omega') = \alpha - j\beta$ then $G(-j\omega') = \alpha + j\beta$. This results in a negative frequency response plot which is a mirror image, about the real axis, of the positive frequency response plot. As an example consider the response plot shown in Fig. 5.17, which becomes that shown in Fig. 5.18.

As a final extension to the plots considered thus far, it is also necessary to completely close the contour by connecting up any remaining unconnected contour parts at infinity. The closing contour at infinity must always be drawn in a clockwise direction, and must be such that the direction of increasing frequency is retained. The closing contour at infinity for the response plot previously considered in Fig. 5.18 is shown in Fig. 5.19.

Once the closing contour at infinity has been drawn, the *complete* plot forms a continuous closed contour, such that the following Nyquist stability criterion can be applied.

Nyquist stability criterion for stable open-loop systems

If an open-loop system, $G(j\omega)$, is stable and its complete frequency response plot does not encircle the $-1 + j0$ point (known as the -1 point) in a clockwise direction, then the corresponding unity feedback closed-loop system is stable. Conversely if the -1 point is encircled in a clockwise direction then the corresponding unity feedback closed-loop system is unstable. When the plot passes through the -1 point in a clockwise direction and does not encircle it then the corresponding unity feedback closed-loop system is critically stable.

The criterion given only deals with open-loop systems which are stable. If the open-loop system is unstable then $G(s)$ will contain N_p poles which lie in the right half of the s-plane. These poles are in fact also poles of the function $1 + G$. Further, if there are any right-half plane closed-loop poles, then these will also be zeros of the function $1 + G$. Let N_z be the number of right-half plane closed-loop poles, such that:

N_p = Number of right-half plane poles of $1 + G$

N_z = Number of right-half plane zeros of $1 + G$ (these are also right-half plane closed-loop poles)

If N is the number of clockwise encirclements of the -1 point, then:

$$N = N_p - N_z \tag{5.4.1}$$

and it is required, for the closed-loop system to be stable, that $N_z = 0$. Hence the Nyquist stability criterion can be generalized to read as follows.

Nyquist stability criterion

If the complete frequency response plot of an open-loop system $G(j\omega)$ encircles the -1 point in a clockwise direction, N times, then the corresponding unity feedback closed-loop system is stable if $N = N_p$, i.e. if the number of encirclements is equal to the number of right-half plane open-loop poles.

Conversely if the -1 point is encircled N times in a clockwise direction and $N_p > N$, then the corresponding unity feedback closed-loop system is unstable.

Example 5.4.2

Consider the system open-loop transfer function:

$$G(s) = \frac{K}{(s + 1)(s + 2)(s + 3)}$$

in which the scalar gain $K > 0$ can be varied. Using the Nyquist stability criterion find the range of values for K over which the unity feedback closed-loop system will be stable.

To investigate the system's frequency response, set $s = j\omega$, such that:

$$G(j\omega) = \frac{K}{(j\omega + 1)(j\omega + 2)(j\omega + 3)}$$

then

$$|G(j\omega)| = \frac{K}{(\omega^2 + 1)^{1/2}(\omega^2 + 4)^{1/2}(\omega^2 + 9)^{1/2}}$$

and

$$\underline{/G(j\omega)} = -\tan^{-1}\left[\frac{\omega}{1}\right] - \tan^{-1}\left[\frac{\omega}{2}\right] - \tan^{-1}\left[\frac{\omega}{3}\right]$$

By employing the four important points of interest, as shown in Example 5.4.1, the complete frequency response plot for the open-loop system of Example 5.4.2 is found to be that shown in Fig. 5.20, where it can be noted that no closing contour at infinity is required.

In order to obtain real and imaginary axis crossing points, $G(j\omega)$ can be reconsidered in terms of real and imaginary parts, i.e.

$$\text{real } [G(j\omega)] = \frac{6K(1 - \omega^2)}{(\omega^2 + 1)(\omega^2 + 4)(\omega^2 + 9)}$$

and

$$\text{imaginary } [G(j\omega)] = \frac{K\omega(\omega^2 - 11)}{(\omega^2 + 1)(\omega^2 + 4)(\omega^2 + 9)}$$

Although the imaginary axis crossing point is not immediately of significance, by setting the real part of $G(j\omega)$ to zero it is revealed that the plot crosses the imaginary axis when $\omega = 1$ rad/s, which produces a magnitude at that point of $|G(j\omega)| = K/10$.

Of much greater significance, as far as the Nyquist stability criterion is concerned, is the real axis crossing point, which can be found by setting the imaginary part of $G(j\omega)$ to zero. Hence for the real axis crossing, $\omega = +\sqrt{11} = 3.317$ rad/s, and this results in a magnitude at that point of $|G(j\omega)| = K/60$. It follows that if $[K/60] < 1$, then the complete frequency response plot will not encircle the point at $-1 + j0$, and thus the closed-loop system will be stable.

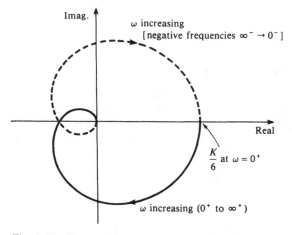

Fig. 5.20 Frequency response plot for Example 5.4.2

However if $[K/60] > 1$, then there will be a clockwise encirclement of the -1 point, and the closed-loop system will be unstable.

So, for a positive gain K, the closed-loop system will be stable as long as: $0 < K < 60$.

If a frequency response plot is also required for negative open-loop gain values K, then this can be obtained in exactly the same way as the method described for positive gain values. No modification is in fact found as far as the open-loop magnitude $|G(j\omega)|$ is concerned, however the phase $\underline{/G(j\omega)}$ is shifted through $180°$ for each frequency ω: see Problem 5.4.

5.4.3 Relative stability

The concepts of gain and phase margin are introduced in Section 5.3.1, with their implementation described in terms of Bode analysis. Both concepts, which relate to the frequency response characteristics of a system, are equally applicable as far as the Nyquist plot is concerned, and are considered here in that context.

In terms of a frequency response plot, if a unity feedback closed-loop system is shown to be stable, then we wish to know by how much the open-loop system gain can be increased before the closed-loop system becomes unstable. The gain margin is defined as the amount (in dB) by which the magnitude of $G(j\omega)$ must be increased in order to be equal to 0 dB, when $\underline{/G(j\omega)} = \pm 180°$. Because the gain margin is uniformly measured in dB, any gain/magnitude values of $G(j\omega)$ obtained from the Nyquist plot must therefore be converted to dB, by means of finding $20 \log_{10}|G(j\omega)|$. The 0 dB, $\underline{/\pm 180°}$ point corresponds to $|G(j\omega)| = 1$, $\underline{/G(j\omega)} = \pm 180°$ or $-1 + j0$. Thus, in terms of

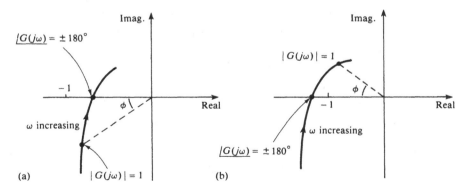

Fig. 5.21   Frequency response plots for stable and unstable closed-loop systems

    (a) Stable
    (b) Unstable

the Nyquist criterion, the gain margin is used as a measure of how far the frequency plot is from the -1 point, when the plot crosses the negative real axis.

Note: The gain margin is negative for unstable systems, see Fig. 5.21.

The phase margin is defined as the additional phase lag required to make $\underline{/G(j\omega)} = \pm 180°$ at the frequency for which the magnitude of $G(j\omega)$ is equal to 0 dB. The phase margin is therefore taken as the phase difference between the negative real axis and the phase of the frequency response plot at the frequency for which $|G(j\omega)| = 1$.

Note: The phase margin is negative for unstable systems, see Fig. 5.21.

Example 5.4.3

Consider the open-loop transfer function of Example 5.4.2:

$$G(j\omega) = \frac{K}{(j\omega + 1)(j\omega + 2)(j\omega + 3)}$$

1. When $K = 20$ the closed-loop system is stable and for $\underline{/G(j\omega)} = \pm 180°$, $|G(j\omega)| = K/60 = 1/3$. So the gain can be increased by a factor of 3, i.e. $1/|G(j\omega)|$, in order to reach the -1 point. The gain margin is therefore $20 \log_{10}(3) = 20 \log_{10}(1/|G(j\omega)|) = 9.542$ dB. Also, when $|G(j\omega)| = 1$ then

$$\frac{20}{(\omega^2 + 1)^{1/2}(\omega^2 + 4)^{1/2}(\omega^2 + 9)^{1/2}} = 1$$

and from this, $\omega_2 = 1.838$ rad/s. At this frequency, $\underline{/G(j\omega_2)} = -135.54°$, such that the phase margin is $44.46°$.

2. When $K = 80$ the closed-loop system is unstable and for $\underline{/G(j\omega)} = \pm 180°$, $|G(j\omega)| = K/60 = 4/3$. So the gain must be 'increased' by a factor of $3/4$, i.e. $1/|G(j\omega)|$, in order to reach the -1 point. The gain margin is therefore $20 \log_{10}(3/4) = -2.5$ dB.

Note: The gain must be 'increased negatively' for an unstable system in order to make it stable; this corresponds to a decrease or attenuation, as indicated by the negative gain margin sign. Also, when $|G(j\omega)| = 1$, then

$$\frac{80}{(\omega^2 + 1)^{1/2}(\omega^2 + 4)^{1/2}(\omega^2 + 9)^{1/2}} = 1$$

and from this, $\omega_2 = 3.766$ rad/s. At this frequency, $\underline{/G(j\omega_2)} = -188.62°$, such that the phase margin is $-8.62°$.

Example 5.4.4

Consider the system open-loop transfer function:

$$G(s) = \frac{K}{s(s + \alpha)}$$

in which the scalar gain $K > 0$ can be varied, and $\alpha > 0$.

Using the Nyquist plot, find the value for K which results in a phase margin of 30°.

To investigate the system's frequency response, set $s = j\omega$, such that

$$G(j\omega) = \frac{K}{j\omega(j\omega + \alpha)}$$

then

$$|G(j\omega)| = \frac{K}{\omega(\omega^2 + \alpha^2)^{1/2}}$$

and

$$\underline{/G(j\omega)} = -90° - \tan^{-1}\left[\frac{\omega}{\alpha}\right]$$

It can be observed directly that the frequency response plot for this system will not cross the negative real axis ($\underline{/G(j\omega)} \rightarrow -180°$ as $\omega \rightarrow \infty$), and so the gain margin is undefined for this system. This can be considered to imply that the gain margin is, to all intents and purposes, ∞, i.e. the gain K can be increased by an infinite amount and the closed-loop system will not become unstable. It is worth mentioning, though, that in practice a plant may well not be an 'exact' second-order system, and that a negative real axis crossing, probably very close to the origin, does in fact exist.

A sketch of the frequency response plot for the system in Example 5.4.4 is shown in Fig. 5.22, and it can be seen that although the gain margin is undefined for this system, the phase margin can be obtained by finding a solution to $|G(j\omega)| = 1$, i.e. when

$$K = \omega(\omega^2 + \alpha^2)^{1/2} \tag{5.4.2}$$

But, it is required that the phase margin $\phi = 30°$, i.e. $\underline{/G(j\omega)} = -150°$, and

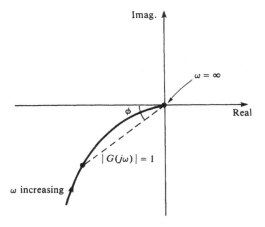

Fig. 5.22   Frequency response plot for Example 5.4.4

this occurs when

$$-150° = -90° - \tan^{-1}\left[\frac{\omega}{\alpha}\right]$$

such that $\omega = \sqrt{3}\ \alpha$.

On substituting for ω in (5.4.2), it is found that an open-loop gain of

$$K = 2\sqrt{3}\ \alpha^2$$

will achieve the desired phase margin.

The value of phase margin selected here is up at the higher end of the values usually chosen. Typical values for reasonable design are 30° to 60° for the phase margin; the gain margin, however, is usually in the range 3 to 12 dB.

5.5   Compensation using the Nyquist plot

Applying compensation to a system by means of the Nyquist plot involves a modification of the original open-loop frequency response characteristics, in order to satisfy some previously specified performance objectives. The objectives can be stated simply in terms of gain and phase margins, and it is this type of requirement which is considered here. In practice, however, it may well be that further desirable performance measurements are given such as bandwidth and pole locations and the Nyquist compensation procedure may well need to be combined with such as Bode analysis and root locus techniques. Indeed this could result in several designs and redesigns being carried out before a final compromise scheme is obtained.

It is considered that the compensation network is included in cascade with the original open-loop system, within a unity feedback loop. This makes it possible to plot firstly the original open-loop system frequency response and secondly the compensated system frequency response as a new open-loop system. The closed-loop stability and relative stability properties of each system can then be witnessed. When feedback compensation is used as well as or instead of cascade compensation, the Nyquist design can be treated either by finding the equivalent sole cascade controller or simply by considering the feedback compensator to be in series with the original plant, such that a Nyquist plot is made of $G(j\omega)H(j\omega)$: this is discussed further in Appendix 4.

5.5.1   Gain compensation

For straightforward and simple performance specifications it is often easiest to meet the requirements by merely adjusting the system open-loop gain K, within a unity feedback loop. In practical terms this will most likely be implemented by means of an amplifier in cascade with the open-loop plant.

Where the plant is of the form $G(j\omega) = KB(j\omega)/A(j\omega)$, if the gain K is increased then this will increase the magnitude of the frequency response plot, without affecting the phase, uniformly over all frequencies. In particular, the frequency at which the phase of $G(j\omega)$ is $\pm 180°$ will not be affected by a variation in K, however the gain of the plot at that frequency and its phase margin will be. Hence stability of a closed-loop system can be ensured by the selection of an appropriate gain value and within limits, both phase and gain margin specifications can be met, as shown in the following example.

Example 5.5.1

Consider the system open-loop transfer function:

$$G(s) = \frac{20(s + 0.5)K}{s(s + 3)(s + 1)}$$

whose frequency response plot for $K = 1$ is considered in detail in Example 5.4.1, and is sketched here again in Fig. 5.23. Using solely the gain $K > 0$, find a value which results in a closed-loop system which has a gain margin of at least 5 dB and a phase margin which lies between $30° < \phi < 50°$.

The original frequency response plot (only positive frequencies are shown in Fig. 5.23) reveals that the -1 point is encircled in a clockwise direction and hence the unity feedback closed-loop system obtained when $K = 1$ is unstable.

While $K = 1$ the gain margin is $20 \log_{10}(3/5) = -4.437$ dB, and in order to obtain a gain margin of at least 5 dB, it is required that the open-loop system gain is $|G(j\omega)| \leqslant 0.5623$ when the phase of $\underline{/G(j\omega)}$ is $\pm 180°$. This value range can be achieved by selecting a gain K such that $0 < K \leqslant 0.3374$.

For a gain $K = 0.3374$, the frequency response plot is equal to unity at a

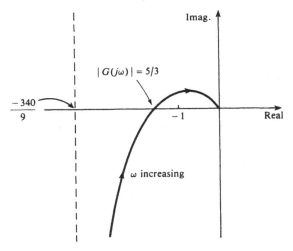

Fig. 5.23 Response plot for Example 5.5.1

frequency found from:

$$| G(j\omega) | = \frac{20(\omega^2 + 25)^{1/2} K}{\omega(\omega^2 + 9)^{1/2}(\omega^2 + 1)^{1/2}} = 1$$

which is satisfied when $\omega = 2.968$ rad/s.

At this frequency the phase $\underline{/G(j\omega)}$ is:

$$\underline{/G(j\omega)} = \tan^{-1}\left[\frac{\omega}{5}\right] - 90° - \tan^{-1}\left[\frac{\omega}{3}\right] - \tan^{-1}[\omega] = -175.38°$$

Hence the phase margin is only $4.62°$.

It can subsequently be found (by trial and error for instance) that by employing a gain $K = 0.05$, the frequency response plot is equal to unity at $\omega = 1.086$ rad/s and this results in a phase of $\underline{/G(j\omega)} = -145.01°$, which gives a phase margin of $34.99°$. So a phase margin which lies within the desired range has been found by the selection of a suitable gain value, K. Further, with $K = 0.05$ the gain margin is $20 \log_{10}(12) = 21.584$ dB, which is certainly more than the required minimum value of 5 dB. Thus, by means of simple gain compensation, the gain and phase margin stipulations have been satisfied.

Unfortunately it is not always possible to satisfy both the gain and phase margin requirements, by only making use of gain compensation. For instance the desire for a gain margin of at least 5 dB but no more than 10 dB could not be achieved simply by gain compensation on the system of Example 5.5.1 if a phase margin within the range $40° < \phi < 50°$ was also required. Hence for a more general type of compensation network it is necessary to employ a slightly more complicated structure, such as the lag/lead compensation considered in the following section.

5.5.2   Lag/lead compensation

The application of a cascade lag/lead compensator is considered in terms of Bode analysis in Section 5.3.3, with a relevant block diagram given in Fig. 5.11. Essentially, the controller adds one pole and one zero to the original open-loop system characteristics, and selection of the positions of these singularities, coupled with a choice of open-loop gain, allows for a great deal more flexibility than is provided by means of gain compensation alone. If the compensator transfer function is given by:

$$D(j\omega) = \frac{j\omega + b}{j\omega + a} \tag{5.5.1}$$

then the frequency response of the cascaded functions $D(j\omega)G(j\omega)$ must be plotted, such that the stability of the unity feedback closed-loop system can be tested. In terms of the Nyquist plot, this means that at each frequency ω, the effect of the cascade compensator is to modify, by a certain amount, the magnitude and phase of the transfer function. The effect of a lead compensator is to make the transfer function phase more positive at each frequency, whereas the effect of a lag compensator is to make the transfer function phase more negative at each frequency.

For a *lag* compensator, the phase angle of the compensator itself is negative in the mid-frequency range, but is $0°$ at the extremes when $\omega = 0$ and $\omega = \infty$. This can be seen by noting that:

$$\underline{/D(j\omega)} = \tan^{-1}\left[\frac{\omega}{b}\right] - \tan^{-1}\left[\frac{\omega}{a}\right] \qquad (5.5.2)$$

and when $a < b$ (for a lag compensator), it follows that the phase will never be positive definite. In fact as

$$|D(j\omega)| = \left[\frac{\omega^2 + b^2}{\omega^2 + a^2}\right]^{1/2} \qquad (5.5.3)$$

the frequency response plot for a lag compensator is a semi-circle, as shown in Fig. 5.24. There are no axis crossing points and plot end conditions are found to be $|D(j\omega)| = b/a$ when $\omega = 0$ and $[D(j\omega)| = 1$ when $\omega = \infty$, with the phase equal to $0°$ in both cases.

So, for any particular frequency value ω, which lies between 0 and ∞, the effect of the lag compensator is

1. To multiply the gain of the original plot by a value which is greater than unity but less than b/a; and
2. To shift the phase by an angle which is less than $0°$ but greater than $-90°$ (for finite a and b).

When the system open-loop gain $K > 0$ can also be varied, then this will not affect the phase angle in any way, but will multiply the gain at all frequencies by a factor K.

If a phase shift of larger negative characteristics is required, then several lag

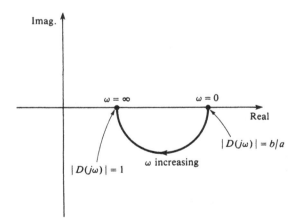

Fig. 5.24 Nyquist plot for a lag compensator

174 Frequency response analysis

compensators can be used in cascade. Restricting ourselves to a single compensator however, an example of its employment is given by the following.

Example 5.5.2

Consider the open-loop transfer function:

$$G(j\omega) = \frac{20K}{j\omega(j\omega + 2)(j\omega + 1)}$$

A frequency response plot for the system, when $K = 1$, with no compensation applied as shown in Fig. 5.25(a), reveals a negative phase margin of $-30.15°$ (when $\omega = 2.532$ rad/s) and a negative gain margin of -10.46 dB (when $\omega = 1.414$ rad/s). Such phase and gain results show that a unity feedback closed-loop system, with no compensation, would be unstable. Note that gain and phase margin results can be obtained either by exact calculations or as approximate values measured from the sketched Nyquist plot.

A phase lag compensator can be used to obtain a gain margin of at least 8 dB and a phase margin of at least $30°$, without affecting the steady-state properties of the system when $K = 1$. This can be achieved by reducing the overall gain at higher frequencies whilst retaining the same value of gain at low frequencies.

The gain margin is found to be -10.46 dB when $\omega = 1.414$ rad/s, for the uncompensated system. If it is ensured that, when the system is compensated, the gain margin will be at least 8 dB, this can be realized by reducing the gain at $\omega = 1.414$ rad/s by approximately 20 dB, on condition that the phase is not also affected to any great extent at that frequency. Such a decrease in gain is obtained by selecting $K = 0.1$. If it is then ensured that $b/a = 10$, this will mean

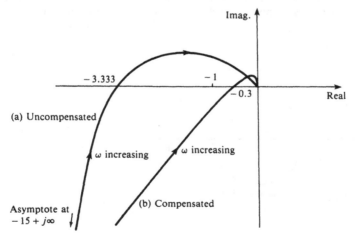

Fig. 5.25 Uncompensated and (b) compensated frequency response plots for the system of Example 5.5.2

that the overall gain at low frequencies will not be affected, i.e.

$$D(j\omega) = \frac{b}{a} = 10 = \frac{1}{K}$$

$$\omega = 0$$

Thus the gain increase, K, is counteracted at low frequencies by the compensator gain, whereas at higher frequencies, no such counteraction takes place. So, at higher frequencies a gain decrease of 20 dB is apparent. This means that, as long as little or no phase change takes place at the frequency at which the overall gain was 20 dB, with $K = 1$, this will become the frequency at which the overall gain is 0 dB, with $K = 0.1$.

Solving

$$|G(j\omega)| = \frac{20}{\omega(\omega^2 + 4)^{1/2}(\omega^2 + 1)^{1/2}} = 10 \ [\,= 20 \text{ dB}]$$

$$[K = 1]$$

gives $\omega = 0.749$ rad/s and at this frequency the phase is:

$$\underline{/G(j\omega)} = -90^\circ - \tan^{-1}\left[\frac{\omega}{2}\right] - \tan^{-1}[\omega] = -147.387^\circ$$

such that a phase margin of 32.61° would be obtained, which is above the minimum required.

A choice of $10a = b < 0.0749$ would ensure that almost no phase change, due to the compensator, occurs at $\omega = 0.749$, and hence a choice of $b = 0.05$, $a = 0.005$ will suffice. The resultant effect, obtained by employing a gain $K = 0.1$, along with the lag compensator:

$$D(j\omega) = \frac{j\omega + 0.05}{j\omega + 0.005}$$

is shown in Fig. 5.25(b). The frequency at which the gain is 0 dB (unity magnitude) is now approximately $\omega = 0.75$ rad/s (as opposed to $\omega = 2.532$ rad/s for the uncompensated system). Further, a gain margin of $+9.5$ dB and a phase margin of approximately 32° are obtained, thus meeting the performance requirements. Employment of a lag compensator has resulted in a lower frequency at which the gain is 0 dB, i.e. the bandwidth has been reduced.

Reduction of bandwidth is a property usually obtained by the use of a lag compensator, and can be looked at in another way, in terms of slowing down the system's response. If the steady-state error constant is held at a steady value, as was done in the example, then both the gain and phase margins will normally be increased. Conversely, by holding the gain and phase margins steady, error constant values can be increased by means of a lag compensator.

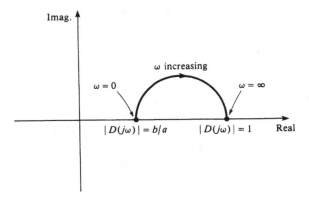

Fig. 5.26   Nyquist plot for a lead compensator

For a *lead* compensator, the phase angle of the compensator itself is positive in the mid-frequency range, but is $0°$ at the extremes when $\omega = 0$ and $\omega = \infty$. This can be seen by noting that (5.5.2) still holds, however now $b < a$ for a lead compensator. It follows that the phase will never be negative definite. In fact as (5.5.3) also holds, the frequency response plot for a lead compensator is a semi-circle, as shown in Fig. 5.26. There are no axis crossing points and plot end conditions are found to be identical to those for the lag compensator, i.e. $|D(j\omega)| = b/a$ when $\omega = 0$ and $|D(j\omega)| = 1$ when $\omega = \infty$, with the phase equal to $0°$ in both cases. So for any particular frequency value ω, which lies between 0 and ∞, the effect of a lead compensator is:

1. To multiply the gain of the original plot by a value which is less than unity but greater than b/a; and
2. To shift the phase by an angle which is greater than $0°$ but less than $+90°$ (for a finite a and b).

When the system open-loop gain $K > 0$ can also be varied, this does not affect the phase angle in any way, but will multiply the gain at all frequencies by a factor K.

If a phase shift of large positive characteristics (more than $90°$) is required, then several lead compensators can be used in cascade. Conversely, if a more complicated compensator structure, which is more selective about the frequency ranges amplified, is required, this can often be achieved by cascading a number of lag and lead compensators. For a single lead compensator, the standard design procedure involves firstly selecting the system open-loop gain value, K, in order to meet gain margin requirements (a decrease in gain margin will be required). The ratio $b : a$ can then be found in order to produce necessary steady-state conditions, such as specified error coefficients to relevant input signals. At higher frequencies the overall gain is thereby increased, the tendency being towards a destabilization of the system. The phase shifting properties of the compensator can then be selected, such that they lie in the mid-frequency range and therefore do not affect the specifications already met by the compensator, see Problem 5.5.

Phase lead compensators are usually employed to improve the transient system response, decreasing the phase and gain margins from those corresponding to a sluggish response to those apparent for a specified, acceptable but faster response. In the majority of cases the frequency at which the overall gain is 0 dB, is increased by a phase lead compensator, thus the bandwidth is increased.

5.6   The Nichols chart

Frequency response analysis of a system involves the use of system gain and phase information obtained over a wide range of frequencies, in theory over all frequencies. Bode and Nyquist plots both deal with the same data type, but present it in a distinctly different way. The Nichols chart, another frequency response technique, deals once more with the same data type but presents it in yet another different way.

Much of the procedure employed with a Nichols chart is similar to that used for Bode and Nyquist design methods. The Nichols chart is therefore not considered here in the same depth as the other two methods, which are in many ways easier to understand and use. A common feature with all three methods is that the open-loop system transfer function response is plotted, while results such as relative stability found from the method, pertain to the closed-loop system when connected in unity feedback mode. Where the closed-loop system in question actually employs a feedback term $H(j\omega)$ which is other than unity, then either the equivalent unity feedback system must be found and the effective open-loop transfer function extracted, or the function $G(j\omega)H(j\omega)$ can be plotted. This is discussed in Appendix 4.

Initially, consider solely the system open-loop transfer function, $G(j\omega)$. The gain in dB of this function can be plotted against the corresponding phase angle values over the range of frequencies $\omega = 0$ to $\omega = \infty$, as shown in the following example.

Example 5.6.1

The open-loop system transfer function,

$$G(j\omega) = \frac{20}{j\omega(j\omega + 1)}$$

produces a gain (dB)–v–phase plot as depicted in Fig. 5.27. The gain for this system is:

$$\text{gain (dB)} = 20 \log_{10} \frac{20}{\omega(\omega^2 + 1)^{1/2}}$$

whereas the phase is:

$$\text{phase} = -90° - \tan^{-1}[\omega]$$

Although for the transfer function shown above it is quite straightforward to

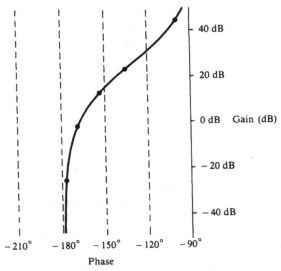

Fig. 5.27 Gain-v-phase plot for Example 5.6.1

obtain the gain-v-phase plot by simply finding the required values for a number of frequencies, because the gain is plotted in dB, terms can be calculated separately and merely added, for both gain and phase, at each frequency. For the example in question we therefore have:

$$\text{gain (dB)} = 20 \log_{10}(20) - 20 \log_{10}(\omega) - 10 \log_{10}(\omega^2 + 1)$$

Table 5.2 Table of gain and phase values for Example 5.6.1

ω	0.1	1	2	5	20
$20 \log_{10} 20$ [dB]	26.02	26.02	26.02	26.02	26.02
$20 \log_{10}(\omega)$ [dB]	-20	0	6.02	13.98	26.02
$10 \log_{10}(\omega^2 + 1)$ [dB]	0.04	3.01	6.99	14.15	26.03
Gain [dB]	45.98	23.01	13.01	-2.11	-26.03
$-90°$	$-90°$	$-90°$	$-90°$	$-90°$	$-90°$
$-\tan^{-1}(\omega)$	$-5.71°$	$-45°$	$-63.43°$	$-78.69°$	$-87.13°$
Phase(°)	$-95.71°$	$-135°$	$-153.43°$	$-168.69°$	$-177.13°$

5.6.1 Relative stability

Gain and phase margins, described earlier in this chapter for unity feedback systems, in terms of Bode and Nyquist plots, can also be obtained by reference to the gain-v-phase plot.

The gain margin is found when the plot of $G(j\omega)$ has a phase angle of $-180°$. At this frequency, stability of the unity feedback closed-loop system is, for the majority of systems, ensured as long as the gain is less than 0 dB (i.e. $|G(j\omega)| < 1$). The gain margin is the amount by which the open-loop gain can be increased until it is equal to 0 dB. So, if ω_1 is the frequency at which the phase of the open-loop transfer function is $-180°$, the gain margin can be found from:

$$\text{gain margin} = 20 \log_{10} \left| \frac{1}{G(j\omega_1)} \right| \text{ dB}$$

or

$$\text{gain margin} = -20 \log_{10} |G(j\omega_1)| \text{ dB}$$

where $|G(j\omega_1)|$ can be either calculated exactly or measured on the gain-v-phase plot.

The phase margin is found when the plot of $G(j\omega)$ has a gain of 0 dB. At this frequency, stability of the unity feedback closed-loop system is, for the majority of systems, ensured as long as the phase is more positive than $\pm 180°$. The phase margin is the amount by which the open-loop phase can be made more negative until it is equal to $\pm 180°$. So, if ω_2 is the frequency at which the gain of the open-loop transfer function is 0 dB, the phase margin can be found from:

$$\text{phase margin} = 180° + \underline{/G(j\omega_2)}$$

where $\underline{/G(j\omega_2)}$ can be either calculated exactly or measured on the gain-v-phase plot.
Note: Although stability can be claimed for the majority of systems if they exhibit positive gain and phase margins, a stricter check must be made using such as the Nyquist criterion to ensure absolute stability, particularly for more complex systems.

Example 5.6.2

Consider the gain-v-phase plot shown in Fig. 5.28. The gain margin, shown as

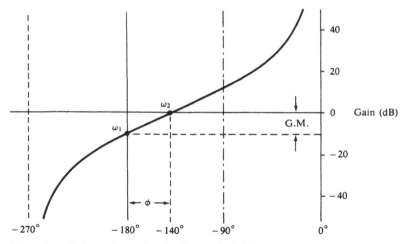

Fig. 5.28 Gain-v-phase plot for Example 5.6.2

G. M. in Fig. 5.28, is found to be 10 dB. This is because the gain is equal to -10 dB when the phase is $-180°$, i.e. at the frequency ω_1. Also, the phase margin, shown as ϕ in Fig. 5.28, is found to be $40°$. This is because the phase is equal to $-140°$ when the gain is 0 dB.

5.6.2 Closed-loop frequency response

In a similar fashion to the Bode and Nyquist methods, the frequency response of the open-loop system $G(j\omega)$ is plotted, and the relative stability of the unity feedback closed-loop system $G(j\omega)/[1 + G(j\omega)]$, is found from the plot.

Defining $G_M(j\omega)$ as

$$G_M(j\omega) = \frac{G(j\omega)}{1 + G(j\omega)} \tag{5.6.1}$$

it follows that for $\omega = \omega'$, then for any particular $G(j\omega')$ there will be one, and only one, unity feedback closed-loop magnitude and phase value, given by $|G_M(j\omega')|$ and $\underline{/G_M(j\omega')}$. However, if a particular magnitude $|G_M(j\omega)|$ is selected, with no constraint placed on phase $\underline{/G_M(j\omega)}$, then this can be achieved by a variety of $G(j\omega)$ values, in terms of magnitude and phase contributions. As an example, consider the requirement that $20 \log_{10} |G(j\omega)| = 2$ dB. The plot in Fig. 5.29 shows that the open-loop gain and phase combinations which produce a closed-loop system with a gain of 2 dB, form a locus. Note that the vertical gain and horizontal phase axes apply to the open-loop system *only*, no closed-loop phase values are shown.

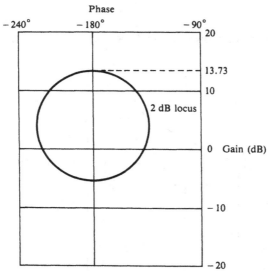

Fig. 5.29 Gain-v-phase plot, showing 2 dB closed-loop gain locus (constant $|G_M(j\omega)|$)

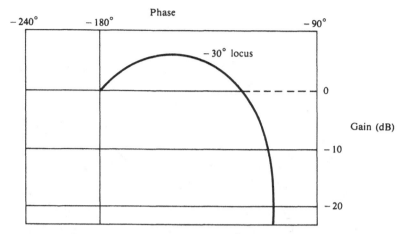

Fig. 5.30   Gain-v-phase plot, showing $-30°$ closed-loop phase locus (constant $\underline{/G_M(j\omega)}$)

Taking one point on the locus, say 2 dB $\underline{/0°}$, this means that

$$\frac{G(j\omega)}{1 + G(j\omega)} = -1.259$$

the solution to which is

$$|G(j\omega)| = 4.861 = 13.734 \text{ dB}$$

and

$$\underline{/G(j\omega)} = -180°$$

So, if the point 13.734 dB $\underline{/-180°}$ corresponds to the gain and phase of the open-loop system $G(j\omega')$ for some frequency ω', then the unity feedback closed-loop gain and phase will be 2 dB $\underline{/0°}$, at the frequency ω'.

A similar analysis can be carried out by selecting a particular phase $\underline{/G_M(j\omega)}$, with no constraint placed on the magnitude $|G_M(j\omega)|$. This phase can then be achieved by a variety of different $G(j\omega)$ values, in terms of magnitude and phase, each different combination producing a different magnitude $|G_M(j\omega)|$ but a constant phase $\underline{/G_M(j\omega)}$. As an example, consider the requirement that $\underline{/G_M(j\omega)} = -30°$. The plot in Fig. 5.30 shows that the open-loop gain and phase combinations, which produce a closed-loop system with a phase of $-30°$, form a locus. Note that the vertical gain and horizontal phase axes apply to the open-loop system *only*, no closed-loop gain values are shown.

Taking one point on the locus, let $G(j\omega) = 0$ dB $\underline{/-60°}$, then

$$G_M(j\omega) = \frac{1\underline{/-60°}}{1 + 1\underline{/60°}} = \frac{1}{\sqrt{3}}\underline{/-30°} = -4.771 \text{ dB}\underline{/-30°}$$

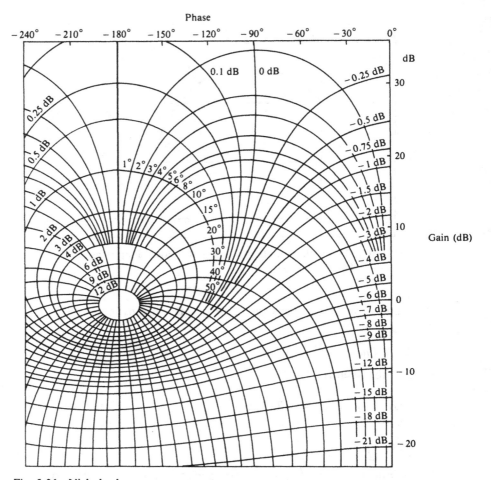

Fig. 5.31 Nichols chart

So, if the point 0 dB $\underline{/-60°}$ corresponds to the gain and phase of the open-loop system $G(j\omega')$ for some frequency ω', then the unity feedback closed-loop gain and phase will be -4.771 dB $\underline{/-30°}$, at the frequency ω'.

A Nichols chart is formed by superimposing unity feedback closed-loop (a) constant magnitude loci and (b) constant phase loci on the open-loop frequency response chart, as shown in Fig. 5.31. The horizontal phase and vertical gain axes correspond to open-loop system values, whereas the curvilinear contours correspond to the closed-loop system.

Note: All closed-loop phase contours shown are those of negative phase angles, e.g. on the curvilinear contour, 40° signifies a closed-loop phase angle of $-40°$.

On the Nichols chart, the frequency response of an open-loop system can be plotted in the normal way by employing the horizontal phase and vertical gain axes, as described in

Example 5.6.2. It is then possible to read, from the curvilinear contours, the closed-loop gain and phase values that the system will produce when connected within a unity feedback loop.

Example 5.6.3

Consider the open-loop transfer function:

$$G(j\omega) = \frac{20}{(j\omega + 1)(j\omega + 2)(j\omega + 3)}$$

The gain (dB) and phase of this open-loop system, over a range of frequencies ω rad/s is tabulated in Table 5.3 and plotted on the Nichols chart in Fig. 5.32.

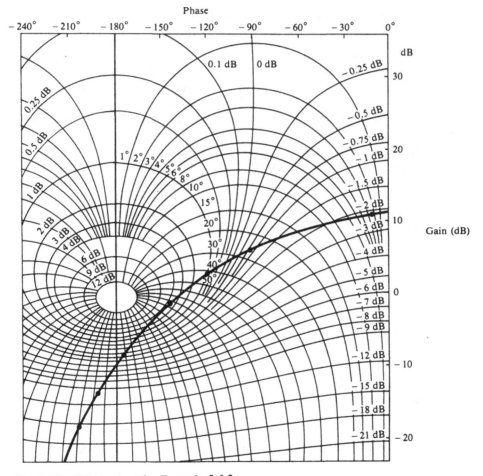

Fig. 5.32   Nichols chart for Example 5.6.3

Table 5.3 Gain and phase values for the system in Example 5.6.3

ω	0.1	1	1.5	2	3	4	5
$\mid G(j\omega)\mid$	10.4	6.02	2.43	-1.14	-7.67	-13.27	-18.07
$\underline{/G(j\omega)}$	$-10.48°$	$-90°$	$-119.74°$	$-142.13°$	$-172.87°$	$-192.53°$	$-205.92°$

The closed-loop gain and phase can then be read directly off the chart, for any chosen value of open-loop gain and phase. In particular the plot can be seen to intersect with the closed-loop $-180°$ phase locus, when the closed-loop gain locus is equal to -6 dB, i.e. when $\mid G_M(j\omega)\mid = 1/2$.

5.7   Compensation using the Nichols chart

Employing the Nichols chart in order to compensate a system response results in a similar process to Bode and Nyquist compensation, that is selecting a compensator in order to reshape the Nichols plot, of an open-loop system, into a more desirable form. It may well take several attempts before a particularly desirable shape is obtained. However, in many cases it is only steady-state and relative stability conditions which are most important, and they can very often be met with little difficulty.

In common with Bode analysis, Nichols chart compensation has the advantage that logarithmic gain values are plotted. Hence any alteration to the system gain can be regarded merely in terms of an addition or subtraction to the original.

As with the sections on Bode and Nyquist compensation, it is considered here that the compensation is applied in cascade with the open-loop system; a unity feedback closed-loop system then being connected around the cascade combination, Fig. 5.11 refers. When the feedback $H(j\omega)$ is non-unity, and/or is employed as the compensator, then either the closed-loop system must be reposed in a unity feedback form, or a plot of $G(j\omega)H(j\omega)$ can be made (see Appendix 4); stability results are the same, whichever approach is employed.

5.7.1   Gain compensation

For relatively simple performance specifications, it is often possible to provide a satisfactory unity feedback control scheme by merely tailoring the open-loop gain K. If the open-loop system is given by $G(j\omega) = KB(j\omega)/A(j\omega)$, an increase in gain K results in the gain-v-phase plot being shifted up by an appropriate amount over all frequencies, whereas a decrease in gain K results in the gain-v-phase plot being shifted down by an appropriate amount over all frequencies. This is because, for any frequency, a variation in gain K has no effect whatsoever on the corresponding phase angle.

Example 5.7.1

Consider the open-loop system transfer function:

$$G(j\omega) = \frac{200K}{(j\omega + 1)(j\omega + 2)(j\omega + 3)}$$

The gain-v-phase plot, for $K = 1$, is shown as the continuous plot in Fig. 5.33.

It is required that K is chosen in order to provide a phase margin of at least $40°$ and a gain margin of 8 to 12 dB. Without gain compensation, i.e. with $K = 1$, it can be seen from Fig. 5.33 that the phase margin is approximately $-30.6°$ whereas the gain margin is approximately -10.5 dB. Because both the phase and gain margins are negative values, the unity feedback system is initially unstable.

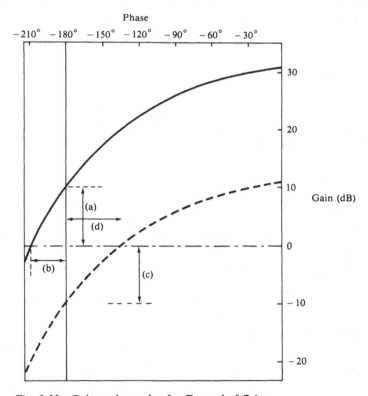

Fig. 5.33 Gain-v-phase plot for Example 5.7.1

(a) Gain margin ⎞
(b) Phase margin ⎠ Uncompensated system, $K = 1$
(c) Gain margin ⎞
(d) Phase margin ⎠ Compensated system, $K = 0.1$

With the phase at $-180°$, the gain is initially 10.5 dB. If this is reduced by 20 dB, by an appropriate choice of gain K, then the overall system gain will be -9.5 dB at a phase angle of $-180°$, thus achieving a gain margin of 9.5 dB, well within the required range. In fact a choice of $K = 0.1$, reduces the gain by 20 dB and therefore provides the desired gain margin. A check can then be made to find that with $K = 0.1$, the phase angle is $-135.5°$ when the system gain is 0 dB. The phase margin is therefore approximately $44.5°$, well above the minimum required.

Although, for this example, it was possible to provide adequate compensation simply by selecting the gain value K, it is not in general possible to achieve even fairly minimal performance objectives by such an elementary procedure. Where gain and phase margin values are required to a much greater tolerance and/or where further performance objectives are specified, a more complicated compensator is necessary, a possible solution to many cases being described in the following section.

5.7.2 Lag/lead compensation

The connection of a lag/lead compensator in cascade with the original open-loop system within a unity feedback closed-loop scheme is shown in Fig. 5.11, where the compensator transfer function is defined by:

$$D(j\omega) = \frac{j\omega + b}{j\omega + a}$$

when the gain of the open-loop system transfer function $G(j\omega)$ is plotted in dB against phase, as on the Nichols chart, so the effect of the cascade compensator $D(j\omega)$ can be merely added in terms of both dB gain and phase angle, at each frequency, in order to obtain the compensated plot.

For either a lag or lead compensator design, as well as a selection of terms a and b, it is normally the case that the open-loop gain K can be specified. This may be carried out in terms of a modification of the actual system gain, or more likely the controller itself may have a gain associated with it. The value K is therefore assumed to be the resultant open-loop gain value, after taking all gains into account.

For a *lag* compensator, the phase angle is negative in the mid-frequency range, and $a < b$. The gain (dB) versus phase plot for a typical lag compensator is shown in Fig. 5.34. In general, the larger is the ratio $b : a$ so the more negative is the peak phase value and the larger in dB magnitude is the steady-state gain value (when $\omega = 0$).

The actual gain value of the compensator cascaded with the plant at any frequency is something which can be selected by K, once a ratio $b : a$ has been chosen. From Fig. 5.34 it can be seen that the gain at lower frequencies will be increased, because of the lag compensator, in relation to the gain at higher frequencies. The lag compensator

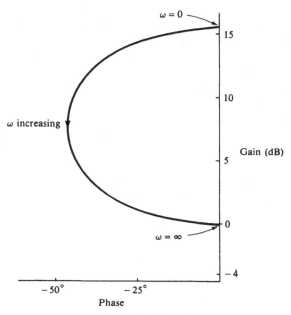

Fig. 5.34   Gain-v-phase plot for a lag compensator with $b = 6$, $a = 1$

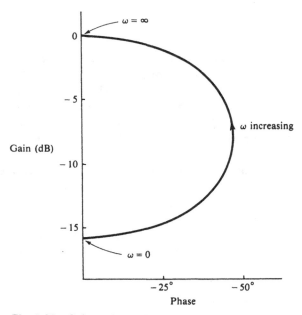

Fig. 5.35   Gain-v-phase plot for a lead compensator with $b = 1$, $a = 6$

therefore emphasizes low-frequency modes at the expense of high-frequency modes, thus slowing down the system response and thereby decreasing the bandwidth.

For a *lead* compensator, the phase angle is positive in the mid-frequency range, and $a > b$. The gain (dB) versus phase plot for a typical lead compensator is shown in Fig. 5.35. In general, the larger is the ratio $a : b$, so the more positive is the peak phase value and the larger in dB magnitude is the steady-state gain value (when $\omega = 0$). By taking into account possible use of the gain K, it can be seen from Fig. 5.35 that the gain at higher frequencies will be increased, because of the lead compensator, in relation to the gain at lower frequencies. The lead compensator therefore emphasizes high-frequency modes at the expense of low-frequency modes, thus speeding up the system response and thereby increasing the bandwidth.

In common with Bode and Nyquist designs, it is possible with a Nichols chart design to apply a compensator which contains both lag and lead networks, should such a more complex form prove necessary or useful.

Example 5.7.2

Consider the open-loop transfer function:

$$G(j\omega) = \frac{20K}{j\omega(j\omega + 2)(j\omega + 10)}$$

with $K = 1$.

A Nichols chart plot for this system, with no compensation applied, is shown in Fig. 5.36(a). This reveals a phase margin of $60.4°$ (when $\omega = 0.907$ rad/s) and a gain margin of 21.6 dB (when $\omega = 4.472$ rad/s). Such phase and gain margin results reveal that a unity feedback closed-loop system, with no compensation, would be rather sluggish. Note that gain and phase margin results can be obtained either by exact calculations or as approximate values measured from the Nichols chart.

Suppose that a requirement is made for the system to be compensated such that the gain margin lies in the range 8 to 12 dB and the phase margin lies in the range $25°$ to $35°$, without steady-state conditions being affected. This design can be achieved by means of a phase lead compensator, along with a suitable choice of gain K. Therefore, if it is ensured that when the system is compensated the gain margin will lie in the required range, this can be realized by increasing the gain at $\omega = 4.472$ rad/s by approximately 12 dB, on condition that the phase is not also affected to any great extent at that frequency. In fact an increase in gain of 12.04 dB is obtained by selecting $K = 4$. With the compensator defined as

$$D(j\omega) = \frac{j\omega + b}{j\omega + a}$$

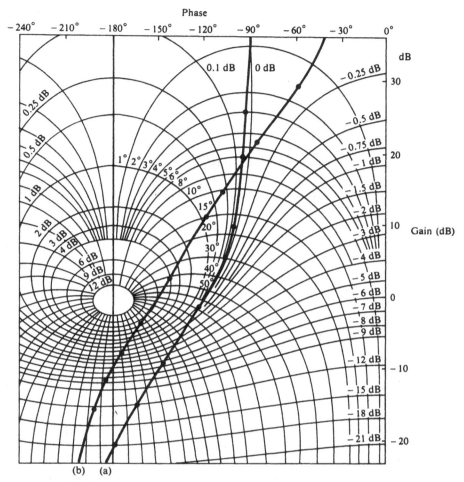

Phase

Fig. 5.36 Nichols chart plot for Example 5.7.2

(a) Uncompensated system
(b) Compensated system

the overall gain at low frequencies will not be affected as long as $a/b = 4$, i.e.

$$D(j\omega) = \frac{b}{a} = \frac{1}{4} = \frac{1}{K}$$

$$\omega = 0$$

Thus the gain increase, K, is counteracted at low frequencies by the compensator gain, whereas at higher frequencies no such counteraction takes place. So, at higher frequencies a gain increase of 12.04 dB is apparent.

As long as little or no phase change takes place at the frequency for which the overall gain was -12.04 dB, with $K = 1$, this will become the frequency at which the overall gain is 0 dB, with $K = 4$.

Solving

$$\frac{|G(j\omega)|}{[K=1]} = \frac{20}{\omega(\omega^2 + 4)^{1/2}(\omega^2 + 100)^{1/2}} = 0.25 \ [= -12.04 \text{ dB}]$$

gives

$$\omega = 2.454 \text{ rad/s}$$

and at this frequency the phase is:

$$\underline{/G(j\omega)} = -90^\circ - \tan^{-1}\left[\frac{\omega}{2}\right] - \tan^{-1}\left[\frac{\omega}{10}\right] = -154.6^\circ$$

such that a phase margin of 25.4° would be obtained, which is within the stated range of 25° to 35°. A choice of $4b = a < 0.245$ would ensure that almost no phase change, due to the compensator, occurs at $\omega = 2.454$ rad/s, and hence a choice of $b = 0.025$, $a = 0.1$ will suffice.

The resultant effect obtained by employing a gain of $K = 4$ along with the lead compensator:

$$D(j\omega) = \frac{j\omega + 0.025}{j\omega + 0.1}$$

is shown in the Nichols chart plot of Fig. 5.36(b). The frequency at which the gain is 0 dB (unity magnitude) is now approximately $\omega = 2.453$ rad/s (as opposed to $\omega = 0.907$ rad/s for the uncompensated system). Further, a gain margin of 9.9 dB and phase margin of approximately 27.2° are obtained, thus meeting the performance requirements. Employment of a lead compensator has resulted in a higher frequency at which the gain is 0 dB, i.e. the bandwidth has been increased.

An increase in bandwidth is a property usually obtained by the use of a lead compensator, and can be seen as speeding up the system's response. If the steady-state error constant is held at a steady value, as was done in the example, then both the gain and phase margins will normally be reduced. Conversely, by holding the gain and phase margins steady, error constant values can be reduced by means of a lead compensator. The phase shifting property of the lead compensator has, in the example, been arranged to occur at frequencies much lower than those at which the gain and phase margins are determined. This can be seen on Fig. 5.36(b) in terms of a phase angle of much less than 90°, whereas for the uncompensated system in Fig. 5.36(a) the phase angle is never less than 90°. The design has therefore been accomplished in terms of the high and low frequency properties of the compensator, the mid-frequency phase shift being placed where it is of no consequence.

The standard design procedure for a single lag compensator, see Problem 5.6, involves firstly selecting the system open-loop gain value, K, in order to meet gain margin requirements (an increase in gain margin will be required). The ratio $b : a$ can then be found in order to produce necessary steady-state conditions, such as specified error coefficients to relevant input signals. At higher frequencies the overall gain is thereby reduced, the tendency being towards a greater extent of relative stability. The phase shifting properties of the compensator can then be selected, such that they lie in the mid-frequency range and therefore do not affect the specifications already met by the compensator.

5.8   Summary

The techniques described in this chapter have all been concerned with analyzing a system in terms of its frequency response. For each of the methods, the basic procedure carried out is essentially the same, that is one of assessing the properties, in particular the stability properties, of a unity feedback closed-loop system, by employing the characteristics belonging to the open-loop system frequency response.

Each of the methods discussed is a different way in which a graphical representation can be given to system frequency response data. If the system transfer function is well known, exact calculations can be made from which both the stability properties and graphical plots are obtained. For many practical real-world systems however, it is unlikely that an 'exact' transfer function will be known. In these cases the frequency response data is collected by experimentation, gain and phase information being found over a wide range of frequencies, to an extent which allows for a good graphical plot to be made. Stability and relative stability closed-loop results can though be taken from the graphical plot, whichever method is used.

Whether the frequency response data has been obtained from real-world experimentation, or from ideal transfer function calculations, this is the same data required for all three of the techniques considered. The different techniques are merely different ways in which the data can be presented. Concepts of stability and relative stability are also common to the methods. Hence if a gain margin of 10 dB is found for a system by employing Bode analysis, the gain margin will still be 10 dB if either Nyquist analysis or the Nichols chart are used instead, i.e. the gain margin and phase margin values are not altered because a different method is used.

For each of the three techniques, the design of a suitable lag or lead compensator in order to meet specified performance objectives was considered. The performance specifications were given as fairly straightforward phase and gain margin requirements, in the examples used, and in most practical cases the requirements will indeed be this fundamental. It may be however that either tighter margin limits must be met, or the compensated frequency response plot is required to pass through further defined points. Although it might be possible to meet these demands by simply fiddling around with the gain K, and the a and b compensator coefficients, it is likely that a number of lag and/or lead compensators must be cascaded in order to provide the degrees of freedom necessary to solve the problem.

Problems

5.1   Find the maximum phase shift due to a phase lead compensator, and the frequency at which it occurs. The compensator is given by

$$D(j\omega) = \frac{j\omega + b}{j\omega + a}$$

where $a > b$.

Similarly, find the minimum phase shift due to a phase lag compensator, and the frequency at which it occurs ($a < b$ for a lag compensator).

5.2   Consider the open-loop transfer function:

$$G(s) = \frac{10K}{s(s + 1)(s + 2)}, \qquad K = 1$$

Employ a Bode plot to find a phase lead cascade compensator, which produces a unity feedback closed-loop system with a minimum gain margin of 10 dB and a minimum phase margin of 30°.

5.3   Show that $G(-j\omega) = G^*(j\omega)$, where $G^*(j\omega)$ denotes the complex conjugate of $G(j\omega)$.

5.4   Consider the open-loop transfer function:

$$G(s) = \frac{K}{(s + 1)(s + 2)(s + 3)}$$

Obtain a frequency response plot for this system when $\infty < K < 0$, and by using the Nyquist criterion, find the range of values for K over which the unity feedback closed-loop system will be stable.

5.5   For the uncompensated open-loop system:

$$G(s) = \frac{20K}{s(s + 2)(s + 10)}, \qquad K = 1$$

By means of a Nyquist design, find the phase lead compensator and gain K which will produce a phase margin in the range 30° to 40° and a gain margin in the range 8 dB to 12 dB.

5.6   For the uncompensated open-loop system in Problem 5.5, with $K = 1$: By means of a Nichols chart design, find the phase lag compensator and gain K which will produce a phase margin in the range 30° to 40° and a gain margin in the range 8 dB to 12 dB.

5.7   Show that when a sinusoidal input signal $u(t) = U \cos \omega t$ is applied to a stable linear system, then the steady-state output signal is also a sinusoid, of magnitude Y, such that $y(t) = Y \cos(\omega t + \phi)$; where $Y = |G(j\omega)| U$ and $\underline{/G(j\omega)} = \phi$.

5.8   Sketch Bode plots for the systems whose open-loop transfer functions are:

(a) $G(s) = s(s + 1)$

(b) $G(s) = \dfrac{(s + 1)(s + 6)}{(s + 2)(s + 3)}$

(c) $G(s) = \dfrac{s + 10}{s^2(s + 1)(s + 100)}$

5.9   A unity feedback system has an open-loop transfer function

$$G(s) = \frac{200K}{s(s + 2)(s + 5)}$$

Obtain a Bode plot for this system when $K = 1$ and find the phase and gain margins. Design a lag compensator in order to achieve a phase margin of approximately $45°$ and a minimum gain crossover frequency of $\omega = 1$ rad/s.

5.10   The open-loop transfer function of a unity feedback system is

$$G(s) = \frac{2K}{(s + 2)(s + 0.4)(s + 0.2)}$$

By means of a Bode plot, when $K = 1$, find the phase and gain margins, and the bandwidth. Also, find the range of values for K, over which the closed-loop system will be unstable.

5.11   Given the open-loop transfer function

$$G(s) = \frac{3(s + 1)}{s(s + \alpha)}$$

By means of a Bode plot, find the value of α which results in a gain crossover frequency of $\omega = 100$ rad/s, when the system is connected within a unity feedback loop.

5.12   The open-loop transfer function for a particular system is

$$G(s) = \frac{30(s + 0.5)}{s^2(s + 5)(s + 3)}$$

For a unity feedback system, find the gain and phase margins by means of a Bode plot, and ascertain the damping ratio.

5.13   When a step input of magnitude 0.04 units is applied to a certain open-loop system, the output response is equal to $[6 - 6 \exp(-t)]$ units. Obtain the Bode plot for this system and find the phase and gain margins.

A phase lead compensator of the form

$$D(s) = \frac{5(s + \beta)}{(s + 5\beta)}$$

is employed such that the phase margin is approximately $45°$. What value of β is required for such a design?

5.14   A unity feedback closed-loop system has an open-loop transfer function of:

$$G(s) = \frac{32K(s + 10)}{s(s + 4)(s + 80)}$$

Also, the Bode gain is approximately -6 dB when $\omega = 60$ rad/s.

Obtain a Bode plot for this system, and find the phase and gain margins, and the bandwidth.

5.15   For the open-loop system transfer function

$$G(s) = \frac{0.5}{s(s + 10)(s + 25)(s + 40)}$$

find a cascade compensator, of the form

$$D(s) = \frac{K(s + b)}{(s + a)}, \qquad K > 0$$

in order to obtain a gain margin of approximately 12 dB and a phase margin of approximately $40°$.

5.16   Find the input to output open-loop transfer functions for the electrical networks shown in Fig. 5.37 and obtain a Bode plot in each case. The input signal $U(s)$ and output signal $Y(s)$ both correspond to voltage transforms, $R_1 = 100$ k$\Omega = R_2$, $C_1 = 2$ μF and $C_2 = 3$ μF.

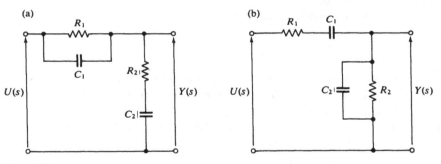

Fig. 5.37

5.17   For a system with open-loop transfer function

$$G(s) = \frac{s + b}{s(s + 1)^2} \qquad 0 < b < \infty$$

find the three different forms which a Nyquist plot can take.

5.18   By means of a frequency response plot, find the values of K for which the unity feedback closed-loop system, whose open-loop transfer function is

$$G(s) = \frac{K}{s(s + 1)(s + 2)}$$

(a) has a gain margin of 6 dB and (b) has a phase margin of $40°$.

5.19 Sketch the frequency response plot for the open-loop system

$$G(s) = \frac{1}{s(s+1)(s+0.5)}$$

and find (a) the frequency of real axis crossing, (b) the frequency at which $|G(j\omega)| = 1$, (c) the magnitude and phase of any asymptotes, (d) the gain margin and (e) the phase margin.

5.20 By means of a frequency response plot, find the phase margin for a unity feedback system whose open-loop transfer function is

$$G(s) = \frac{4\alpha^2}{(s+\alpha)^2}$$

5.21 Find a phase lead cascade compensator which will produce a phase margin of approximately 45° and a gain margin of approximately 10 dB, where the original open-loop system is:

$$G(s) = \frac{40K}{s(s+2)(s+8)}, \qquad K = 1$$

by means of: (a) Nyquist design; (b) Nichols chart design.

5.22 By means of a Nichols chart design, find a suitable phase lag cascade compensator which will produce a gain margin of approximately 10 dB and a phase margin of approximately 45°, where the original open-loop system is:

$$G(s) = \frac{K}{s(s+5)(s+10)}$$

Further reading

Bode, H. W., 'Relations between attenuation and phase in feedback amplifier design', *Bell System Tech. J.*, 1940, pp. 421–54.

Brockett, R. W. and Willems, J. L., 'Frequency domain stability criteria – Part I', *IEEE Trans. on Automatic Control*, **AC-10**, 1965, pp. 255–61.

Brockett, R. W. and Willems, J. L., 'Frequency domain stability criteria – Part II', *IEEE Trans. on Automatic Control*, **AC-10**, 1965, pp. 407–13.

Brogan, W. L., *Modern control theory*, Prentice-Hall, 1991.

DiStefano, J. J., Stubberud, A. R. and Williams, I. J., *Feedback and control systems*, McGraw-Hill (Schaum), 1976.

Douce, J. L., 'A note on frequency-response measurement', *Proc. IEE*, **133**, Pt.D, No. 4, 1986, pp. 189–90.

Eveleigh, V. W., *Control systems design*, Tata McGraw-Hill, 1972.

Healey, M., *Principles of automatic control*, Hodder and Stoughton, 1981.

Hind, E. C., 'Closed-loop transient response from the open-loop frequency response', *Measurement and Control*, 1978, pp. 302–8.

Horowitz, I. M., *Synthesis of feedback systems*, Academic Press, 1963.

Hunt, V. D., *Mechatronics*, Chapman & Hall, 1989.

Jackson, L. B., *Signals, systems and transforms*, Addison-Wesley, 1991.

Jacobs, O. L. R., *Introduction to control theory*, Oxford University Press, 1974.

MacFarlane, A. G. J., *Frequency response methods in control systems*, IEEE Press, 1979.

Maciejowski, J. C., *Multivariable feedback design*, Addison-Wesley, 1990.

Morris, N. M., *Control engineering*, 4th ed., McGraw-Hill, 1991.

Nyquist, H., 'Regeneration theory', *Bell System Tech. J.*, 1932, pp. 126–47.

Phillips, C. and Harbor R., *Feedback control systems*, Prentice-Hall, 1991.

Raven, F. H., *Automatic control engineering*, 2nd ed., McGraw-Hill, 1990.

Schwarzenbach, J. and Gill, K. F., *System modelling and control*, Edward Arnold, 1984.

Westcott, J. H., 'The frequency response method: its relationship to transient behaviour in control system design', *Trans. of Soc. of Instr. Tech.*, **4,** No. 3, 1952, pp. 113–24.

6

State equations

6.1   Introduction

In the previous chapters emphasis has been placed strongly on an input/output s-domain transfer function description of the system. All design has been based on the differential equations describing the relationship between input and output, often these equations have been built up from individual system components. The purpose of this chapter is to introduce a further variable, namely the system state, and to show how the system descriptions obtained when using the state are arrived at.

As an example of the state of a system, consider a football match (American, rugby or any other) in which the system input is effort on the part of the teams. The output is the score in terms of points or goals at the end of the game, whereas the system state could be the score at any point during the game. In general the state is a set of values, often scalar numbers, which if known along with the system input will result in complete knowledge of firstly any future states and secondly the system output, subject to the dynamics of the system also being known.

The state of any particular system is not necessarily unique, in the football match example it could be the team's morale during the match for instance and, as in this case, unlike the system input and output, the state is not always a measurable variable. When considering one system, then, a large number of state–space representations are possible, depending on the state variables chosen. The approach taken in this chapter is to start from the original s-domain transfer function in order to obtain the time domain state–space model for a defined state vector.

Once in the state–space it is possible to transform one representation into another by means of operations on the state variables in question, and hence simpler representations can be found that exhibit certain system properties. It is shown how the roots of the transfer function denominator can be important in these transformations and the possibilities for moving to a state representation with no obvious link to the original transfer function are suggested.

6.2   State equations from transfer functions

Consider a system with one input and one output, a single-input-single-output (SISO) system, whose output depends not only on the system input, but also on the rate of change of the output. That is:

$$y(t) = \frac{1}{a}\dot{y}(t) - \frac{b}{a}u(t) \tag{6.2.1}$$

where a and b are scalar values, $y(t)$ the system output at time t and $u(t)$ the system input at time t. Also $\dot{y}(t) = (\mathrm{d}/\mathrm{d}t)y(t) = \mathrm{D}y(t)$. Equation (6.2.1) can, however, also be written in the form:

$$\dot{y}(t) = ay(t) + bu(t) \tag{6.2.2}$$

Then, defining the state, $x_1(t)$, to be equal to the output $y(t)$, it follows that

$$\dot{x}_1(t) = ax_1(t) + bu(t) \tag{6.2.3}$$

and

$$y(t) = x_1(t)$$

This is a rather simple case to serve as an example of a state–space system description; the description will now be extended to cover more generalized and complicated expressions.

6.2.1   Matrix representations

Assuming an extended version of equation (6.2.2), let

$$\mathrm{D}^3 y(t) + a_2\mathrm{D}^2 y(t) + a_1\mathrm{D}y(t) + a_0 y(t) = b_0 u(t) \tag{6.2.4}$$

Now however, let

$$x_1(t) = y(t)$$
$$x_2(t) = \mathrm{D}x_1(t) = \mathrm{D}y(t)$$
$$x_3(t) = \mathrm{D}x_2(t) = \mathrm{D}^2 y(t)$$

Then, (6.2.4) can be written as

$$\begin{bmatrix} \dot{x}_1(t) \\ \dot{x}_2(t) \\ \dot{x}_3(t) \end{bmatrix} = \begin{bmatrix} 0 & 1 & 0 \\ 0 & 0 & 1 \\ -a_0 & -a_1 & -a_2 \end{bmatrix} \begin{bmatrix} x_1(t) \\ x_2(t) \\ x_3(t) \end{bmatrix} + \begin{bmatrix} 0 \\ 0 \\ b_0 \end{bmatrix} u(t)$$

and

$$y(t) = [1 \quad 0 \quad 0] \begin{bmatrix} x_1(t) \\ x_2(t) \\ x_3(t) \end{bmatrix} \tag{6.2.5}$$

or, in more general terms;

$$\underline{\dot{x}}(t) = A\underline{x}(t) + Bu(t)$$

and   (6.2.6)

$$y(t) = C\underline{x}(t) + Du(t)$$

where A is the coefficient matrix and $\underline{x}(t)$ is the state vector at time t. It will be seen that the matrix D, which gives a direct relationship between system input and output, is often equal to the null matrix, i.e. all its elements are zero.

In the equations considered, one of the states is made equal to the output. This is not necessarily the case in general, as can be seen when

$$D^3 y(t) + a_2 D^2 y(t) + a_1 D y(t) + a_0 y(t) = b_2 D^2 u(t) + b_1 D u(t) + b_0 u(t)$$

                                                                                      (6.2.7)

or

$$(D^3 + a_2 D^2 + a_1 D + a_0) y(t) = (b_2 D^2 + b_1 D + b_0) u(t)$$ (6.2.8)

The variable, $z(t)$, is then introduced such that:

1. $u(t) = (D^3 + a_2 D^2 + a_1 D + a_0) z(t)$
2. $y(t) = (b_2 D^2 + b_1 D + b_0) z(t)$ (6.2.9)

The states can now be defined as:

$$x_1(t) = z(t)$$
$$x_2(t) = D z(t)$$
$$x_3(t) = D^2 z(t)$$

which gives a representation equal to:

$$\underline{\dot{x}}(t) = \begin{bmatrix} 0 & 1 & 0 \\ 0 & 0 & 1 \\ -a_0 & -a_1 & -a_2 \end{bmatrix} \underline{x}(t) + \begin{bmatrix} 0 \\ 0 \\ 1 \end{bmatrix} u(t)$$ (6.2.10)

and

$$y(t) = [b_0 \quad b_1 \quad b_2] \underline{x}(t)$$

in which

$$\underline{x}(t) = \begin{bmatrix} x_1(t) \\ x_2(t) \\ x_3(t) \end{bmatrix}$$

and $\underline{x}(t_0)$, the initial state, is known.

It can be seen from (6.2.8) that a general form of the transfer equation is:

$$(D^n + a_{n-1} D^{n-1} + \cdots + a_1 D + a_0) y(t)$$
$$= (b_m D^m + b_{m-1} D^{m-1} + \cdots + b_1 D + b_0) u(t) \quad (6.2.11)$$

and subject to $m < n$ this will lead to an n-dimensional state vector, i.e.,

$$\underline{x}^{\mathrm{T}}(t) = [x_1(t), x_2(t), ..., x_n(t)]$$

6.2.2 Transfer functions

The differential equation (D-operator) description of the relationship between system input, $u(t)$, and system output, $y(t)$, is merely another way of writing the transfer function from input to output. If it is assumed, when considering the s-domain transfer function, that initial conditions are zero, the two descriptions, D-operator and s-domain, are identical when s^j is replaced by D^j. If nonzero initial conditions are apparent, though, this simple transformation from one description to another does not exist and a return to the D-operator description by means of an inverse Laplace transform is necessary. Hence, assuming zero initial conditions, the function

$$\frac{Y(s)}{U(s)} = \frac{b_{n-1}s^{n-1} + b_{n-2}s^{n-2} + \cdots + b_1 s^1 + b_0}{s^n + a_{n-1}s^{n-1} + \cdots + a_1 s^1 + a_0} \qquad (6.2.12)$$

can result in state equations of the form:

$$
\begin{bmatrix} \dot{x}_1 \\ \dot{x}_2 \\ \cdot \\ \cdot \\ \cdot \\ \dot{x}_{n-1} \\ \dot{x}_n \end{bmatrix}
=
\begin{bmatrix}
0 & 1 & 0 & \cdot & & 0 \\
0 & 0 & 1 & 0 & & 0 \\
\cdot & \cdot & \cdot & & & \cdot \\
\cdot & \cdot & \cdot & & \ddots & 0 \\
0 & \cdot & \cdot & 0 & & 1 \\
-a_0 & -a_1 & \cdot & -a_{n-2} & & -a_{n-1}
\end{bmatrix}
\begin{bmatrix} x_1 \\ x_2 \\ \cdot \\ \cdot \\ \cdot \\ x_{n-1} \\ x_n \end{bmatrix}
+
\begin{bmatrix} 0 \\ \cdot \\ \cdot \\ \cdot \\ \cdot \\ 0 \\ 1 \end{bmatrix} u(t) \qquad (6.2.13)
$$

and

$$
y(t) = [b_0, b_1, ..., b_{n-2}, b_{n-1}]
\begin{bmatrix} x_1 \\ x_2 \\ \cdot \\ \cdot \\ \cdot \\ x_{n-1} \\ x_n \end{bmatrix}
$$

This is, however, just one of an infinite number of possible state–space representations of the transfer function (6.2.12), and follows as a generalization to (6.2.10). The differences between representations are dependent on the method of state definition.

6.2.3 Worked example

Consider the RLC circuit shown in Fig. 6.1. Mesh 1 has a current $i_1(t)$ flowing through it, whereas Mesh 2 has a current $i_2(t)$.

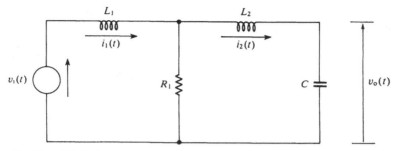

Fig. 6.1   RLC circuit with input $v_i(t)$

Summing voltages in Mesh 1:

$$v_i(t) = L_1 D i_1(t) + R_1 i_1(t) - R_1 i_2(t)$$

and summing voltages in Mesh 2:

$$0 = R_1 i_2(t) + L_2 D i_2(t) + v_o(t) - R_1 i_1(t)$$

and

$$i_2(t) = C D v_o(t)$$

Assuming that the input signal $u(t) = v_i(t)$, and the output signal $y(t) = v_o(t)$, the currents $i_1(t)$ and $i_2(t)$ can be classified as the states $x_3(t)$ and $x_2(t)$ with the output, $v_o(t)$, as the state, $x_1(t)$. The state equations can then be written as:

$$\dot{\underline{x}}(t) = \begin{bmatrix} 0 & 1/C & 0 \\ -1/L_2 & -R_1/L_2 & R_1/L_2 \\ 0 & R_1/L_1 & -R_1/L_1 \end{bmatrix} \underline{x}(t) + \begin{bmatrix} 0 \\ 0 \\ 1/L_1 \end{bmatrix} u(t)$$

and

$$y(t) = [1 \quad 0 \quad 0]\underline{x}(t)$$

where

$$\underline{x}^T(t) = [x_1(t), x_2(t), x_3(t)]$$

In this example the states have physical meanings and can be measured if desired. However, as long as $x_1(t_0) = v_o(t_0)$, $x_2(t_0) = i_2(t_0)$, $x_3(t_0) = i_1(t_0)$ and $u(t) = v_i(t)$ are known for all $t \geqslant t_0$, all states, including the output signal, can be calculated, without being measured, for all $t \geqslant t_0$.

6.3   Solution to the state equations

Consider the state equation

$$\dot{\underline{x}}(t) = A\underline{x}(t) + Bu(t) \tag{6.3.1}$$

The right-hand side of this equation consists of two parts, the first of these, $A\underline{x}(t)$, is known as the homogeneous part whilst the second, $Bu(t)$, is known as the forcing function.

In its present condition, for any A, B and $u(t)$, equation (6.3.1) will not give an explicit state vector value due to $\underline{\dot{x}}(t)$ on one side of the equation and $\underline{x}(t)$ on the other. This equation must therefore be solved such that under these conditions a unique state vector is found.

6.3.1   The state transition matrix

The $n \times n$ state transition matrix, $\phi(t)$, is the solution to equation (6.3.1) with the input $u(t)$ set to zero. That is, it satisfies the homogeneous equation

$$\underline{\dot{x}}(t) = A\underline{x}(t) \tag{6.3.2}$$

With initial conditions at time t_0, the solution to this equation (6.3.2) is

$$\underline{x}(t) = \exp[A(t - t_0)]\underline{x}(t_0) \tag{6.3.3}$$

in which

$$\phi(t - t_0) = \exp[A(t - t_0)] \tag{6.3.4}$$

The state transition matrix is then expressive of the natural system response when the input is zero, and relates the state vector at time t to the state vector at time t_0, the initial time.

Where t_0 is considered to be zero:

$$\phi(t) = \exp[A(t)] \tag{6.3.5}$$

An alternative approach in the calculation of the state transition matrix is to take the Laplace transform of (6.3.2), to give:

$$s\underline{X}(s) - \underline{x}(0) = A\underline{X}(s) \tag{6.3.6}$$

By rearranging this expression:

$$[sI - A]\underline{X}(s) = \underline{x}(0) \tag{6.3.7}$$

On the assumption that $[sI - A]$ is nonsingular and therefore invertible:

$$\underline{X}(s) = [sI - A]^{-1}\underline{x}(0) \tag{6.3.8}$$

Now, if inverse Laplace transforms, denoted by $\mathscr{L}^{-1}\{\cdot\}$, are taken

$$\underline{x}(t) = \mathscr{L}^{-1}\{[sI - A]^{-1}\}\underline{x}(0) \tag{6.3.9}$$

for all $t \geqslant 0$ such that the state transition matrix, $\phi(t)$, becomes:

$$\phi(t) = \mathscr{L}^{-1}\{[sI - A]^{-1}\} = \exp[A(t)] \tag{6.3.10}$$

and the zero-input response with initial time set to zero is therefore:

$$\underline{x}(t) = \phi(t)\underline{x}(0) \tag{6.3.11}$$

6.3.2 Properties of the state transition matrix

Having found, in the previous section, the state transition matrix as a solution to the state equations, it is now shown that this matrix has several important properties; these are as follows:

(a) If A is the null matrix $= 0(n \times n)$, i.e. a matrix in which every coefficient is zero,

$$\phi(0) = \exp[0] = I$$

(b) $\phi(t) \cdot \phi(t) \dots \phi(t) = \phi^m(t) = \phi(tm)$

This holds because:

$$\phi(t) \cdot \phi(t) \dots \phi(t) = e^{At}e^{At} \dots e^{At}$$
$$= e^{At + At + \dots + At} = e^{mAt}$$

and

$$e^{mAt} = (e^{At})^m = \phi^m(t)$$

or

$$e^{mAt} = (e^A)^{tm} = \phi(tm)$$

(c) $\qquad \phi^{-1}(t) = e^{-At} = (e^A)^{-t} = \phi(-t)$

$\Rightarrow \quad \phi^{-1}(t) = \phi(-t)$

(d) $\qquad \phi(t_2 - t_1) \cdot \phi(t_1 - t_0) = e^{A(t_2 - t_1)}e^{A(t_1 - t_0)}$

$$= e^{At_2} \cdot e^{-At_1} \cdot e^{At_1} \cdot e^{-At_0} = e^{At_2}e^{-At_0}$$
$$= e^{A(t_2 - t_0)} = \phi(t_2 - t_0)$$

$\Rightarrow \quad \phi(t_2 - t_1) \cdot \phi(t_1 - t_0) = \phi(t_2 - t_0)$

Further properties follow from those already given, but will not be considered here.

6.3.3 General solution of the state equation

Thus far the solution, via the state transition matrix, of the homogeneous part of the equation, assuming zero input, has been found. In this section it is considered that the forcing function is nonzero, then by taking the Laplace transform of (6.3.1):

$$s\underline{X}(s) - \underline{x}(0) = A\underline{X}(s) + BU(s) \tag{6.3.12}$$

Rearranging this equation gives:

$$\underline{X}(s) = [sI - A]^{-1}\underline{x}(0) + [sI - A]^{-1}BU(s) \tag{6.3.13}$$

where the first part of equation (6.3.13) is identical to (6.3.8).

By taking the inverse Laplace transform of (6.3.13), however, the following is obtained:

$$\underline{x}(t) = \mathscr{L}^{-1}\{[sI - A]^{-1}\}\underline{x}(0) + \mathscr{L}^{-1}\{[sI - A]^{-1}BU(s)\} \tag{6.3.14}$$

and using the representation for the state transition matrix, (6.3.14) can be written as:

$$\underline{x}(t) = \phi(t)\underline{x}(0) + \int_0^t \phi(t - \tau)Bu(\tau)\,d\tau \tag{6.3.15}$$

where the initial time is assumed to be zero. If all the elements in the initial state vector are also zero, then the zero-state response is given by:

$$\underline{x}(t) = \int_0^t \phi(t - \tau)Bu(\tau)\,d\tau \tag{6.3.16}$$

6.3.4 Worked example

A unit step function is applied to the system whose state equation is

$$\underline{\dot{x}}(t) = \begin{bmatrix} 0 & 1 \\ -3 & -4 \end{bmatrix} \underline{x}(t) + \begin{bmatrix} 0 \\ 1 \end{bmatrix} u(t)$$

Find the state transition matrix and hence solve the state transition equation.

The solution is as follows:

$$sI - A = \begin{bmatrix} s & 0 \\ 0 & s \end{bmatrix} - \begin{bmatrix} 0 & 1 \\ -3 & -4 \end{bmatrix} = \begin{bmatrix} s & -1 \\ 3 & s+4 \end{bmatrix}$$

The inverse of this matrix $[sI - A]$ is then:

$$[sI - A]^{-1} = \frac{1}{(s+3)(s+1)} \begin{bmatrix} s+4 & 1 \\ -3 & s \end{bmatrix}$$

The state transition matrix can now be obtained by taking the inverse Laplace transform, $\phi(t) = \mathscr{L}^{-1}\{[sI - A]^{-1}\}$, such that:

$$\phi(t) = \frac{1}{2} \begin{bmatrix} 3e^{-t} - e^{-3t} & e^{-t} - e^{-3t} \\ 3e^{-3t} - 3e^{-t} & 3e^{-3t} - e^{-t} \end{bmatrix}$$

but, a further equation arises in that:

$$[sI - A]^{-1}Bu(s) = \frac{1}{(s+3)(s+1)} \begin{bmatrix} 1 \\ s \end{bmatrix} \frac{1}{s}$$

and taking the inverse Laplace transform of this results in:

$$\mathscr{L}^{-1}\{[sI - A]^{-1}BU(s)\} = \frac{1}{6} \begin{bmatrix} 2 - 3e^{-t} + e^{-3t} \\ e^{-t} - e^{-3t} \end{bmatrix}$$

The overall solution to the state transition equation is therefore:

$$\underline{x}(t) = \frac{1}{2} \begin{bmatrix} 3e^{-t} - e^{-3t} & e^{-t} - e^{-3t} \\ 3e^{-3t} - 3e^{-t} & 3e^{-3t} - e^{-t} \end{bmatrix} \underline{x}(0)$$

$$+ \frac{1}{6} \begin{bmatrix} 2 - 3e^{-t} + e^{-3t} \\ e^{-t} - e^{-3t} \end{bmatrix}: \quad t \geq 0$$

Hence, if the initial values of the elements in the state vector at time $t_0 = 0$ are known, the value of these elements at any subsequent time t can be found.

6.4   The characteristic equation

In the previous section it was shown how the state of a system at any time t can be found in terms of the input function and the initial state values. In fact, the term $[sI - A]$ in its inverse form appears as the basis for the transfer functions found. This matrix is also apparent in the overall transfer function relating system input to output as can be seen from the following.

Taking the Laplace transform of the initial output state equation (6.2.6):

$$Y(s) = C\underline{X}(s) + DU(s) \tag{6.4.1}$$

The state vector, $\underline{X}(s)$, was determined in (6.3.13) to be dependent on initial state values and the system input; this equation can therefore be substituted into (6.4.1), and this leads to:

$$Y(s) = C[sI - A]^{-1}\underline{x}(0) + \{C[sI - A]^{-1}B + D\}U(s) \tag{6.4.2}$$

To obtain the final transfer function relating input to output, though, initial conditions must be set to zero, i.e. $\underline{x}(0) = 0$, this follows from the transfer function definition. The input/output relationship is therefore given by:

$$Y(s) = \{C[sI - A]^{-1}B + D\}U(s) \tag{6.4.3}$$

This can, however, also be written in the form of (6.2.12) as:

$$\frac{Y(s)}{U(s)} = \frac{C\{\text{adj}[sI - A]\}B + |sI - A|D}{|sI - A|} \tag{6.4.4}$$

Hence, the denominator of the expression (6.4.4) is:

$$|sI - A| = s^n + a_{n-1}s^{n-1} + \cdots + a_1 s + a_0 \tag{6.4.5}$$

and this is called the *characteristic polynomial* of the matrix A.

For the existence of a finite transfer function the matrix $(sI - A)$ must be nonsingular, where this is not so:

$$|sI - A| = 0 \tag{6.4.6}$$

and this (6.4.6) is termed the *characteristic equation* of the matrix A, i.e. the denominator of the input to output transfer function is set to zero in order to obtain the characteristic equation.

6.4.1   Eigenvalues and eigenvectors

The roots of the characteristic equation are referred to as the eigenvalues of the matrix A or the poles of the input to output transfer function. The roots of the numerator of the input to output transfer function are, on the other hand, called the zeros of the transfer function.

Eigenvectors are associated with the previously obtained eigenvalues such that if λ_i denotes the ith eigenvalue of matrix A, its associated $n \times 1$ dimensional eigenvector, v_i, is found to be such that

$$(\lambda_i I - A)v_i = 0 \tag{6.4.7}$$

where

$$v_i^T = [v_{i1}, v_{i2}, \ldots, v_{in}] \tag{6.4.8}$$

Example 6.4.1

Let the $n = 2$ dimensional matrix A be defined as:

$$A = \begin{bmatrix} 3 & 2 \\ 1 & 4 \end{bmatrix}$$

The characteristic equation is therefore obtained from:

$$|sI - A| = \begin{vmatrix} s-3 & -2 \\ -1 & s-4 \end{vmatrix} = 0$$

or

$$s^2 - 7s + 10 = 0$$

The eigenvalues are thus, $\lambda_1 = 5$, $\lambda_2 = 2$.

These can now be used to obtain the eigenvector. Firstly the eigenvector corresponding to λ_1, using equation (6.4.7);

$$\begin{bmatrix} \lambda_1 - 3 & -2 \\ -1 & \lambda_1 - 4 \end{bmatrix} \begin{bmatrix} v_{11} \\ v_{12} \end{bmatrix} = 0$$

or, substituting for $\lambda_1 = 5$

$$\begin{bmatrix} 2 & -2 \\ -1 & 1 \end{bmatrix} \begin{bmatrix} v_{11} \\ v_{12} \end{bmatrix} = 0$$

This is satisfied if $v_{11} = v_{12} = 1$. Secondly, the eigenvector corresponding to λ_2 must be found from:

$$\begin{bmatrix} -1 & -2 \\ -1 & -2 \end{bmatrix} \begin{bmatrix} v_{21} \\ v_{22} \end{bmatrix} = 0$$

and this is satisfied when $v_{21} = 2$, $v_{22} = -1$. Hence the eigenvalues are $\lambda_1 = 5$, $\lambda_2 = 2$, with associated linearly independent eigenvectors

$$v_1 = \begin{bmatrix} 1 \\ 1 \end{bmatrix}, \qquad v_2 = \begin{bmatrix} 2 \\ -1 \end{bmatrix}$$

It will be seen in the following section how the eigenvectors are of great importance in matrix diagonalization. Firstly however a result fundamental to the study of linear systems must be introduced.

6.4.2 Cayley–Hamilton theorem

The Cayley-Hamilton theorem states that every square matrix A satisfies its own characteristic equation. In other words, if A is an $n \times n$ dimensional square matrix resulting in a characteristic polynomial;

$$a(s) = |sI - A| = s^n + a_{n-1}s^{n-1} + \cdots + a_1 s + a_0$$

then

$$a(A) \triangleq A^n + a_{n-1}A^{n-1} + \cdots + a_1 A + a_0 I = 0$$

Proof: The inverse of $(sI - A)$ can be obtained as

$$(sI - A)^{-1} = \frac{\text{adj}(sI - A)}{|sI - A|} \tag{6.4.9}$$

where the adjoint of $(sI - A)$ is a matrix whose elements are all polynomials in s of degree less than n, i.e. $(n = 1)$ or less, such that the matrices B_i can be defined as:

$$\text{adj}(sI - A) = B_0 + B_1 s + B_2 s^2 + \cdots + B_{n-1}s^{n-1} \tag{6.4.10}$$

It follows though, that by rearranging (6.4.9)

$$|sI - A| \cdot I = (sI - A) \cdot \text{adj}(sI - A) \tag{6.4.11}$$

or by using (6.4.10)

$$(s^n + a_{n-1}s^{n-1} + \cdots + a_1 s + a_0)I = (sI - A) \cdot (B_0 + B_1 s + \cdots + B_{n-1}s^{n-1}) \tag{6.4.12}$$

Hence, by equating like powers of s in (6.4.12), the following relationships result:

$$B_{n-1} = I$$
$$B_{n-2} - AB_{n-1} = a_{n-1}I$$
$$B_{n-3} - AB_{n-2} = a_{n-2}I$$
$$\vdots \qquad \vdots$$
$$B_0 - AB_1 = a_1I$$
$$- AB_0 = a_0I$$

The first equation can now be multiplied throughout by A^n, the second equation by A^{n-1} and so on, the final (bottom) equation being multiplied by I. This results in

$$A^n + a_{n-1}A^{n-1} + a_{n-2}A^{n-2} + \cdots + a_1A + a_0I$$
$$= A^n B_{n-1} + A^{n-1}(B_{n-2} - AB_{n-1}) + A^{n-2}(B_{n-3} - AB_{n-2}) + \cdots$$
$$+ A(B_0 - AB_1) - AB_0 = 0 \quad (6.4.13)$$

The Cayley–Hamilton theorem is thus proved.

Example 6.4.2

To show that the matrix A in Section 6.4.1 satisfies its own characteristic equation, and to find A^{-1} using the Cayley–Hamilton theorem.

$$A = \begin{bmatrix} 3 & 2 \\ 1 & 4 \end{bmatrix}$$

and its characteristic equation is

$$a(s) = s^2 - 7s + 10 = 0$$

Substituting A for s leads to

$$a(A) = A^2 - 7A + 10I = 0$$

or

$$a(A) = \begin{bmatrix} 11 & 14 \\ 7 & 18 \end{bmatrix} - \begin{bmatrix} 21 & 14 \\ 7 & 28 \end{bmatrix} + \begin{bmatrix} 10 & 0 \\ 0 & 10 \end{bmatrix} = 0$$

Also

$$A^2 - 7A + 10I = 0$$

Therefore, multiplying both sides of this equation by A^{-1} gives:

$$A - 7I + 10A^{-1} = 0$$

or

$$A^{-1} = \tfrac{1}{10}\{7I - A\}$$

Hence

$$A^{-1} = \frac{1}{10} \left\{ \begin{bmatrix} 7 & 0 \\ 0 & 7 \end{bmatrix} - \begin{bmatrix} 3 & 2 \\ 1 & 4 \end{bmatrix} \right\} = \begin{bmatrix} 0.4 & -0.2 \\ -0.1 & 0.3 \end{bmatrix}$$

These properties show that this method can be used to obtain A^k where k is a large number and successive multiplications of the A matrix would prove time consuming.

Further operations on the system matrix will now be considered in the following section.

6.5   Canonical forms

In Section 6.2 it was commented that for any particular transfer function there are an infinite number of possible state–space representations, these being dependent on the definition of the state vector itself. It can be the case, therefore, that the state representation obtained results in a complicated expression, especially with regard to the A matrix. In order to simplify any subsequent analysis and design it is preferable to transform the state vector into a condition such that the A matrix corresponds with one of a number of simplified forms which exhibit specific properties. This section describes firstly the method of transforming a state vector and secondly a number of possible simplified A matrix forms, these being termed *canonical forms*.

6.5.1   State variable transformation

Let the n-dimensional system be characterized by the state equations:

$$\dot{\underline{x}}(t) = A\underline{x}(t) + Bu(t)$$

and   (6.5.1)

$$y(t) = C\underline{x}(t) + Du(t)$$

where A is an $n \times n$ coefficient matrix. The state vector $\underline{x}(t)$ can then be transformed to the state vector $\underline{x}'(t)$ by means of the relationship

$$\underline{x}(t) = Q\underline{x}'(t) \tag{6.5.2}$$

By substitution of (6.5.2) into (6.5.1) a new state–space representation of the same system is obtained (6.5.3). Either

$$Q\dot{\underline{x}}'(t) = AQ\underline{x}'(t) + Bu(t)$$

or

$$\dot{\underline{x}}'(t) = Q^{-1}AQ\underline{x}'(t) + Q^{-1}Bu(t) \tag{6.5.3}$$

and

$$y(t) = CQ\underline{x}'(t) + Du(t)$$

These equations hold on condition that Q is an $n \times n$ nonsingular matrix. Summarizing (6.5.3), the equations can also be written as:

$$\underline{\dot{x}}'(t) = A'\underline{x}'(t) + B'u'(t) \tag{6.5.4}$$

and

$$y'(t) = C'\underline{x}'(t) + D'u'(t)$$

where

$$A' = Q^{-1}AQ, \qquad B' = Q^{-1}B$$

in which A' and A are termed similar matrices. Also

$$C' = CQ \qquad \text{and} \qquad D' = D$$

Finally it must be noted that as $y'(t) = y(t)$, and $u'(t) = u(t)$, the system output and input are unaffected by any state variable transformation.

6.5.2   Phase variable canonical form

A single input, time-invariant, linear system is said to be in phase variable canonical form if the state equations are written such that:

$$\underline{\dot{x}}(t) = \begin{bmatrix} 0 & 1 & 0 & \cdot & 0 \\ 0 & 0 & 1 & \cdot & 0 \\ \cdot & \cdot & \cdot & & 0 \\ \cdot & \cdot & \cdot & \cdot & \\ 0 & 0 & 0 & \cdot & 1 \\ -a_0 & -a_1 & -a_2 & \cdot & -a_{n-1} \end{bmatrix} \underline{x}(t) + \begin{bmatrix} 0 \\ \cdot \\ \cdot \\ \cdot \\ 0 \\ 1 \end{bmatrix} u(t) \tag{6.5.5}$$

and

$$y(t) = C\underline{x}(t)$$

The form of the C matrix and the existence of the D matrix are not, therefore, important for this definition.

It can be seen that the phase variable form is in fact the same as that described in Section 6.2.2 for the representation in the state–space of an s-domain system transfer function, the parameters $a_0, a_1, ..., a_{n-1}$ being the coefficients of the characteristic polynomial. This phase variable form, simple because of its sparsity, can thus be obtained directly from the system transfer function with numerator coefficients appearing in the C matrix, as described in 6.2.2.

6.5.3 Diagonalization

For any n-dimensional diagonal matrix, the n associated eigenvalues $\lambda_1, \lambda_2, ..., \lambda_n$ are also the elements of the main diagonal, this follows from the definition of eigenvalues given in Section 6.4.1. The state transition matrix will, therefore, also be diagonal, such that:

$$\exp \Lambda t = \exp \begin{bmatrix} \lambda_1 & & & 0 \\ & \lambda_2 & & \\ & & \ddots & \\ 0 & & & \lambda_n \end{bmatrix} t = \begin{bmatrix} \exp \lambda_1 t & & & 0 \\ & \exp \lambda_2 t & & \\ & & \ddots & \\ 0 & & & \exp \lambda_n t \end{bmatrix}$$

(6.5.6)

To find a diagonal matrix which is similar to the original matrix A, a matrix $\bar{Q}$ must be obtained which satisfies:

$$\Lambda = A' = \bar{Q}^{-1} A \bar{Q}$$

(6.5.7)

where Λ is a diagonal matrix and $\bar{Q}$ is that value of Q which when used as an operator on A causes A' to be diagonal. It follows that:

$$\bar{Q}\Lambda = A\bar{Q}$$

(6.5.8)

Let the n columns of $\bar{Q}$ be denoted by q_i where $i = 1, ... n$. Then (6.5.8) must satisfy:

$$q_i \lambda_i = A q_i = \lambda_i q_i \qquad i = 1, ..., n$$

(6.5.9)

which is true when λ_i is an eigenvalue of the matrix A, with an associated eigenvector $q_i (= v_i)$, this is a consequence of (6.4.7). The matrix $\bar{Q}$ can therefore be found by means of the eigenvectors of A, resulting in a diagonal matrix, Λ, which consists of the corresponding eigenvalues.

Hence

$$\bar{Q} = [q_1, q_2, ..., q_n]$$

in which

$$q_i = v_i \qquad i = 1, 2, ..., n$$

From (6.5.9) it is apparent that a nonsingular matrix $\bar{Q}$ can be obtained only when matrix A has n linearly independent eigenvectors, which has to be true for (6.5.7) to hold. Only eigenvectors corresponding to distinct eigenvalues are independent, thus where A has multiple order eigenvalues it may not be possible to calculate a diagonal similar matrix, but when A has n distinct eigenvalues a diagonal transform can always be found.

6.5.4 Jordan canonical form

A matrix, as described in Section 6.5.3 with eigenvalues along the main diagonal is a special case of the Jordan canonical form. This form copes with the possibility of multiple order eigenvalues by means of a set of Jordan blocks, the total number of which is equal to the number of linearly independent eigenvectors, each eigenvector being associated with one block only. In general, if λ_i is an $\bar{n}$th order eigenvalue, the $(\bar{n} \times \bar{n})$ dimensional Jordan block associated with this eigenvalue is given by:

$$\begin{bmatrix} \lambda_i & 1 & & 0 \\ & \lambda_i & 1 & \\ & & \ddots & 1 \\ 0 & & & \lambda_i \end{bmatrix}$$

Where $\bar{n} = 1$, however, the Jordan block reduces to the eigenvalue λ_i.

The overall matrix in Jordan canonical form is then made up of Jordan blocks along its 'main' diagonal. As an example, assume matrix A has three eigenvalues, λ_1, λ_2, λ_3, where λ_1 and λ_3 are single ordered but λ_2 is triple ordered. The Jordan canonical form matrix is found to be

$$\begin{bmatrix} \lambda_1 & 0 & 0 & 0 & 0 \\ 0 & \lambda_2 & 1 & 0 & 0 \\ 0 & 0 & \lambda_2 & 1 & 0 \\ 0 & 0 & 0 & \lambda_2 & 0 \\ 0 & 0 & 0 & 0 & \lambda_3 \end{bmatrix}$$

As an example, consider the matrix

$$A = \begin{bmatrix} 2 & 1 & 0 \\ 3 & 1 & -2 \\ -2.5 & 2 & 4 \end{bmatrix}$$

The characteristic equation is found from $|\lambda I - A|$ as:

$$|\lambda I - A| = \lambda^3 - 7\lambda^2 + 15\lambda - 9 = 0$$

which has roots equal to $(\lambda - 3)^2(\lambda - 1)$, and hence eigenvalues $\lambda_1 = 3$ (a double eigenvalue) and $\lambda_2 = 1$. The Jordan canonical form matrix is therefore

$$\Lambda = \begin{bmatrix} 3 & 1 & 0 \\ 0 & 3 & 0 \\ 0 & 0 & 1 \end{bmatrix}$$

There are thus two Jordan blocks, an off-diagonal unity term resulting in the block corresponding to the multiple eigenvalue.

6.6   Decomposition

Having described the Jordan canonical form in the previous section, its relationship with the original transfer function, in terms of multiple order eigenvalues, will be made use of in describing the various methods available with which a state–space description can be obtained from the original transfer function. One or two of these methods have already been considered in Section 6.2 during the introduction of state–space techniques. Here, however, they are looked at more formally such that the advantages and disadvantages of each method can be considered.

6.6.1   Partial fraction decomposition

This particular method, also known as parallel decomposition, is designed to result in a Jordan canonical form representation, a special case being when all eigenvalues are distinct, whereby a diagonal A matrix will result. Let the transfer function to be decomposed be defined as:

$$\frac{Y(s)}{U(s)} = \frac{b_{n-1}s^{n-1} + b_{n-2}s^{n-2} + \cdots + b_1 s + b_0}{s^n + a_{n-1}s^{n-1} + \cdots + a_1 s + a_0} \tag{6.6.1}$$

in which signal initial conditions are assumed to be zero. This can, however, also be represented in partial fraction form as:

$$\frac{Y(s)}{U(s)} = \frac{c_1}{(s-\lambda_1)^m} + \cdots + \frac{c_{m-1}}{(s-\lambda_1)^2} + \frac{c_m}{s-\lambda_1} + \frac{c_{m+1}}{s-\lambda_2} + \cdots + \frac{c_n}{s-\lambda_{n-m+1}} \tag{6.6.2}$$

where only one multiple eigenvalue, λ_1, is present and is of order m.
Note: When m is unity all the eigenvalues are distinct.

The sum obtained in (6.6.2), when multiplied throughout by $U(s)$, can also be written as a linear combination:

$$Y(s) = \sum_{i=1}^{n} c_i X_i(s) \tag{6.6.3}$$

such that the state vector can be defined as

$$\underline{x}^T(t) = [x_1(t), x_2(t), \ldots, x_n(t)] \tag{6.6.4}$$

in which the individual states are found as follows:

$$X_1(s) = \frac{U(s)}{(s-\lambda_1)^m} \tag{6.6.5}$$

but also

$$X_2(s) = \frac{U(s)}{(s-\lambda_1)^{m-1}} \tag{6.6.6}$$

and substituting for this in (6.6.5);

$$X_1(s) = \frac{1}{s - \lambda_1} X_2(s)$$

similarly

$$X_2(s) = \frac{1}{s - \lambda_1} X_3(s)$$

until

$$X_m(s) = \frac{1}{s - \lambda_1} U(s) \tag{6.6.7}$$

then

$$X_{m+1}(s) = \frac{1}{s - \lambda_2} U(s)$$

and finally

$$X_n(s) = \frac{1}{s - \lambda_{n-m+1}} U(s)$$

Taking the first equation of (6.6.7) as an example, this can also be expressed as

$$sX_1(s) - \lambda_1 X_1(s) = X_2(s) \tag{6.6.8}$$

or

$$sX_1(s) = \lambda_1 X_1(s) + X_2(s)$$

In the time domain, this becomes, by taking the inverse Laplace transform:

$$\dot{x}_1(t) = \lambda_1 x_1(t) + x_2(t) \tag{6.6.9}$$

The same transform can be taken for the remaining $n - 1$ states. As an example consider the mth state, which becomes:

$$\dot{x}_m(t) = \lambda_1 x_m(t) + u(t) \tag{6.6.10}$$

Writing the states in vector form results in the standard n-dimensional state characterization of a system (6.5.1) in which the matrices are defined by:

$$A = \left[\begin{array}{cccc:cccc} \lambda_1 & 1 & 0 & \cdot & 0 & & & \\ & \cdot & & & & & & \\ 0 & & \lambda_1 & & 0 & & 0 & \\ \vdots & & & & 1 & & & \\ 0 & \cdot & \cdot & 0 & \lambda_1 & & & \\ \hdashline & & & & & \lambda_2 & 0 & \cdot \cdot & 0 \\ & & 0 & & & 0 & \cdot & & 0 \\ & & & & & \vdots & & & \\ & & & & & 0 & \cdot \cdot & 0 & \lambda_p \end{array}\right]; \quad B = \left[\begin{array}{c} 0 \\ \cdot \\ \cdot \\ 0 \\ 1 \\ \cdot \\ 1 \end{array}\right\} (m-1) \text{ zeros} \tag{6.6.11}$$

Also

$$C^{T} = [c_1, c_2, ..., c_n] \quad \text{and} \quad D = 0$$

where

$$p = n - m + 1$$

The A matrix is hence of the Jordan canonical form, and will reduce to a diagonal form only when no multiple eigenvalues are present.

6.6.2 Direct decomposition

The previous description relies on the fact that it is possible, without too much difficulty, to obtain the eigenvalues (poles) of the transfer function. This may not always be the case and so a more direct approach, as was carried out in Section 6.2, must be taken.

Let the top and bottom of equation (6.6.1) be multiplied by $X(s)$, then if numerators and denominators are equated, this results in

$$Y(s) = (b_{n-1}s^{n-1} + b_{n-2}s^{n-2} + \cdots + b_1s + b_0)X(s)$$

and (6.6.12)

$$U(s) = (s^n + a_{n-1}s^{n-1} + \cdots + a_1s + a_0)X(s)$$

Multiplying out the second of these equations leads to:

$$U(s) = s^n X(s) + a_{n-1}s^{n-1}X(s) + \cdots + a_1sX(s) + a_0 X(s) \qquad (6.6.13)$$

Then by letting $X_1(s) = X(s)$ and $X_i(s) = sX_{i-1}(s)$; $i = 2, ..., n$ equation (6.6.13) becomes:

$$U(s) = sX_n(s) + a_{n-1}X_n(s) + \cdots + a_1 X_2(s) + a_0 X_1(s) \qquad (6.6.14)$$

or by rearranging

$$sX_n(s) = -a_{n-1}X_n(s) - \cdots - a_1 X_2(s) - a_0 X_1(s) + U(s)$$

When a time domain description is required, the state–space representation can be built up such that

$$A = \begin{bmatrix} 0 & 1 & 0 & \cdot & 0 \\ \cdot & \cdot & \cdot & & \cdot \\ \cdot & \cdot & \cdot & & 0 \\ 0 & \cdot & \cdot & 0 & 1 \\ -a_0 & -a_1 & \cdot & -a_{n-2} & -a_{n-1} \end{bmatrix} ; \quad B = \begin{bmatrix} 0 \\ \cdot \\ \cdot \\ 0 \\ 1 \end{bmatrix}$$

and

$$C^{T} = [b_0, b_1, ..., b_{n-2}, b_{n-1}]$$

in which the matrix C is formulated directly from the first of the equations defined in (6.6.12). The direct method of decomposition therefore results in a phase variable canonical form realization.

There are various other methods of decomposition dependent on the conditions applicable to both system numerator and denominator. In this section two methods have been considered, firstly a method which relies upon all the factors of the system denominator being available, and secondly a method in which no assumptions are made concerning factorization.

6.7   Summary

The fundamental concept of the state–space as a basis for controller design, rather than the more standard input/output transfer function description, has been considered. It has been shown how, by starting with the transfer function parameters, a state–space model can be obtained directly. Canonical matrix forms, which much simplify further analysis, have been formulated by means of state transformations such that where a particular state vector gives rise to a problematic description, through complicated parameter structures, this can be converted to a canonical form in order to clarify the situation. By locating the roots, eigenvalues, of the transfer function denominator, diagonal forms of state matrix, A, can be obtained, although where multiple roots are found certain modifications must be made using Jordan forms.

The state–space framework built up in this chapter will now be employed in the control of systems. Emphasis is placed on specific important aspects such as closed-loop stability via assignment of the transfer function denominator roots.

Problems

6.1   Consider the electrical circuit shown in Fig. 6.2 with an input signal $v_i(t)$. Find a state–space model for this system by taking the two mesh currents and the output voltage as states.

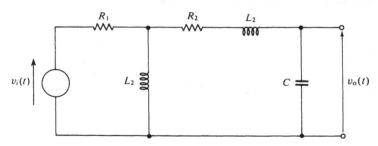

Fig. 6.2

6.2   The transfer function of a position control Ward–Leonard unit is given by:

$$\frac{\theta(s)}{V(s)} = \frac{K_1 K_2}{s(R + Ls)(B + Js)}$$

Find a state–space model for this system if the states are given by:

$$x_1(s) = \frac{V(s)}{s}, \qquad x_2(s) = \frac{V(s)}{R + Ls}, \qquad x_3(s) = \frac{V(s)}{B + Js}$$

6.3   Given the state–space description

$$\underline{\dot{x}}(t) = \begin{bmatrix} 0 & 1 & 0 \\ 0 & 0 & 1 \\ -2 & -6 & -5 \end{bmatrix} \underline{x}(t) + \begin{bmatrix} 6 \\ 3 \\ 4 \end{bmatrix} u(t)$$

and

$$y(t) = [1 \quad 0 \quad 0]\underline{x}(t)$$

find the input $u(t)$ to output $y(t)$ transfer function.

6.4   Consider the two tanks shown in Fig. 6.3.

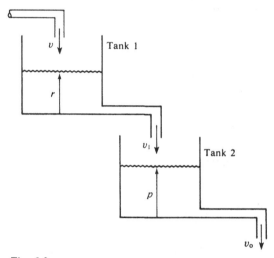

Fig. 6.3

Water flows into tank 1 at a rate v (the input) and this tank, which has unit cross-sectional area, has a head of water equal to height r. Water flows out of tank 1 and into tank 2 at a rate v_1, where $v_1 = 2r$. Tank 2 has unit cross-sectional area and a head of water equal to height p. The output, which is the rate of water flow out of tank 2, is given by $v_o = 3p$.

Describe the tank system by means of a state–space model, the two states being the tank heads r and p.

6.5   Consider the system transfer function:

$$(D^3 + 6D^2 + 11D + 6)y(t) = (D^3 + 9D^2 + 14D + 10)u(t)$$

relating system input $u(t)$ to output $y(t)$.

Reconsider this transfer function in terms of an s-domain representation and find an equivalent state–space model in which:

$$\dot{\underline{x}}(t) = A\underline{x}(t) + B\underline{u}(t)$$
$$y(t) = C\underline{x}(t) + Du(t)$$

6.6   Consider the state–space system model:

$$\dot{\underline{x}}(t) = \begin{bmatrix} 0 & 1 & 0 \\ 0 & 0 & 1 \\ 0 & -1 & -2 \end{bmatrix} \underline{x}(t) + \begin{bmatrix} 0 \\ 1 \\ 1 \end{bmatrix} u(t)$$

$$y(t) = [1 \quad 0 \quad 1]\underline{x}(t)$$

Find the input $u(t)$ to output $y(t)$ transfer function for this system.

6.7   For the state–space model of Problem 6.6, obtain a Jordan canonical form state description by carrying out an appropriate state transformation.

6.8   Find the state transition matrix for

$$\dot{\underline{x}}(t) = \begin{bmatrix} 3 & 4 \\ 2 & 1 \end{bmatrix} \underline{x}(t) + \begin{bmatrix} 0 \\ 1 \end{bmatrix} u(t)$$

and solve the state transition equation.

6.9   Find the state transition matrix for

$$\dot{\underline{x}}(t) = \begin{bmatrix} 0 & 1 \\ -6 & -7 \end{bmatrix} \underline{x}(t) + \begin{bmatrix} 2 \\ 1 \end{bmatrix} u(t)$$

and solve the state transition equation.

6.10   Consider the state–space model

$$\dot{\underline{x}}(t) = \begin{bmatrix} -4 & 3 \\ \alpha & -2 \end{bmatrix} \underline{x}(t) + \begin{bmatrix} 1 \\ 0 \end{bmatrix} u(t)$$

Reconfigure the state equations into phase variable canonical form if $\alpha = 2$ and the same input $u(t)$ is applied. Can a phase variable canonical form be obtained if $\alpha = 0$?

6.11   Consider the state–space model

$$\dot{\underline{x}}(t) = \begin{bmatrix} 0 & 1 & 0 \\ 3 & 0 & 2 \\ -12 & -7 & -6 \end{bmatrix} \underline{x}(t) + \begin{bmatrix} 0 \\ 0 \\ 1 \end{bmatrix} u(t)$$

Find a matrix Q such that $\underline{x}(t) = Q\underline{x}'(t)$ transforms the system to phase variable canonical form.

6.12   Consider the state–space model

$$\underline{\dot{x}}(t) = \begin{bmatrix} -1 & 4 \\ 1 & -1 \end{bmatrix} \underline{x}(t) + \begin{bmatrix} 1 \\ 1 \end{bmatrix} u(t)$$

with $u(t) = 0$, for all t. After 1 second it is found that $x_1(t) = 2$ and $x_2(t) = 1$.
What is the value of $\underline{x}(t)$ when $t = 0$?
    What conclusions can be drawn from your results?

6.13   By means of the Cayley–Hamilton theorem, calculate the state transition matrix
for

$$\underline{\dot{x}}(t) = \begin{bmatrix} -1 & 3 \\ 2 & -2 \end{bmatrix} \underline{x}(t) + \begin{bmatrix} 0 \\ 1 \end{bmatrix} u(t)$$

$$y(t) = [1 \quad 0]\underline{x}(t)$$

Find $y(t)$ if $u(t) = 1$ for $t \geqslant 0$, $x_1(0) = 1$ and $x_2(0) = 0$.

6.14   Find the eigenvalues and eigenvectors for the matrix

$$A = \begin{bmatrix} 0 & 1 \\ -2 & -3 \end{bmatrix}$$

Calculate $\exp(At)$ and write $(D^2 + 3D + 2)x(t) = \exp(-3t)$, $x(0) = 0$, $\dot{x}(0) = 1$
in state–space form.

6.15   A state–space model of a particular system is given by:

$$\underline{\dot{x}}(t) = \begin{bmatrix} -1 & 1 \\ 0 & -2 \end{bmatrix} \underline{x}(t) + \begin{bmatrix} 0 \\ 1 \end{bmatrix} u(t)$$

$$y(t) = [1 \quad 1]\underline{x}(t)$$

By noting that the state transition matrix is equal to the inverse Laplace
transform of $[sI - A]^{-1}$, see (6.3.10), show that this is equal to $\exp[A(t)]$.

6.16   For the system in Problem 6.15, find $y(t)$ if $u(t) = 1$ for $t \geqslant 0$, $x_1(0) = 1$ and
$x_2(0) = 0$. Also, find $y(t)$ if $u(t) = 1$ for $t \geqslant 0$, but $x_1(0) = -1$, $x_2(0) = 0$.

6.17   For the system shown in Fig. 6.4, find a state–space description by employing
two states and calculate the state transition matrix. Also, find $y(t)$ if $u(t) = 1$
for $t \geqslant 0$ and $x_1(0) = x_2(0) = 0$.

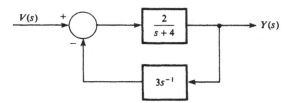

Fig. 6.4

6.18   Obtain the input $u(t)$ to output $y(t)$ transfer function for the system described by:

$$\dot{\underline{x}}(t) = \begin{bmatrix} 0 & 1 & 0 \\ 0 & 0 & 1 \\ 6 & -1 & 4 \end{bmatrix} \underline{x}(t) + \begin{bmatrix} -1 \\ -4 \\ 26 \end{bmatrix} u(t)$$

and

$$y(t) = \begin{bmatrix} 1 & 1 & 1 \end{bmatrix} \underline{x}(t)$$

6.19   For the system shown in Fig. 6.5, find a state–space description by employing two states and calculate the state transition matrix.

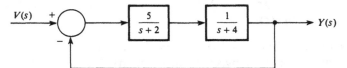

Fig. 6.5

6.20   Consider the system transfer function:

$$G(s) = \frac{6}{(s+3)^3} + \frac{7}{(s+3)^2} + \frac{2}{s+3} + \frac{5}{s+4}$$

Show that the Jordan form state–space description is:

$$\dot{\underline{x}}(t) = \begin{bmatrix} -3 & 1 & 0 & 0 \\ 0 & -3 & 1 & 0 \\ 0 & 0 & -3 & 0 \\ 0 & 0 & 0 & -4 \end{bmatrix} \underline{x}(t) + \begin{bmatrix} 0 \\ 0 \\ 1 \\ 1 \end{bmatrix} u(t)$$

when

$$y(t) = \begin{bmatrix} 6 & 7 & 2 & 5 \end{bmatrix} \underline{x}(t)$$

$$Y(s) = G(s)U(s)$$

Further reading

Blackman, P. F., *Introduction to state-variable analysis*, Macmillan, 1977.

Brockett, R. W., *Finite dimensional linear systems*, John Wiley, 1970.

Delchamps, D. F., *State-space and input-output linear systems*, Springer-Verlag, 1988.

DeRusso, P. M., Roy, R. J. and Close, C. M., *State variables for engineers*, Wiley, 1965.

Doyle, J., Francis, B., and Tannenbaum, A., *Feedback control theory*, Macmillan, 1992.

Elloy, J.-P. and Piasco, J.-M., *Classical and modern control*, Pergamon Press, 1981.

Godfrey, K. and Jones, P. (eds.), *Signal processing for control*, Springer-Verlag, 1986.

Healey, M., 'Study of methods of computing transition matrices', *Proc. IEE*, **120**, No. 8, 1973, pp. 905–12.

Kailath, T., *Linear systems*, Prentice-Hall, 1980.

Kalman, R. E., 'On the general theory of control systems', *Proc. IFAC 1st Int. Congress Aut. Control*, Moscow, 1960, pp. 481–93.

Kucera, V., *Analysis and design of discrete linear control systems*, Prentice-Hall, 1991.

Ogata, K., *State space analysis of control systems*, Prentice-Hall, 1967.

Olsson, G. and Piani, G., *Computer systems for automation and control*, Prentice-Hall, 1992.

O'Reilly, J. (ed.), *Multivariable control for industrial applications*, Peter Peregrinus Ltd, 1987.

Prime, H. A., *Modern concepts in control theory*, McGraw-Hill, 1969.

Rohrer, R., *Circuit analysis: an introduction to the state variable approach*, McGraw-Hill, New York, 1970.

Rosenbrock, H. H., *State space and multivariable theory*, Nelson, 1970.

Rubin, W. B., 'A simple method for finding the Jordan form of a matrix', *IEEE Trans. on Automatic Control*, **AC-17**, 1972, pp. 145–6.

Sinha, N. K., *Control systems*, Wiley Eastern, 1994.

Wiberg, D. W., *Theory and problems of state space and linear systems*, McGraw-Hill, New York, 1971.

Wolovich, W. A., *Linear multivariable systems*, Springer-Verlag, 1974.

7

State variable feedback

7.1   Introduction

In Chapter 6 we reconsidered a system in terms of three variable types, namely input, output and state. Relationships between input-output transfer functions were shown and various manipulations possible when in a state–space format were discussed. However, in the analysis carried out thus far the advantages and flexibility provided by a state–space basis have, as yet, not really been uncovered. In fact it is only this chapter in which the ideas of providing a controller scheme and in particular a feedback control system based on a state–space framework are introduced.

In general the only control we have over a given system is provided by the control input $u(t)$ which we can choose; both the system output and the state variables are resultant values. What we would really like is to achieve a control over the system, such that we can alter a state variable by taking it from its present value to a value we desire, a certain time later, simply by an appropriate choice of control input. In fact we would like to be able to do this for all the state variables, i.e. we would have complete control of the system by being able to choose an input $u(t)$ which would move any state vector from one value to some other arbitrary value after a certain time. Then the system is said to be completely state controllable, although often it is simply called completely controllable.

One further important factor is whether or not we can actually see all the states of the system, usually this means by some direct or indirect measurement. If we cannot measure directly certain of the states, it must be possible to reconstruct each of them simply in terms of the available information, i.e. the system inputs and outputs. Consider for example a bank balance for which no bank account statement, i.e. no direct state information, is given, then the state of the account in terms of the present balance can be calculated by noting the withdrawals (output) and deposits (input) made. The theory behind this is that if the standing of the balance is known (probably this means how much it is in the red) then an appropriate deposit can be made. Purchasers of this book will be able to feel pleased in that they have helped to prove the practicality of this theory as far as my own state variable is concerned. However, by considering a general state vector, when, over any finite time period, the complete vector can be obtained from the system input and output information available over that same time period, the system is said to be completely state observable.

In this chapter the concepts of state controllability and observability are presented in terms of the state–space system descriptions previously introduced; following this, tests for controllability and observability which can be carried out on a particular system are described by making use of the state matrices available.

The overall design for a controller is, by means of a state–space description, split up into two parts, (a) the control, dependent on the control action required and (b) the observer, obtaining a reconstruction of the state vector. In the later sections of this chapter, details are given of some of the different types of controller design possible with emphasis on the approaches taken for pole assignment and optimal control, and finally methods are described in which the state vector can be estimated or observed by employing the input and output variable data.

7.2   Controllability and observability

The problems of controllability and observability are very often treated together and indeed this seems sensible in that both are tests applied to a feedback control system in order to ascertain whether or not we can actually dictate exactly system performance. Controllability is concerned with whether or not we can control a system, whereas observability is concerned with whether or not we can see, or work out, the result of the control applied. In some cases the importance of one of these two measures may well dominate, in that it may be required to find out what has happened even if we cannot do much about it, or vice versa.

7.2.1   Concepts of controllability and observability

General expressions for a state–space system description were given in the previous chapter, in (6.2.6); these are repeated here as:

$$\dot{\underline{x}}(t) = A\underline{x}(t) + Bu(t)$$
$$y(t) = C\underline{x}(t) + Du(t)$$

(7.2.1)

A schematic diagram of this basic system is shown in Fig. 7.1, where the input $u(t)$ and output $y(t)$ are connected by a middle man, the state $\underline{x}(t)$.

The loop can be closed by deriving the input in terms of system response information, in this case in terms of the state variable vector $\underline{x}(t)$. Let the control input be defined as:

$$u(t) = - F\underline{x}(t)$$

(7.2.2)

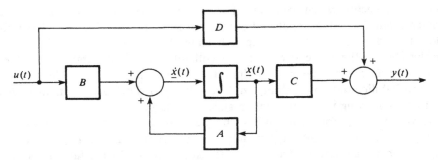

Fig. 7.1 Block schematic of a state–space system description

where F is a $1 \times n$ feedback vector; for a single input system with an n-dimensional state vector.

On substitution of the control input from (7.2.2) into the state equation of (7.2.1) this results in the equation

$$\dot{\underline{x}}(t) = [A - BF]\underline{x}(t) \qquad (7.2.3)$$

A schematic of this is given in Fig. 7.2.

A solution to the equation (7.2.3) can be found in a similar way to the procedure carried out in the previous chapter, see Section 6.3.1, and this results in

$$\underline{x}(t) = \exp[(A - BF)(t - t_0)]\underline{x}(t_0) \qquad (7.2.4)$$

which shows that for a given system $A, B,$ and an initial state condition $\underline{x}(t_0)$, we would like to be able to select a vector F such that the state vector $\underline{x}(t)$ settles down to a steady value. In this case it is desired that all elements in the state vector tend to zero; this is the case when the input is derived solely by state feedback. The addition of a nonzero external input as shown in Section 7.4 results in nonzero steady-state state variables. But, can we choose an F such that $\underline{x}(t)$ settles down? This is in fact the same as asking the question: 'is the system controllable?'

Definition of controllability: a system defined by (7.2.1) is said to be state controllable

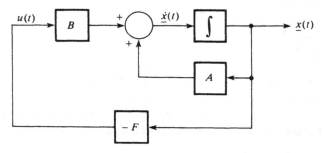

Fig. 7.2 Use of feedback on the state equation

if a control $u(t)$ exists which will change the state $\underline{x}(t_0)$ to any desired state $\underline{x}(t)$ over the finite time period t_0 to t: where $t > t_0$.

This definition generalizes the idea of controllability such that from any initial state value $\underline{x}(t_0)$ we wish to be able to take the state to a particular value $\underline{x}(t)$, not necessarily zero. However, where only state feedback is employed, i.e. no external input is present as shown in Fig. 7.2, then the final state value $\underline{x}(t)$ is required to be zero. By means of state feedback only, controllability is concerned with whether the state variables will settle down or not, whereas by applying an external input, one not generated by feedback, this will dictate the steady values themselves: zero external input, as considered thus far, means zero steady-state requirements for a controllable state vector. A system for which the definition of controllability is shown not to apply is said to be uncontrollable.

Observability of a system is dependent on whether or not information from all of the state variables can be obtained at the system output. When a system has more than one output then every state must have an effect on at least one of the outputs. If one or more states of a particular system cannot be observed at the output then that system is said to be unobservable.

The concept of observability is extremely important in that for many state–space system descriptions the state vector is chosen to be some mathematical tool required to meet specific needs, hence leading to a quite complex relationship for each state variable in terms of any physically real measurements. It is important therefore that we can obtain the state vector by means of system input and output information only; if indeed this is, as we would hope, possible then the system is observable.

Definition of observability: a system defined by (7.2.1) is said to be state observable if the system state vector $\underline{x}(t)$ can be obtained over any finite time period, e.g. t_0 to t_1, from complete knowledge of the system input and output over the same time period.

The inclusion of system input knowledge in a definition of observability is in a sense a red herring, in that the system input does not in any way determine itself whether a system is observable or unobservable. In order to calculate the state vector we do need to know the value of the system input, however; just what the input is makes no difference to whether or not it is actually possible to find the vector. In order to simplify things the input $u(t)$ can be set to zero, resulting in a state variable solution from (7.2.1) of

$$\underline{x}(t) = \exp[A(t - t_0)]\underline{x}(t_0) \tag{7.2.5}$$

On substitution of this into the output equation of (7.2.1), remembering that $u(t) = 0$, we have

$$y(t) = C \exp[A(t - t_0)]\underline{x}(t_0) \tag{7.2.6}$$

What is then desired is to be able to obtain $\underline{x}(t)$ over any time period, starting at time t_0, from (7.2.5) given that $\underline{x}(t_0)$ is dependent in some way on $y(t)$. It is fairly straightforward to show (see Problem 7.1) that by premultiplying both sides of (7.2.6) by $\exp[A^{\mathrm{T}}(t - t_0)]C^{\mathrm{T}}$ and then integrating, it is possible for $\underline{x}(t_0)$ to be found from $y(t)$, and thence $\underline{x}(t)$ from the same source, subject to certain conditions on the A and C matrices. Further, as the actual input value does not affect the observability of a system, it will still be possible to calculate $\underline{x}(t)$ for any known input as long as it is possible to do so for the zero-input conditions described.

It can be noted that the concept of controllability concerns the input $u(t)$ and state $\underline{x}(t)$ along with matrices A and B, whereas the concept of observability involves the output $y(t)$ and state $\underline{x}(t)$ along with matrices A and C. These associations will now be further developed in terms of tests that can be carried out in order to ascertain whether or not a system is controllable and/or observable.

7.2.2 Tests for controllability and observability

Although several tests are possible in order to determine controllability and observability conditions of a system, those most commonly encountered and perhaps therefore simplest are described here. Examples are given of how to apply the tests, however no proofs are shown as these are considered to be a little too detailed for this text and indeed can be found in several of the referenced works.

Test for controllability: a system defined by (7.2.1) is completely state controllable if the $n \times n$ controllability matrix has rank n, i.e. if it is nonsingular.

The controllability matrix V_c is defined as:

$$V_c = [B, AB, ..., A^{n-1}B] \tag{7.2.7}$$

Note: The nonsingularity condition does not apply to systems with more than one input. As we are only considering single input systems here, this is of no direct concern.

Example 7.2.1

Consider the $n = 2$ state variable system

$$A = \begin{bmatrix} 0 & 1 \\ -2 & -3 \end{bmatrix}; \quad B = \begin{bmatrix} 0 \\ 1 \end{bmatrix}, \quad C = \begin{bmatrix} 1 & 1 \end{bmatrix}$$

such that

$$AB = \begin{bmatrix} 1 \\ -3 \end{bmatrix}$$

so

$$V_c = [B, AB] = \begin{bmatrix} 0 & 1 \\ 1 & -3 \end{bmatrix}$$

which is nonsingular, i.e. it has rank 2.

Hence the system given in this example is completely controllable.

Example 7.2.2

Consider the $n = 2$ state variable system

$$A = \begin{bmatrix} 0 & -2 \\ 1 & -3 \end{bmatrix}; \qquad B = \begin{bmatrix} 1 \\ 1 \end{bmatrix}; \qquad C = [0 \quad 1]$$

such that

$$AB = \begin{bmatrix} -2 \\ -2 \end{bmatrix}$$

so

$$V_c = [B, AB] = \begin{bmatrix} 1 & -2 \\ 1 & -2 \end{bmatrix}$$

which is singular, it has rank 1.
  Hence the system given in this example is uncontrollable.

Test for observability: a system defined by (7.2.1) is completely state observable if the $n \times n$ observability matrix has rank n, i.e. it is nonsingular.
  The observability matrix V_0 is defined:

$$V_0 = \begin{bmatrix} C \\ CA \\ \vdots \\ CA^{n-1} \end{bmatrix} \qquad\qquad (7.2.8)$$

Note: The nonsingularity condition does not apply to systems with more than one output. As only single output systems are considered here, this is of no direct concern.

Example 7.2.3

Consider the $n = 2$ state variable system

$$A = \begin{bmatrix} 0 & 1 \\ -2 & -3 \end{bmatrix}; \qquad B = \begin{bmatrix} 0 \\ 1 \end{bmatrix}; \qquad C = [1 \quad 1]$$

such that

$$CA = [-2 \quad -2]$$

so

$$V_0 = \begin{bmatrix} C \\ CA \end{bmatrix} = \begin{bmatrix} 1 & 1 \\ -2 & -2 \end{bmatrix}$$

which is singular, it has rank 1.

Hence the system given in this example is unobservable.

Note: This is the same system as that in Example 7.2.1, where it was shown to be controllable. So this system is controllable but unobservable.

Example 7.2.4

Consider the $n = 2$ state variable system

$$A = \begin{bmatrix} 0 & -2 \\ 1 & -3 \end{bmatrix}; \quad B = \begin{bmatrix} 1 \\ 1 \end{bmatrix}; \quad C = [0 \quad 1]$$

such that

$$CA = [1 \quad -3]$$

so

$$V_0 = \begin{bmatrix} C \\ CA \end{bmatrix} = \begin{bmatrix} 1 & 1 \\ 1 & -3 \end{bmatrix}$$

which is nonsingular, i.e. it has rank 2.

Hence the system given in this example is completely observable.

Note: This is the same system as that in Example 7.2.2, where it was shown to be uncontrollable. So this system is uncontrollable but observable.

The results of the controllability and observability tests are completely separate from each other; as well as the conditions shown in the examples it is quite possible for a system to be both controllable and observable or to be neither controllable nor observable.

7.2.3   Invariant theorems

It was described in the previous chapter how the state variable vector could be transformed to an alternative form, possibly to make use of advantageous properties, such as simplicity, held by the new form.

Consider the original state representation to be (7.2.1) with a new state vector $\underline{x}'(t)$ related to the original $\underline{x}(t)$ by means of

$$\underline{x}(t) = Q\underline{x}'(t) \tag{7.2.9}$$

resulting in the new state description

$$\begin{aligned} \underline{\dot{x}}'(t) &= A'\underline{x}'(t) + B'u(t) \\ y(t) &= C'\underline{x}'(t) + Du(t) \end{aligned} \tag{7.2.10}$$

where

$$A' = Q^{-1}AQ, \ B' = Q^{-1}B \text{ and } C' = CQ$$

with Q nonsingular.

The following two statements can then be made:

1. Invariant theorem of controllability: If the pair $[A,B]$ is completely state controllable then the pair $[A',B']$ is also completely state controllable, i.e. if the system (7.2.1) is controllable then so is (7.2.10).

   It is also true that if $[A,B]$ is uncontrollable then so is $[A',B']$.

   Proof: If V_c, the controllability matrix for a system described by (7.2.1), is of full rank then not only is that particular system controllable but also the controllability matrix V_c' for the system described by (7.2.10) is:

   $$V_c' = [B', A'B', (A')^2B', ..., (A')^{n-1}B']$$

   and as $(A')^i = Q^{-1}A^iQ$ it follows that

   $$V_c' = [Q^{-1}B, Q^{-1}AB, Q^{-1}A^2B, ..., Q^{-1}A^{n-1}B]$$

   or

   $$V_c' = Q^{-1}V_c$$

   Therefore, on condition that Q^{-1} is nonsingular, i.e. of full rank, then so too will be V_c', hence the system described by (7.2.10) is also controllable.

2. Invariant theorem of observability: If the pair $[A,C]$ is completely state observable then the pair $[A',C']$ is also completely state observable, i.e. if the system (7.2.1) is observable then so is (7.2.10).

   It is also true that if $[A,C]$ is unobservable then so is $[A',C']$.

   Proof: Similar to the proof given for the invariant theorem of controllability, see Problem 7.5.

7.3   Controllability and observability in transfer functions

In the previous chapter, Section 6.2 in particular, the idea of state–space system descriptions was introduced by means of a transfer function approach. In this chapter controllability and observability have been introduced as system characteristics with specific dependence on a state–space framework. But any one particular system is the same system irrespective of whether we prefer to think of it in terms of the state–space rather than as a transfer function. So, having uncovered the tests for controllability and observability in the state–space, these must necessarily also have a meaning in terms of a transfer function description. A consideration of this last point is the purpose of this section.

7.3.1   Transfer function realization

Before discussing controllability and observability in terms of a system transfer function it will be shown how a transfer function, considered open-loop in this section, can be simply obtained in terms of the state matrices.

On taking Laplace transforms of (7.2.1), assuming zero initial conditions on the state vector, it follows that

$$sX(s) = AX(s) + BU(s) \tag{7.3.1}$$

or

$$X(s) = [sI - A]^{-1}BU(s) \tag{7.3.2}$$

On substitution of the state vector into the transformed output equation of (7.2.1), we have

$$Y(s) = \{C[sI - A]^{-1}B + D\}U(s) \tag{7.3.3}$$

and thus if the open-loop transfer function relating input to output, defined previously in (6.2.12), is given as

$$Y(s) = G(s) \cdot U(s) \tag{7.3.4}$$

two basic statements can be made.

1. Realization theorem: A state–space system description $\{A, B, C, D,\}$ is a realization of the transfer function system description $G(s)$ if

$$G(s) = C[sI - A]^{-1}B + D \tag{7.3.5}$$

However, for any one $G(s)$ many possible selections of $\{A, B, C, D\}$ exist, depending on the state vector specified. In fact for any $G(s)$ the order n of the state vector chosen is not unique, which leads to the second statement.

2. Minimal realization: A state–space system description $\{A, B, C, D\}$ is a minimal realization of the transfer function system description $G(s)$ if (7.3.5) holds and n, the order of the state vector, is the minimum possible order.

It will be seen in Section 7.3.2 how the idea of a minimal realization is linked with the controllability and observability of that system.

7.3.2   Pole-zero cancellation

The controllability and observability of a state–space system description are dependent on there being no common factors between the numerator and denominator of $G(s)$, i.e. poles and zeros of a transfer function which are identical and can therefore be cancelled to produce a lower order, simpler transfer function. This can best be seen by an example, so consider the transfer function:

$$G(s) = \frac{(s+2)}{(s+2)(s+3)} = \frac{1}{(s+3)} = \frac{s+2}{s^2+5s+6}$$

It can be seen that this transfer function has a common pole-zero at $s = -2$.

Further, by making use of (7.3.5) it can easily be verified that the following are all state–space realizations of $G(s)$. Note D is the null matrix in each case.

(a) $A = \begin{bmatrix} -3 & 0 & 0 \\ 0 & -1 & 0 \\ 0 & 0 & -2 \end{bmatrix}$; $B = \begin{bmatrix} 1 \\ 0 \\ 0 \end{bmatrix}$; $C^T = \begin{bmatrix} 1 \\ 1 \\ 1 \end{bmatrix}$; $n = 3$

(b) $A = \begin{bmatrix} 0 & -6 \\ 1 & -5 \end{bmatrix}$; $B = \begin{bmatrix} 2 \\ 1 \end{bmatrix}$; $C^T = \begin{bmatrix} 0 \\ 1 \end{bmatrix}$; $n = 2$

(c) $A = \begin{bmatrix} -3 & 0 \\ 0 & -2 \end{bmatrix}$; $B = \begin{bmatrix} 1 \\ 0 \end{bmatrix}$; $C^T = \begin{bmatrix} 1 \\ 0 \end{bmatrix}$; $n = 2$

(d) $A = \begin{bmatrix} 0 & 1 \\ -6 & -5 \end{bmatrix}$; $B = \begin{bmatrix} 0 \\ 1 \end{bmatrix}$; $C^T = \begin{bmatrix} 2 \\ 1 \end{bmatrix}$; $n = 2$

(e) $A = [-3]$; $B = [1]$; $C^T = [1] = C$; $n = 1$

After toying around for a while with the above set of realizations it is apparent that it is relatively easy to obtain realizations for higher values of n, provided that the 'extra' poles and zeros produced cancel in the final transfer function, the dimension n and nature of the state vector being different in each case. The only realization in which no pole-zero cancellations occur is (e), which is in fact also a minimal realization of $G(s)$.

By applying the test for controllability and observability introduced in Section 7.2.2, results from the realizations (a) to (e) are as follows:

(a) *un*controllable observable $n = 3$
(b) *un*controllable observable $n = 2$
(c) *un*controllable *un*observable $n = 2$
(d) controllable *un*observable $n = 2$
(e) controllable observable $n = 1$

This list shows that for any transfer function, of the many possible state–space realizations some are controllable, some are not, some are observable, some are not, some are both controllable and observable, some are neither: each particular case being dependent on the state vector, and hence state–space description, selected. It can be seen from the list that although (c) is an uncontrollable, unobservable description a higher order state vector can be found which is either controllable or observable, in the case of (a) this is observable. The only system description which is both controllable and observable is though (e), which is a minimal realization and which defines the transfer function with no pole-zero cancellations. In fact this leads to the general rule for system transfer functions.

1. Complete controllability and observability in a transfer function. A transfer function, $G(s)$, represents a system which is both controllable and observable if no pole-zero cancellations occur.
   Note: A minimal realization of a system is a controllable and observable realization, if it is possible to obtain a controllable and observable realization for that system.

   A second rule follows directly from this:

2. A transfer function, $G(s)$, represents a system which is uncontrollable or unobservable or both if at least one pole-zero cancellation takes place.

The two definitions made, arising from controllability and observability concepts, point to the fact that very often it is not possible to witness everything that is taking place in a system simply from its input-output transfer function. Specifically when pole-zero cancellations occur, this effectively corresponds to information on the system lost somewhere between input and output. By means of a state–space description this information can be retrieved in terms of the state vector and hence a system which cannot be controlled simply by consideration of its transfer function can be controlled, as desired, if the cancelling modes are suitably acted upon in the state–space realization. Several methods which can be employed on a state–space system description in order to achieve a suitable control action will now be discussed.

7.4   State variable feedback

It has been shown thus far in this chapter that the problem of controlling a system in terms of a state–space framework can actually be split up into two separate sub-problems. Firstly, observability: can we obtain enough knowledge about the state vector? This topic is considered more fully in Section 7.5, and is independent of the type of control action selected. Secondly, controllability: can we actually control the system as desired by manipulating the state vector, irrespective of how that vector has been obtained? That is, assuming the state vector to be available, we consider how to provide a state feedback control scheme in order to alter, hopefully improve, the performance of the system under control, and this type of feedback control, as depicted in Fig. 7.3, is the subject of this section.

The diagram shown in Fig. 7.3 is a modification of that in Fig. 7.2, and includes a state

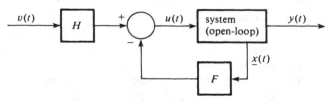

Fig. 7.3 State variable feedback

feedback parameter vector, F. However, also included is a reference input vector $v(t)$, introduced via a feedforward gain H, something which we as the controller designer can choose. In general it is usually required that the output $y(t)$ follows the desired value $v(t)$, the criterion placed on the way this reference following is carried out then defines the type of control action that is needed. In Fig. 7.2 effectively it was assumed that the reference input $v(t)$ was zero, i.e. the output was desired to be zero, this is commonly referred to as the regulator problem.

7.4.1   Pole placement

Pole placement control is concerned with the specification of the poles of the closed-loop system, that is the roots of the characteristic equation, or in other words the closed-loop denominator coefficients. The basic philosophy is that if we know how the closed-loop poles affect the system dynamics and hence performance, then we can tailor that performance as required if we can arbitrarily select the pole positions.

In terms of state variable feedback design the pole placement scheme can be described as one where we are given the desired closed-loop pole positions and must select the state feedback parameters in order to achieve those pole positions. The closed loop in this sense refers to the relationship between the reference value $v(t)$ and the actual output value $y(t)$.

The general state–space description has a state equation

$$\dot{\underline{x}}(t) = A\underline{x}(t) + Bu(t) \tag{7.4.1}$$

where $\underline{x}(t)$ is the $(n \times 1)$ dimensional state vector, and the single control input $u(t)$ is defined as

$$u(t) = - F\underline{x}(t) + Hv(t) \tag{7.4.2}$$

in which F is the $(1 \times n)$ dimensional vector of state feedback parameters. It is in fact these n feedback parameters which we must find in order to achieve the desired pole positions. Further, $v(t)$ is a scalar reference signal and H is a gain which we can select; this latter choice will be considered briefly a little later in this section, but is though not the main subject of this discussion.

On substitution of the control input equation (7.4.2) into the state equation (7.4.1), this results in

$$\dot{\underline{x}}(t) = [A - BF]\underline{x}(t) + BHv(t) \tag{7.4.3}$$

It also follows that if the output equation is defined by

$$y(t) = C\underline{x}(t) + Du(t) \tag{7.4.4}$$

then on substitution of the control input (7.4.2) into this equation, we find that

$$y(t) = [C - DF]\underline{x}(t) + DHv(t) \tag{7.4.5}$$

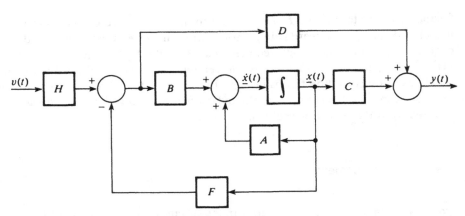

Fig. 7.4 Closed-loop system with state feedback

A diagram showing the state–space system description with feedback connected is shown in Fig. 7.4, this closed-loop system is completely defined by the two equations (7.4.3) and (7.4.5).

Viewing (7.4.3) in terms of its Laplace transform, so

$$X(s) = [sI - \bar{F}]^{-1}BHV(s) \tag{7.4.6}$$

where $\bar{F} = A - BF$, and the state vector is assumed to be initially zero.

On substitution of (7.4.6) into the output equation transformed from (7.4.5), it follows that

$$Y(s) = \{[C - DF][sI - \bar{F}]^{-1}B + D\}HV(s) \tag{7.4.7}$$

and it can be seen from this that the closed-loop poles are in fact the roots of the equation:

$$|sI - \bar{F}| = |sI - A + BF| = 0 \tag{7.4.8}$$

Equation (7.4.8) shows the possibility that all of the closed-loop poles can be arbitrarily placed by selection of F and that this is dependent on the state matrices A and B only, which leads to the following.

Pole placement by state variable feedback: The n state feedback parameters contained within the vector F can be selected in order to obtain an arbitrary set of n closed-loop poles if the pair $\{A,B\}$ is controllable.

In order to show the operation of pole placement by example, assume that the desired closed-loop response ot a second-order system is given by the denominator

$$s^2 + 5s + 4 = (s + 4)(s + 1)$$

Example 7.4.1

Consider the $n = 2$ dimensional, controllable system with

$$A = \begin{bmatrix} 0 & 1 \\ -2 & -3 \end{bmatrix}; \qquad B = \begin{bmatrix} 0 \\ 1 \end{bmatrix}$$

Then we have also that the state feedback parameter vector is given by

$$F = [f_1 \quad f_2]$$

where f_1 and f_2 are to be found in order to achieve pole placements. But

$$[A - BF] = \begin{bmatrix} 0 & 1 \\ -2 - f_1 & -3 - f_2 \end{bmatrix} = \bar{F}$$

and

$$[sI - \bar{F}] = \begin{bmatrix} s & -1 \\ f_1 + 2 & s + f_2 + 3 \end{bmatrix}$$

such that

$$|sI - \bar{F}| = s^2 + (f_2 + 3)s + (f_1 + 2)$$

In order to obtain the desired closed-loop pole positions it is required that

$$f_2 + 3 = 5, \quad \text{i.e.} \quad f_2 = 2$$

and

$$f_1 + 2 = 4, \quad \text{i.e.} \quad f_1 = 2$$

which results in $F = [2 \quad 2]$.

Example 7.4.2

Consider the $n = 2$ dimensional, uncontrollable system with

$$A = \begin{bmatrix} 0 & -2 \\ 1 & -3 \end{bmatrix}; \quad B = \begin{bmatrix} 1 \\ 1 \end{bmatrix}$$

and

$$F = [f_1 \quad f_2]$$

so

$$[sI - \bar{F}] = \begin{bmatrix} s + f_1 & f_2 + 2 \\ f_1 - 1 & s + f_2 + 3 \end{bmatrix}$$

and

$$|sI - \bar{F}| = s^2 + s(f_1 + f_2 + 3) + f_1 + f_2 + 2$$

In order to achieve the desired closed-loop pole positions it is required that

$$f_1 + f_2 + 3 = 5 \quad \text{or} \quad f_1 + f_2 = 2$$
$$f_1 + f_2 + 2 = 4 \quad \text{or} \quad f_1 + f_2 = 2$$

which result in one equation but two unknowns.

Hence, for this uncontrollable system it is not possible to obtain a state feedback parameter vector F which will achieve the desired positions.

Note: The selection of the gain H is usually carried out in order to ensure that the actual output $y(t)$ is exactly equal to the reference $v(t)$ once steady-state conditions are reached, i.e. H is chosen to counteract the inherent steady-state gain of the closed-loop system. So it is therefore straightforward, with reference to (7.4.7), and assuming a step input, to choose H as:

$$H = \{ [C - DF] [sI - \bar{F}]^{-1} B + D \}^{-1} |_{s=0} \qquad (7.4.9)$$

This calculation of H therefore must come after the selection of the vector F and hence $\bar{F}$ also.

7.4.2   Optimal control

In Section 7.4.1 a controller was designed with the intention of ensuring that the closed-loop system response is of a form that we want, irrespective of the open-loop characteristics of the system, and this is done by forcing the closed-loop denominator to assume a specified polynomial. But this is just one particular criterion on which a controller design can be based. In fact many requirements can be made for the closed-loop system, including fairly straightforward and perhaps obvious limitations such as maximum and minimum possible input values, e.g. due to the type of amplifier used, or the allowable settling time, i.e. the time taken for a response to settle to approximately its final value. Perhaps one of the most commonly encountered and widely researched overall control performance objectives is that of optimal control, and in this section a glimpse is given of the possibilities open to the designer by making use of the state variable feedback ideas previously introduced.

An optimal controller is one which is best or most favorable, and this automatically implies that it is best in terms of a particular sense, i.e. in terms of a defined performance requirement or specification. So the statement that a certain controller is optimal is, on its own, fairly meaningless unless it is also stated in which sense the control is optimal. Very often the performance specification is made in terms of a mathematical function, the optimal control then being the one which minimizes (or maximizes if appropriate) this function. For a single input system a cost function defining performance could be simply:

$$CF = \int_{t_i}^{t_f} u^2(t) \, dt \qquad (7.4.10)$$

where the function is considered over the time inverval t_i = initial time to t_f = final time, and the initial and final time state vectors are assumed to be available. A minimization of this particular function by a selection of the control input to be zero over the time period specified, results in a cost function which is also zero. A straightforward requirement that the controller be designed in order to minimize the magnitude of the control input therefore results in a rather obvious solution, i.e. no control input at all. The input will be a signal, e.g. voltage or velocity, and hence the cost function (7.4.10) is then proportional to the energy expended in controlling the system; we can therefore minimize the energy expended, as required from (7.4.10), by not expending any at all.

An alternative cost function could be one in which the state vector is seen as the important variable and in particular the error between the state at the initial time t_i and the final time t_f; note that the n-dimensional vector at time t_f, $\underline{x}(t_f)$ is termed the origin of the state–space. This cost function can then be defined as

$$CF = \int_{t_i}^{t_f} \underline{x}^T(t)\underline{x}(t) \; dt \tag{7.4.11}$$

such that if this is minimized by an appropriate choice of control input, the vector $\underline{x}(t_i)$ will approach the vector $\underline{x}(t_f)$ with very little overshoot. However it will most likely be the case that certain states are considered to be more important than others, e.g. a particular velocity or acceleration might be critical whereas another might be of relatively little importance. Weightings can therefore be placed on certain states and/or certain combinations of states, which means that the cost function (7.4.11) becomes

$$CF = \int_{t_i}^{t_f} \underline{x}^T(t)Q\underline{x}(t) \; dt \tag{7.4.12}$$

where Q is an $n \times n$ weighting matrix and is usually symmetric with all positive, real coefficients.

An interesting and commonly encountered special case of (7.4.12) arises when $Q = C^T C$ and $D = 0$, see (7.4.4). In which case the cost function described (7.4.12) refers to the selection of an input signal which minimizes the error with respect to time, between the actual output, with initial conditions at time t_i, and the final desired output at time t_f. However, whether in terms of the output or the states of (7.4.12), choosing a control input $u(t)$ which minimizes this cost function may well result in a set of high magnitude values for $u(t)$; $t_i \leqslant t \leqslant t_f$, with large and rapid variations occurring during the time interval, and in practice this may well result in actuator saturation and hence nonlinear characteristics.

It appears appropriate therefore to combine the cost functions derived in order to put emphasis on both the control input and the states, such that an overall function can be described as

$$CF = \int_{t_i}^{t_f} [\underline{x}^T(t)Q\underline{x}(t) + u^T(t)Ru(t)] \; dt \tag{7.4.13}$$

in which (a) a multi-input system is allowed for, and (b) the real and usually symmetric and positive definite weighting matrix R is used to put emphasis on different inputs and on the input(s) with respect to the states. This latter point refers to the relative costing placed on the inputs as opposed to the states, i.e. the importance of (7.4.10) as opposed to (7.4.12), and points to the fact that this latest cost function, when minimized by appropriate choice of control input, results in an action which is a compromise between the two extremes considered.

The cost function (7.4.13) is termed a quadratic performance index and when it is wished to find the control input $u(t)$ from a linear system (7.4.1) as a solution to the cost

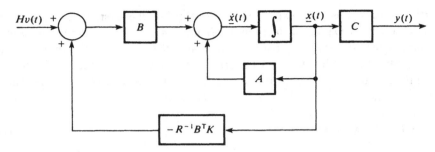

Fig. 7.5   LQ feedback control

function minimization for $t_i = 0$, $t_f = \infty$, then this is referred to as the linear quadratic control problem (*LQ*). Note R is a scalar for a single input.

In order to consider an overall closed-loop transfer function, an external reference input signal, $v(t)$, is required, and this is generally introduced to the control input in the form of (7.4.2). The external input does however not affect the optimality of the control employed; this is obtained by the choice of the state feedback parameters within the vector F. In particular, the optimal control which minimizes the cost function (7.4.13) is given by (7.4.2) in which

$$F = R^{-1}B^T K \tag{7.4.14}$$

where K is a positive definite matrix and the unique solution to the algebraic matrix Riccati equation

$$Q = KBR^{-1}B^T K - A^T K - KA \tag{7.4.15}$$

A block diagram of the overall system is shown in Fig. 7.5.

Unfortunately it is not a simple matter to obtain the solution (7.4.14) and (7.4.15) from the cost function (7.4.13), this problem is however dealt with in several texts (see D'Azzo and Houpis (1983), for example) and is not considered further here.

Finally it must be noted that the closed-loop poles of the optimal system are the roots of the equation:

$$|sI - A + BF| = |sI - A + BR^{-1}B^T K| = 0 \tag{7.4.16}$$

and these lie in the left half of the s-plane. Further, the minimum value of the cost function is then $x^T(t_i)Kx(t_i)$.

7.5   State estimation

The state vector has thus far been assumed to be either known or directly measurable, the control input being formed as some multiple of this known vector. The state vector is though very often selected merely as a mathematical entity with very little obvious physical relevance, and is therefore generally not measurable. Hence, an alternative scheme must be employed in order to calculate or more likely estimate the value of each state vector element. The control input can then be formed as a multiple of the estimated state vector which is used in place of its true but unavailable value, hopefully the estimate being a reasonably good one.

The mathematical tool employed in order to estimate the state vector is called an observer and, as will be described in this section, the measurable system inputs and outputs are used in the estimation procedure such that the estimated state vector is linearly related to the actual, but unknown, vector. This means that an error will exist between the actual and the estimated state values, one property desired of an observer then being that this error tends towards zero. It will though be shown that it is not simply a case of trying to reduce the error to zero as quickly as possible because this can lead to problems due to poor noise rejection. The selection of observer characteristics is therefore an involved task and it must be remembered that if the observer is part of a feedback path, which will very likely be the case, then the observer characteristics will directly affect the closed-loop system characteristics.

7.5.1 Asymptotic estimation

The linear system under consideration is defined by

$$\dot{\underline{x}}(t) = A\underline{x}(t) + Bu(t)$$
$$y(t) = C\underline{x}(t) \tag{7.5.1}$$

which is identical to (7.2.1) except that $D = 0$.

The matrices A, B and C are assumed to be known, as are the system input and output, $u(t)$ and $y(t)$ respectively. However the state vector $\underline{x}(t)$ is unknown, all that we do have is an estimate of this vector denoted by $\hat{\underline{x}}(t)$, which means that by considering the output equation in (7.5.1) we can in fact write that as

$$\hat{y}(t) = C\hat{\underline{x}}(t) \tag{7.5.2}$$

in which $\hat{y}(t)$ is an approximation to the output signal but, as it can be calculated from $\hat{\underline{x}}(t)$, is a known value. The error between the two outputs, actual and approximate, is then

$$l(t) = y(t) - \hat{y}(t) \tag{7.5.3}$$

and by referring to the state equation of (7.5.1) it seems reasonable that $\hat{x}(t)$ can be derived from

$$\dot{\hat{\underline{x}}}(t) = A\hat{\underline{x}}(t) + Bu(t) + Gl(t) \tag{7.5.4}$$

where the estimated state vector is subject to the same system characteristics as the actual state vector, the term $Gl(t)$ accounting for any error. By an appropriate choice of the parameters in the vector G so the state estimate can be made to vary significantly even for only small errors $l(t)$, when G is large, or to vary only slightly even for large errors, when G is small. So how should G be chosen? A further insight into the problem is given by considering the error between the actual and the estimated state, defined here by

$$\varepsilon(t) = \underline{x}(t) - \underline{\hat{x}}(t) \tag{7.5.5}$$

If (7.5.4) is subtracted from the state equation of (7.5.1) this results in

$$\underline{\dot{x}}(t) - \underline{\dot{\hat{x}}}(t) = A\underline{x}(t) + Bu(t) - A\underline{\hat{x}}(t) - Bu(t) - Gl(t) \tag{7.5.6}$$

and by including (7.5.5) and the output equation of (7.5.1) this becomes

$$\dot{\varepsilon}(t) = (A - GC)\varepsilon(t) \tag{7.5.7}$$

which has a solution

$$\varepsilon(t) = \exp\{[A - GC](t - t_0)\}\varepsilon(t_0) \tag{7.5.8}$$

where t_0 is the time at which initial conditions are taken.

It is therefore only the error present at the initial time t_0, $\varepsilon(t_0)$, which causes the remaining error $\varepsilon(t)$ for all following time. Further it can be seen from (7.5.8) that the response of the error $\varepsilon(t)$ is dictated by the eigenvalues of the matrix $(A - GC)$, these being called the observer poles, such that if there are any roots which lie in the right half of the s-plane, the error equation will be unstable resulting in the error heading off to infinity. For an observable system though, the observer poles can be arbitrarily chosen and hence the observer can not only be selected as stable but can also have a specified rate of convergence to zero, i.e. the exponential decay can be chosen. The error will therefore asymptotically tend to zero, which means that the estimated state vector will become equal to the actual state vector, and this will occur whatever the system input and output values are.

7.5.2   Observer and system stability

Irrespective of the type and nature of the control action in operation, if an observer is introduced such that an estimated state vector is used in place of its unknown actual value then the closed-loop system no longer depends on the actual value but rather is affected by the characteristics associated with the estimated version. It must be noted though that for an asymptotic observer, which reduces the estimation error to zero as $t \to \infty$, then under steady-state conditions there will in fact be no difference whatsoever in the closed-loop system.

The most important question that must be asked is this: Assuming a particular control system design to be otherwise stable, will the closed-loop system remain stable when an observer, which is itself stable, is incorporated?

In order to obtain an answer to this problem, consider again the system equation of (7.5.1) along with the state feedback equation

$$u(t) = - F\underline{x}(t) \tag{7.5.9}$$

with vector F containing the state feedback parameters. If it is initially assumed that the state vector $\underline{x}(t)$ is known, then on substitution of the control input of (7.5.9) into the state equation of (7.5.1) it follows that

$$\underline{\dot{x}}(t) = (A - BF)\underline{x}(t) \tag{7.5.10}$$

However, if the estimated state vector $\underline{\hat{x}}(t)$ is used in place of $\underline{x}(t)$, its actual value, in order to obtain the control input from (7.5.9), then on substituting $u(t)$ defined as

$$u(t) = -F\underline{\hat{x}}(t) \tag{7.5.11}$$

into the state equation of (7.5.1), we have that

$$\underline{\dot{x}}(t) = A\underline{x}(t) - BF\underline{\hat{x}}(t) \tag{7.5.12}$$

Also, if the observer equation (7.5.4) is considered, when $u(t)$ is substituted it follows that

$$\underline{\dot{\hat{x}}}(t) = GC\underline{x}(t) + (A - BF - GC)\underline{\hat{x}}(t) \tag{7.5.13}$$

and in matrix form this is

$$\begin{bmatrix} \dot{x}(t) \\ \dot{\hat{x}}(t) \end{bmatrix} = \begin{bmatrix} A & -BF \\ GC & A - BF - GC \end{bmatrix} \begin{bmatrix} x(t) \\ \hat{x}(t) \end{bmatrix} \tag{7.4.14}$$

This means that the characteristic equation for the overall closed-loop system can be calculated from

$$\text{characteristic eqn} = \det\begin{bmatrix} sI - A & BF \\ -GC & sI - (A - BF - GC) \end{bmatrix} \tag{7.5.15}$$

Note: See Section 6.3.1 for details.

By elementary matrix operations (see Problem 7.7) the determinant is more simply reconfigured as:

$$\text{characteristic eqn} = \det\begin{bmatrix} sI - (A - BF) & BF \\ 0 & sI - (A - GC) \end{bmatrix} \tag{7.5.16}$$

or

$$\text{characteristic eqn} = \det(sI - A + BF) \cdot \det(sI - A + GC)$$

and this means that the characteristic polynomial of the overall system is found to be the characteristic polynomial of the controller, see (7.2.4), multiplied by the characteristic polynomial of the observer. By appropriate selection of the vector gains F and G therefore, a stable closed-loop denominator, characteristic polynomial, can be obtained. Further, if the system is completely controllable and completely observable then all the roots of the characteristic polynomial can be arbitrarily chosen.

The fact that the characteristic equation of the closed-loop system is obtained simply as a combination of the controller and observer polynomials is important in that the controller design can be carried out regardless of the nature of the state vector, e.g. it can be assumed that we know its actual value, and also the observer can be designed irrespective of the type of control action to be applied. This is known as the separation principle.

As a final note to this section it must be pointed out that no reference signal was included in the control input defined in (7.5.9) and (7.5.11). By a simple addition to the input $u(t)$, as was done in (7.4.2), such a reference can be introduced and this results in only a trivial extension to the equations (7.5.12) and (7.5.13): see Problem 7.8.

Example 7.5.1

To show how the design of an observer is carried out consider the system described by the equations (7.5.1) in which:

$$A = \begin{bmatrix} -3 & 1 \\ 0 & -2 \end{bmatrix}; \qquad B = \begin{bmatrix} 0 \\ 1 \end{bmatrix}; \qquad C^T = \begin{bmatrix} 1 \\ 0 \end{bmatrix}$$

An observer, as defined by (7.5.4) can then be employed in order to obtain an estimate of the unknown and unmeasurable state. For this system the gain vector is therefore

$$G = \begin{bmatrix} g_1 \\ g_2 \end{bmatrix}$$

where the selection of the values g_1 and g_2 is the object of this example.

The poles of the observer are obtained from the characteristic equation given as the determinant of

$$sI - (A - GC) = \begin{bmatrix} s + 3 + g_1 & -1 \\ g_2 & s + 2 \end{bmatrix} = 0$$

and the determinant of this matrix is

$$s^2 + (5 + g_1)s + (6 + 2g_1 + g_2)$$

The gain values g_1 and g_2 can now be selected in order to obtain a required form for this polynomial, i.e. in order to place the observer poles in their desired positions. Let us assume that the requirement is for the poles to lie at $s = -6$, and $s = -1$. This is satisfied by choosing $g_1 = 2$ and $g_2 = -4$ which results in a characteristic equation

$$s^2 + 7s + 6 = 0$$

Finally, the observer can be then written as

$$\hat{\underline{x}}(t) = \begin{bmatrix} -5 & 1 \\ 4 & -2 \end{bmatrix} \hat{\underline{x}}(t) + \begin{bmatrix} 0 \\ 1 \end{bmatrix} u(t) + \begin{bmatrix} 2 \\ -4 \end{bmatrix} y(t)$$

remembering that because

$$\hat{\underline{x}}(t) = A\hat{\underline{x}}(t) + Bu(t) + Gl(t)$$

this is also, by means of (7.5.3)

$$\dot{\hat{\underline{x}}}(t) = (A - GC)\hat{\underline{x}}(t) + Bu(t) + Gy(t)$$

7.6   Summary

In this chapter some of the basic facilities provided when a state–space system description is employed have been described. In the first instance the ideas of controllability and observability were introduced and it was shown how simple tests can be made by using the system matrices, in order to find out whether the system under consideration is controllable and/or observable.

Any state–space description of a system can also be reconfigured in terms of a transfer function polynomial relationship between input and output. The terms controllability and observability therefore have a meaning in so far as system transfer functions are concerned and this particular topic was discussed in Section 7.3.

Tests for controllability and observability merely point to whether or not a system is controllable and/or observable, the actual design procedures employed for controllers and observers providing the subject matter for Sections 7.4 and 7.5. In Section 7.4 controller analysis was carried out by means of state variable feedback, making the assumption that the state vector was actually available to be fed back. The methods described for controller design were firstly pole placement and secondly LQ optimal, two very commonly encountered techniques. In Section 7.5 it was however pointed out that the state vector is very often not available; an estimation of the vector is therefore made and this estimate is employed, instead of the actual but unknown value, for state variable feedback. It is required though that the state estimate be as near as possible to its actual value, i.e. that it is a good estimate, and that it does not show signs of instability; these points were also covered in Section 7.5.

Problems

7.1   If the input to a particular system is zero, i.e. $u(t) = 0$, for all t, and the system is described by the equations:

(a) $\underline{x}(t) = \exp[A(t - t_0)]\underline{x}(t_0)$
(b) $y(t) = C\underline{x}(t)$

where t_0 is the initial time considered, show that $\underline{x}(t_0)$ can be found in terms of $y(t)$, by premultiplying (b) by $\exp[A^T(t - t_0)C^T]$ and then integrating both sides of the equation. What conditions must be assumed to hold for this calculation to be possible?

7.2   Repeat Problem 7.1 for a general input signal $u(t)$ and show that the same conditions must hold for the calculation of $\underline{x}(t_0)$ to be possible.

7.3   Consider the state–space problem model:

$$\underline{\dot{x}}(t) = \begin{bmatrix} 0 & 1 \\ -2 & -3 \end{bmatrix} \underline{x}(t) + \begin{bmatrix} 1 \\ 1 \end{bmatrix} u(t)$$

$$y(t) = [1 \quad 2]\underline{x}(t)$$

Show that the system is completely state controllable and completely state observable.

7.4   Consider the state–space model:

$$\underline{\dot{x}}(t) = \begin{bmatrix} 1 & 0 & 0 \\ \alpha & -2 & 1 \\ -2 & 0 & -1 \end{bmatrix} \underline{x}(t) + \begin{bmatrix} 1 \\ 1 \\ 0 \end{bmatrix} u(t)$$

$$y(t) = [-1 \quad 1 \quad 0]x(t)$$

Find the value of α that will cause the system to be unobservable. With this value of α, find the minimal order canonical state–space representation.

7.5   If the pair $[A,C]$ for a given system is completely state observable, then prove that the transformed pair $[A',C']$ is also completely state observable. What conditions must be assumed?

7.6   The state–space equations for a particular system are:

$$\underline{\dot{x}}(t) = \begin{bmatrix} 1 & 0 \\ 1 & -1 \end{bmatrix} \underline{x}(t) + \begin{bmatrix} 1 \\ 0 \end{bmatrix} u(t)$$

and it is required that the quadratic cost function

$$CF = \int_0^\infty [\underline{x}^T(t)Q\underline{x}(t) + u(t)Ru(t)] \, dt$$

must be minimized, where

$$Q = \begin{bmatrix} 1 & 0 \\ 0 & 0 \end{bmatrix} \quad \text{and} \quad R = 1$$

The optimal LQ problem is solved by using the matrix Riccati equation:

$$Q = -A^T K - KA + KBR^{-1}B^T K$$

where

$$u(t) = -R^{-1}B^T K\underline{x}(t)$$

Find the optimal state feedback and obtain the open-loop and closed-loop eigenvalues.

7.7   For a system described by

$$\underline{\dot{x}}(t) = A\underline{x}(t) + Bu(t)$$

$$y(t) = C\underline{x}(t)$$

with a control input $u(t) = -F\underline{\hat{x}}(t)$ and the observer vector G defined in

(7.5.4), the overall closed-loop system characteristic equation is:

$$\text{char. eqn} = \det \begin{bmatrix} sI - A & BF \\ -GC & sI - (A - BF - GC) \end{bmatrix}$$

Show, by elementary matrix operations, that this can be written more simply as:

$$\text{char. eqn} = \det(sI - A + BF) \cdot \det(sI - A + GC)$$

7.8   For a system described by

$$\dot{\underline{x}}(t) = A\underline{x}(t) + Bu(t)$$
$$y(t) = C\underline{x}(t)$$

the control input is $u(t) = -F\hat{\underline{x}}(t) + Hv(t)$ where $v(t)$ is a scalar reference signal and H is a gain. What effect does the inclusion of a reference signal have on observer equation (7.5.4) and, hence, on the characteristic equation (7.5.15)?

7.9   Show that the state–space system described by:

$$\dot{\underline{x}}(t) = \begin{bmatrix} 1 & 2 \\ 0 & -4 \end{bmatrix} \underline{x}(t) + \begin{bmatrix} 1 \\ 1 \end{bmatrix} u(t)$$

has at least one eigenvalue which lies in the right half of the s-plane, i.e. it is an open-loop unstable system. Also, show that by a suitable state feedback design $u(t) = -F\underline{x}(t)$, the system can be made closed-loop stable. Choose F such that both closed-loop eigenvalues (poles) lie at $s = -4$.

7.10   Test the system described by

$$\dot{\underline{x}}(t) = \begin{bmatrix} 1 & 3 & 2 & 4 \\ 2 & 1 & 5 & 6 \\ 0 & 0 & -1 & 2 \\ 0 & 0 & 0 & -2 \end{bmatrix} \underline{x}(t) + \begin{bmatrix} 1 \\ 2 \\ 0 \\ 0 \end{bmatrix} u(t)$$

for controllability and observability properties. Find the state feedback vector F, where $u(t) = -F\underline{x}(t)$, such that the four closed-loop eigenvalues lie at $s = -4, -3, -2$ and -1.

7.11   Find the values of α which mean that an observer cannot be constructed for the system

$$\dot{\underline{x}}(t) = \begin{bmatrix} 1 & 1 \\ 2 & 0 \end{bmatrix} \underline{x}(t) + \begin{bmatrix} 0 \\ 1 \end{bmatrix} u(t)$$
$$y(t) = [\alpha \quad 1]\underline{x}(t)$$

When $\alpha = 0$ find an observer vector G, defined in (7.5.4), such that the estimation error decays with eigenvalues at $s = -2$ and -1.

7.12   A system is modeled in the state–space by:

$$\dot{\underline{x}}(t) = \begin{bmatrix} -4 & -4 & -2 \\ 1 & 0 & 1 \\ 6 & 9 & 2 \end{bmatrix} \underline{x}(t) + \begin{bmatrix} -1 \\ 0 \\ 2 \end{bmatrix} u(t)$$

Show that this system is completely state controllable, but has at least one eigenvalue which lies in the right half of the *s*-plane, i.e. it is open-loop unstable.

By a suitable state transformation, repose the system description in phase variable canonical form. Find the state variable feedback vector F, $u(t) = - F\underline{x}(t)$, which moves the right half plane eigenvalue to $s = - 3$, whilst not affecting the other two eigenvalues.

7.13   A state–space system description is given as:

$$\dot{\underline{x}}(t) = \begin{bmatrix} 3 & 1 \\ 2 & 4 \end{bmatrix} \underline{x}(t) + \begin{bmatrix} 1 \\ 3 \end{bmatrix} u(t)$$

Find Q, such that if $\underline{x}(t) = Q\underline{x}'(t)$ then

$$B' = \begin{bmatrix} 0 \\ 1 \end{bmatrix}$$

where $B' = Q^{-1}B$ and $\dot{\underline{x}}'(t) = A'\underline{x}'(t) + B'u(t)$.

Also, find the state variable feedback $u(t) = - F\underline{x}(t)$ such that the closed-loop system has eigenvalues at $s = - 2$ and $- 3$.

7.14   Show that the system described by

$$\dot{\underline{x}}(t) = \begin{bmatrix} 0 & 1 \\ -2 & -3 \end{bmatrix} \underline{x}(t) + \begin{bmatrix} 0 \\ 1 \end{bmatrix} u(t)$$

is completely state controllable.

Find a control input $u(t) = \alpha$, $t \geqslant 0$, where α is a constant, such that the input is driven to $\underline{x}^T(\log 2) = [0 \quad 0]$ from $\underline{x}^T(0) = [-1 \quad 4]$.

7.15   A system described by

$$\dot{\underline{x}}(t) = \begin{bmatrix} 0 & 1 \\ 0 & 0 \end{bmatrix} \underline{x}(t) + \begin{bmatrix} 0 \\ 1 \end{bmatrix} u(t)$$

is subject to the initial conditions $\underline{x}^T(0) = [2 \quad -2]$. The state must be driven to $\underline{x}^T(1) = [0 \quad 0]$ subject to the cost function

$$CF = \frac{1}{2} \int_0^1 u^2(t) \, dt$$

Find the minimizing control and minimum cost.

7.16   For the state–space model of Problem 7.3, assuming that a state feedback of the form $u(t) = - F\underline{x}(t) + v(t)$ is applied, find a vector F which will cause the closed-loop system to be unobservable.

7.17   For the system:

$$\dot{\underline{x}}(t) = \begin{bmatrix} 0 & 1 \\ 4 & 0 \end{bmatrix} \underline{x}(t) + \begin{bmatrix} 0 \\ 1 \end{bmatrix} u(t)$$

$$y(t) = [1 \quad 0]\underline{x}(t)$$

design an asymptotic observer with eigenvalues at $s = - 5$ and $- 6$.

If, instead, we have that

$$y(t) = [1 \quad \alpha]\underline{x}(t)$$

find the values of α for which an asymptotic observer cannot be obtained.

7.18   For the system:

$$\underline{\dot{x}}(t) = \begin{bmatrix} -2 & 2 \\ -1 & -1 \end{bmatrix} \underline{x}(t) + \begin{bmatrix} 0 \\ 1 \end{bmatrix} u(t)$$

test for controllability and open-loop stability (eigenvalue positions must all be in the left half of the s-plane). Find the state variable feedback which produces closed-loop eigenvalues which both lie at $s = -1$.

7.19   Consider the system with open-loop transfer function:

$$G(s) = \frac{2}{(s+1)(s+2)(s+5)}$$

The system is connected in the closed-loop configuration in Fig. 7.6 where $H(s) = Ks + 1$.

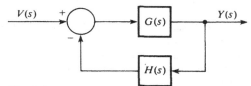

Fig. 7.6

Find an equivalent state–space representation, and use this to obtain the closed-loop transfer function. Select a value for K such that the closed-loop eigenvalues lie at $s = -4$, -3 and -1. Find a matrix Q such that a phase variable canonical form can be obtained for the system by making a state transformation.

7.20   Consider the system equation:

$$(a_2\mathrm{D}^2 + a_1\mathrm{D})y(t) = b_0 u(t)$$

relating system input $u(t)$ to output $y(t)$.

Find a state–space representation for the system.

The LQ problem is to find the optimal control which minimizes the cost function.

$$\mathrm{CF} = \frac{1}{2} \int_0^\infty (y^2(t) + qb_0 u^2(t))\, \mathrm{d}t$$

for some value q, by means of the control $u(t) = -R^{-1}BK\underline{x}(t)$. Find $u(t)$ in terms of a_2, a_1, b_0 and q; and obtain $R^{-1}BK$ when $b_0 = 0.8a_2$, $a_1 = 5a_2$ and $q = 10^{-5}$. Finally, calculate the system closed-loop eigenvalues.

Further reading

Bonnell, R. D., 'An observability criterion for a class of linear systems', *IEEE Trans. on Automatic Control*, **AC-II**, 1966, p. 135.

Brockett, R. W., 'Poles, zeros and feedback: state space interpretation', *IEEE Trans. on Automatic Control*, **AC-10**, 1965, pp. 129–35.

Chen, C. T., *Introduction to linear system theory*, Holt, Rinehart and Winston, 1970.

Chen, G., *Approximate Kalman filtering*, World Scientific Publishing Co., 1993.

D'Azzo, J. J. and Houpis, C. H., *Linear control system analysis and design*, McGraw-Hill, 1983.

Delchamps, D. F., *State-space and input-output linear systems*, Springer-Verlag, 1988.

Doyle, J., Francis, B., and Tannenbaum, A., *Feedback control theory*, Macmillan, 1992.

Godfrey, K. and Jones, P. (eds.), *Signal processing for control*, Springer-Verlag, 1986.

Kailath, T., *Linear systems*, Prentice-Hall, 1980.

Kalman, R. E., Ho, Y. C. and Narendra, K. S., 'Controllability of linear dynamical systems', in LaSalle *et al.* (eds.), *Contributions to differential equations*, **1**, 1962, Interscience Publ.

Kucera, V., *Analysis and design of discrete linear control systems*, Prentice-Hall, 1991.

Kwakernaak, H. and Sivan, R., *Linear optimal control systems*, Wiley-Interscience, 1972.

Luenberger, D. G., 'Observing the state of a linear system', *IEEE Trans. on Military Electronics*, **MIL-8**, 1964, pp. 74–80.

Luenberger, D. G., 'An introduction to observers', *IEEE Trans. on Automatic Control*, **AC-16**, No. 6, 1971, pp. 596–602.

Olsson, G. and Piani, G., *Computer systems for automation and control*, Prentice-Hall, 1992.

O'Reilly, J., *Observers for linear systems*, Academic Press, 1983.

Owens, D. H., *Multivariable and optimal systems*, Academic Press, 1981.

Schultz, D. G. and Melsa, J. L., *State functions and linear control systems*, McGraw-Hill, 1967.

Tropper, A. M., *Linear algebra*, Nelson, 1979.

Willems, J. C. and Mitter, S. K., 'Controllability, observability, pole allocation and state reconstruction', *IEEE Trans. on Automatic Control*, **AC-16**, 1971, pp. 582–95.

Wolovich, W. A., *Linear multivariable systems*, Springer-Verlag, 1974.

8

Digital control

8.1   Introduction

Once an analog controller has been designed and built it is generally a permanent structure. Change in operation must be accomplished by adjustment of a potentiometer or by arrangement of switch settings to arrive at a desired gain value. This can even result in a certain amount of rewiring.

By employing a computer in the control loop, the controller action is obtained partly by means of software operations. To change a particular setting only one line of the program needs to be altered or acted upon. This results in a great deal more flexibility and speed. It also allows for more adaptive controller schemes in which control parameter tuning can take place.

However, the change in mode of controller operation results in another theoretical base which is necessary to deal with understanding the principles of controller design procedures. Essentially, the computer can only take brief glimpses, or samples, of the real-world system under control. In general this merely means discrete views of system input and output signals, although in certain cases the system state may also be available.

In Section 8.2 the computer's eye view of the system to be controlled is introduced leading to z-domain representations of discrete time systems. Stability of systems in the z-domain and the effects of sampling are considered in the following section. The later sections, meanwhile, deal with forms of controller design in the discrete time case, Section 8.5 being based on state variable methods in particular. Section 8.6, however, concentrates on choosing closed-loop denominator coefficients or closed-loop pole locations, the controller therefore being formed with this as the desired objective.

8.2   Sampled data systems

Most digital computer systems are designed as that of Fig. 8.1.

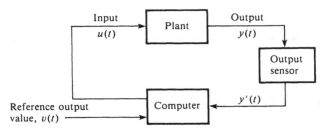

Fig. 8.1 Block diagram showing a computer applied to a continuous time plant

The plant itself is a real-world system operating in continuous time; this indeed is the case for virtually all control applications. Its input is defined here as $u(t)$, at time t, and its output as $y(t)$, at time t. However, due to measurement errors, because of the sensors applied, the system output becomes $y'(t)$ when eventually sampled by the computer. A reference output value, $v(t)$, is fed into the computer by software means, i.e. it is typed into the computer. The program itself must then obtain the error, $e(t)$, between the reference output, $v(t)$, and the measured output value, $y'(t)$. The way in which $e(t)$ is used in order to obtain the plant input, $u(t)$, being the control algorithm held within the rest of the software.

The computer shown in Fig. 8.1 itself consists of several smaller functional parts. The system (plant) to which the computer is connected is a real-world analog system, whereas the program carrying out the control operations requires signal values to be in a digital form such that it can work with discrete events. The analog value $y'(t)$ must therefore be converted into a digital (possibly binary representation) form on which the computer can operate. This conversion is carried out in hardware using an analog-to-digital (A/D) converter, see Fig. 8.2, a finite time being taken before a particular sampled analog value can be represented in digital form.

Conversely, once the correct plant input has been calculated from the programmed algorithm, it will initially be in digital form. Fig. 8.2 therefore contains a digital-to-analog (D/A) converter on the computer output side

The standard procedure is to use a real-time clock such that $y'(t)$ is sampled periodically, resulting in a digital representation of its analog value, every T seconds. Once the digital value has been obtained, the program is used to calculate a new plant input value $u(t)$, and this is immediately fed into the D/A converter resulting in a new

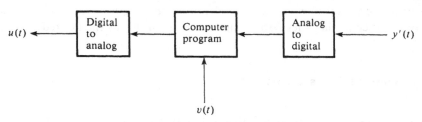

Fig. 8.2 Sampling operation of the computer

analog value being passed to the plant. The program then waits until T seconds have expired from the time of the previous sampling, whereupon the process is repeated. The A/D time along with the program time plus the D/A time must in total then be less than T seconds.

8.2.1   Delay operator representation

It is assumed here that once a sample of $y'(t)$ has been taken its digital value is subtracted from the reference input, $v(t)$, in order to obtain the error between actual and desired outputs. In T seconds time a further error value will be obtained, and the same will be true T seconds later still. The program can not only make use of these values, but also previous values it has calculated for the plant input. At one particular time instant, when $t = k$, let the error signal at that time be $e(k)$ and the new control (plant) input be $u(k)$. Also, let the error signal obtained one sample period (T seconds) prior to this, be $e(k - 1)$ with a control input at the same time of $u(k - 1)$. This series can continue until $e(0)$ and $u(0)$, if it is supposed that the initial time $t = t_0 = 0$. At time $t = k$, therefore, the new control input $u(k)$ can be found as a function of past input values and past and present error values, that is:

$$u(k) = f(u(k - 1), u(k - 2), ..., u(0); e(k), e(k - 1), ..., e(0)) \qquad (8.2.1)$$

In practice, however, only a relatively small and therefore finite number of $u(.)$ and $e(.)$ terms are actually required. Assuming the number of $u(.)$ terms to be n and the number of $e(.)$ terms to be $m + 1$, equation (8.2.1) can be reduced to:

$$\begin{aligned} u(k) = {} & f_1 u(k - 1) + f_2 u(k - 2) + \cdots + f_n u(k - n) \\ & + g_0 e(k) + g_1 e(k - 1) + \cdots + g_m e(k - m) \end{aligned} \qquad (8.2.2)$$

in which f_i: $i = 1, ..., n$ and g_i: $i = 0, 1, ..., m$; are scalar coefficients.

This method of representation is termed a difference equation and can be used, often with only a few parameters (four to five), to represent many physical systems. The exact means by which difference equations can be employed in a transfer function description will now be investigated.

8.2.2   Z-transform analysis

Let $\delta(t - kT)$ be the unit impulse function at time $t = kT$. When a particular continuous time signal, $f(t)$, is sampled, it follows that the sampler output consists of a series of impulses of varying magnitude, at a specific time kT the value of this sampled signal is therefore $f(kT) \cdot \delta(t - kT)$, where T is the sampling period. If it is considered that the time origin, $t_0 = 0$, the overall sampled version, $f'(t)$, of $f(t)$ can hence be written as:

$$f'(t) = \sum_{k=0}^{\infty} f(kT) \cdot \delta(t - kT) \qquad (8.2.3)$$

or, by taking the Laplace transform of this:

$$\mathcal{L}\{f'(t)\} = \sum_{k=0}^{\infty} f(kT) \cdot e^{-sTk} \qquad (8.2.4)$$

The term z is now defined as

$$z = e^{sT} \qquad (8.2.5)$$

such that the z-transform of $f(t)$, denoted by the operation $Z\{.\}$, is defined by

$$Z\{f(t)\} = \sum_{k=0}^{\infty} f(kT)z^{-k} = F(z) \qquad (8.2.6)$$

or

$$F(z) = f(0) + f(T)z^{-1} + f(2T)z^{-2} + \cdots$$

Tables of certain z-transforms are to be found in the Appendices. It is not intended to investigate the underlying mathematics of this operation further here however, as many works cover the underlying theory, such as Shannon's sampling theorem.

8.2.3   Properties of the z-transform

Three properties associated with the z-transform are considered, firstly the final value theorem which is useful later on for steady-state response calculations, secondly the initial value theorem and finally the effect of a time shift on the original signal.

Final value theorem

$$\lim_{t \to \infty} \{f(t)\} = \lim_{z \to 1} \{(z - 1)F(z)\}$$

on condition that $F(z)$ converges when $|z| > 1$ and all the denominator roots of $(1 - z)F(z)$ have magnitude less than unity.

A special case of the final value theorem occurs when a unit step input is applied to the transfer function $F(z)$, the z-transform of a unit step function being $z/(z - 1)$. This means that the output signal, obtained as the multiple of $F(z)$ and the unit step, can be found in terms of its steady-state (final) value simply by summing the coefficients of $F(z)$. This is due to the fact that the $(z - 1)$ term from the final value operation and the $(z - 1)$ term in the unit step function cancel each other, leaving $F(1)$ as the steady-state value, where $z = 1$. This mode of expressing the steady-state value, i.e. $F(1)$, in which the z is replaced by unity, is often encountered in digital control literature.

Initial value theorem

$$\lim_{t \to 0} \{f(t)\} = \lim_{z \to \infty} \{F(z)\}$$

$$= \lim_{z \to \infty} \{f(0) + f(1)z^{-1} + f(2)z^{-2} + \cdots\}$$

$$= f(0)$$

This condition applies only when the limit exists.

Time shifting theorem

$$Z\{F(t - nT)\} = z^{-n}F(z)$$

This follows from:

$$Z\{f(t - nT)\} = \sum_{k=0}^{\infty} f(kT - nT)z^{-k}$$

$$= \sum_{k=0}^{\infty} f(k'T)z^{-(k'+n)}$$

$$= F(z)z^{-n}$$

in which $k' = k - n$.

8.3   Stability analysis

Perhaps the most important characteristic of a system is its stability or lack of stability. Just as with continuous time systems, analysis to determine whether or not a discrete time system is stable is carried out at the initial design stage, further controller design being dependent on the results obtained.

This section will initially introduce the stable region for system poles, as an equivalent to the stability region in the s-plane. Subsequently transforming systems represented in the s-domain to their z-domain equivalent will be considered when the input to the system is sampled.

8.3.1   The unit circle

A linear continuous time system, characterized by its s-domain representation, is stable if all its poles (denominator roots) lie in the left half of the s-plane. This is true whether the system is in open-loop or closed-loop mode. In the previous section (8.2) it was defined that the s-domain is related to the z-domain by means of the equality

$$z = e^{sT} \qquad\qquad (8.3.1)$$

which is also

$$z = e^{(\sigma + j\omega)T} = e^{\sigma T} e^{j\omega T}$$

Stability in the s-plane is given in the region represented by $\sigma < 0$, i.e. a negative real part of s. The value $e^{\sigma T}$ then determines the magnitude of z, whereas $e^{j\omega T}$ determines the phase. Therefore the corresponding stability region in the z-domain is such that $|z| = e^{\sigma T}$ where $\sigma < 0$, i.e. the magnitude of z must always be less than unity, no restriction being placed on phase.

Critical stability in the s-plane is found when roots lie on the imaginary axis, that is when $\sigma = 0$. In the z-domain, therefore, critical stability occurs when roots are such that $|z| = e^0 = 1$, this means that critical stability occurs in the z-plane when a root is of magnitude 1, with any phase angle. Equivalent stability regions between s- and z-planes are drawn in Fig. 8.3, showing stability occurring within the unit circle of the z-plane, this area being equivalent to the left half of the s-plane.

To summarize, a discrete time system, represented in the z-plane, is stable if all its poles lie within the unit circle of the z-plane.

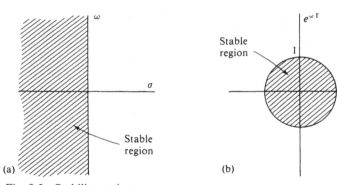

Fig. 8.3   Stability regions

(a) s-plane
(b) z-plane

8.3.2   Zero order hold (ZOH)

When a computer is used in the control of a system, the system output has to be fed into an analog-to-digital converter, see Fig. 8.4, before being acceptable in digital form to the

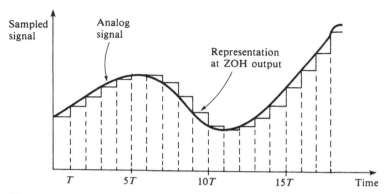

Fig. 8.4 Staircase effect of a ZOH on a continuous time signal

computer. The analog signal is sampled, and this value is held, prior to being passed through to the A/D converter and used in program calculations. While these calculations are being carried out, although the actual analog signal value can vary, it is the sampled and held value that is still being processed, see Fig. 8.4.

The ZOH has the effect of multiplying the analog signal at a particular sampling instant, by unity. If the sampled period is T seconds, then T seconds later the new analog signal multiplied by unity minus the old value becomes the new ZOH output. The ZOH can therefore be regarded as a unit step function with Laplace transform:

$$G_{ZOH}(s) = \frac{1}{s} - \frac{e^{-Ts}}{s} = \frac{1 - e^{-Ts}}{s} \tag{8.3.2}$$

where $1/s$ is the unit step and e^{-Ts}/s is the unit step delayed by T seconds.

The overall output signal value is thus obtained by multiplying the ZOH by the system transfer function and the input signal.

8.3.3 Worked example for ZOH system

A system is considered to be sampled every 0.5 seconds, by means of a zero order hold, as shown in Fig. 8.5, in which the system transfer function is given by:

$$G'(s) = \frac{1}{s(s+2)}$$

The z-transform of the overall transfer function, given by $G'(s)$ and the zero order hold, is denoted by $G(z)$, that is

$$G(s) = \frac{(1 - e^{-sT})}{s^2(s+2)}$$

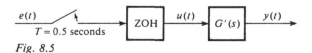

Fig. 8.5

such that

$$G(z) = (1 - z^{-1}) . Z\left\{\frac{1}{s^2(s + 2)}\right\}$$

where $Z\{1 - e^{-sT}\} = 1 - z^{-1}$. Then

$$G(z) = (1 - z^{-1})\left\{\frac{T}{2}\frac{z}{(z - 1)^2} - \frac{(1 - e^{-2T})z}{2^2(z - 1)(z - e^{-2T})}\right\}$$

and $T = 0.5$ seconds, hence

$$G(z) = \frac{1}{4}\left\{\frac{1}{(z - 1)} - \frac{1 - e^{-1}}{(z - e^{-1})}\right\}$$

or

$$G(z) = \frac{1}{4}\left\{\frac{ze^{-1} + 1 - 2e^{-1}}{(z - 1)(z - e^{-1})}\right\}$$

which is

$$G(z) = \frac{0.09197z + 0.06606}{(z - 1)(z - 0.3679)}$$

or

$$G(z) = \frac{(9.197z + 6.606)100^{-1}}{z^2 - 1.3679z + 0.3679}$$

From the analysis given in the previous section, this particular system is critically stable, having a pole actually on the unit circle (disc).

A table of further common z-transforms can be found in the Appendices.

8.4   Digital compensation

A computer can be thought of as perhaps the most flexible, due to programming possibilities, of a group of digital compensators. In general, however, a digital compensator can be regarded as any device which samples its own input signal and provides a sampled version of its output. Hence resistor-capacitor networks, for example, can also fit the digital compensator definition. Although, as in the continuous time case, closed-loop stability remains of great importance, digital compensation is generally used for obtaining desirable transient performance, that is the response to a change in plant input signal.

In this section the time response, especially under steady-state conditions, of a discrete time transfer function is considered. It is then shown how this transfer function can be used as the starting point in the design of a compensator.

8.4.1   Transfer function time response

Consider the second-order discrete time transfer function given in equation (8.4.1).

$$y(k) = \frac{(b_1 z + b_2)}{(z^2 + a_1 z + a_2)} u(k) \qquad (8.4.1)$$

in which a_1, a_2, b_1, b_2 are scalar coefficient gain values. This equation can, however, also be written as:

$$(z^2 + a_1 z + a_2) \cdot y(k) = (b_1 z + b_2) \cdot u(k)$$

or

$$y(k + 2) + a_1 y(k + 1) + a_2 y(k) = b_1 u(k + 1) + b_2 u(k)$$

However, dependent on exactly where in time the specific time instant k is chosen; this is the same as:

$$y(k) + a_1 y(k - 1) + a_2 y(k - 2) = b_1 u(k - 1) + b_2 u(k - 2) \qquad (8.4.2)$$

in which the time of the most recent output value is considered to be k rather than $k + 2$ as was the previous case. This procedure could be carried out by defining a new variable j as $j = k + 2$, hence $j - 1 = k + 1$ and $j - 2 = k$, but inclusion of a new variable is thought to complicate matters. Two forms of equation (8.4.2) will now be considered. Firstly rewrite the equation as:

$$(1 + a_1 z^{-1} + a_2 z^{-2}) y(k) = (b_1 z^{-1} + b_2 z^{-2}) u(k)$$

or

$$y(k) = \frac{(b_1 z^{-1} + b_2 z^{-2})}{(1 + a_1 z^{-1} + a_2 z^{-2})} u(k)$$

This form of equation (8.4.2) is obtained directly from equation (8.4.1) simply by multiplying top and bottom of the equation by z^{-2}. An nth order transfer function can therefore be written in either of the forms shown, simply by multiplying through the transfer function, when in the form of equation (8.4.1), by z^{-n}.

The second mode in which equation (8.4.2) can be regarded is such that:

$$y(k) = - a_1 y(k - 1) - a_2 y(k - 2) + b_1 u(k - 1) + b_2 u(k - 2) \qquad (8.4.3)$$

By this method, at time instant k, the value of the system output can be calculated by multiplying past output and input values (regressors) by their associated scalar coefficients and summing the results. This is the main means by which a system response to a specified input can be calculated, as is shown by the worked example of Section 8.4.2.

It suffices in this section merely to generalize the statements made regarding the second-order transfer function specifically detailed. If the transfer function relating system input, $u(k)$, to system output, $y(k)$, is termed $G(z)$ then this can be defined as:

$$G(z) = \frac{b_1 z^{n-1} + \cdots + b_n}{z^n + a_1 z^{n-1} + \cdots + a_n} \tag{8.4.4}$$

where certain of the b_i and a_i; $i = 1, \ldots, n$; terms can be zero to match the definition given. Significantly, if $b_1, \ldots, b_j$ terms are all zero, a delay of $j + 1$ sample periods is required before a change of input value has any effect on the output. As will be seen shortly, digital compensation cannot really cancel out this inherent delay in a system.

8.4.2   Time response – worked example

Consider the second-order system defined by:

$$y(k) = \frac{(2z + 1)}{(z^2 + 1.4z + 0.6)}\, u(k) \tag{8.4.5}$$

As described in the previous section, this can also be written as:

$$y(k) = -1.4y(k - 1) - 0.6y(k - 2) + 2u(k - 1) + u(k - 2)$$

Now, let a step occur at the input from an initial value zero up to unity, such that $u(0) = $ unity. It follows then that:

$$y(1) = -1.4y(0) - 0.6y(-1) + 2u(0) + u(-1)$$

where $k = 1$, i.e. the output value one sampling period after the step input has occurred is considered. As the input was zero up until $k = 0$, it is assumed that all outputs up until $k = 0$ are also zero. However, as $y(0)$ relies only on values up to and including $k = -1$, all of which are zero, $y(0)$ is also zero. The above equation therefore reduces to:

$$y(1) = 2u(0) = 2$$

when $k = 2$, we have:

$$y(2) = -1.4y(1) - 0.6y(0) + 2u(1) + u(0)$$

such that this reduces to:

$$y(2) = -1.4 \times 2 + 2 \times 1 + 1 = 0.2$$

By this means a table can be produced showing the new output value at each sampling instant, see Table 8.1. In this particular example it is apparent that once transient conditions have dissipated, the steady-state output value is unity, equal to the magnitude of step input. It is not generally the case, however, that the system output becomes unity under steady-state conditions, as will be seen in the next section.

Table 8.1 Table of output values for the worked example of Section 8.4.2

k	$u(k-2)$	$u(k-1)$	$u(k)$	$y(k-2)$	$y(k-1)$	$y(k)$
0	0	0	1	0	0	0
1	0	1	1	0	0	2
2	1	1	1	0	2	0.2
3	1	1	1	2	0.2	1.52
4	1	1	1	0.2	1.52	0.75
5	—	—	—	1.52	0.75	1.04
6	—	—	—	0.75	1.04	1.10
7	—	—	—	1.04	1.10	0.84
8	—	—	—	1.10	0.84	1.16
9	—	—	—	0.84	1.16	0.87
10	—	—	—	1.16	0.87	1.09
11	—	—	—	0.87	1.09	0.96
12	1	1	1	1.09	0.96	1.01
13	1	1	1	0.96	1.01	1.01
14	1	1	1	1.01	1.01	0.97

8.4.3   Compensator design

Consider the computer-controlled system shown in Fig. 8.1: assuming that $y(t) = y'(t)$, the computer program can be such that it provides a rational transfer function in the z-domain, $D(z)$ as a multiple of the error between actual and reference outputs. This is shown in Fig. 8.6, where $G(z)$ is the multiple of the zero-order hold and the z-transform of $G(s)$, its continuous time form.

It can be seen from Fig. 8.6 that the closed-loop transfer function is:

$$y(t) = \frac{D(z)G(z)}{1 + D(z)G(z)} \, v(t) \qquad (8.4.6)$$

such that if a certain second-order system is defined by:

$$G(s) = \frac{3}{s(s+1)}$$

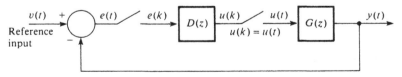

Fig. 8.6   Compensator $D(z)$ within a sampled data control system

when the sampling period, $T = 0.5$ seconds, its z-domain representation becomes:

$$G(z) = (1 - z^{-1})Z\left[\frac{3}{s^2(s+1)}\right]$$

or

$$G(z) = \frac{3[(e^{-0.5} - 0.5)z + (1 - 1.5e^{-0.5})]}{(z-1)(z - e^{-0.5})}$$

Hence

$$G(z) = \frac{0.3196(z + 0.8467)}{(z-1)(z - 0.6065)}$$

As an example of compensator design, assume this particular problem to be (a) to provide a compensator, $D(z)$, such that the open-loop pole at $z = 0.6065$ is replaced by an open-loop pole at $z = \lambda$, and (b) to choose a gain K such that along with a correct choice of λ, the closed-loop denominator becomes

$$z^2 + 0.3z + 0.1$$

Let the compensator be:

$$D(z) = \frac{K(z - 0.6065)}{(z + \lambda)}$$

The open-loop gain therefore is:

$$D(z)G(z) = \frac{0.3196K(z + 0.8467)}{(z-1)(z + \lambda)}$$

and part (a) of the design problem has been accomplished.

The closed-loop expression relating $v(t)$ to $y(t)$ is found from (8.4.6) to be:

$$y(t) = \frac{0.3196K(z + 0.8467)}{(z^2 + z(\lambda - 1) - \lambda) + 0.3196K(z + 0.8467)} v(t)$$

such that the closed-loop denominator is then:

$$z^2 + z(0.3196K + \lambda - 1) + (0.2706K - \lambda)$$

and this must be equated with the desired denominator $z^2 + 0.3z + 0.1$, resulting in the matrices

$$\begin{bmatrix} 0.3196 & 1 \\ 0.2706 & -1 \end{bmatrix} \begin{bmatrix} K \\ \lambda \end{bmatrix} = \begin{bmatrix} 1.3 \\ 0.1 \end{bmatrix}$$

which give values of $K = 2.3721$, $\lambda = 0.5419$, satisfying part (b) of the design problem. The final compensator is thus:

$$D(z) = \frac{2.3721(z - 0.6065)}{(z + 0.5419)}$$

8.5   State variable forms

In the two previous chapters a state–space framework was firstly introduced and then employed in controller design work. This was done entirely with continuous time systems in mind. In this chapter, however, discrete time systems have been discussed, in particular with reference to an input/output discrete time polynomial transfer function. This section gives details of a state–space framework for use with discrete time systems, specifically concerning itself with a transform into the state–space from the original polynomial format. The state variables are firstly defined such that subsequently the characteristic equation and a return to the polynomial description can be obtained.

8.5.1   State variables for discrete time systems

It has previously been mentioned that for any one system there are an infinite number of possible state–space representations, dependent on the state vector chosen. This is no less true about discrete time state–space descriptions, therefore one representation only will be considered, other state–space forms being obtainable from this by means of state variable transforms.

Consider the discrete-time transfer function (8.4.4), in which:

$$y(k) = \frac{b_1 z^{n-1} + \cdots + b_n}{z^n + a_1 z^{n-1} + \cdots + a_n} u(k) \tag{8.5.1}$$

or

$$y(k) = G(z)u(k) = \frac{B(z)}{A(z)} u(k)$$

This can also be written as:

$$(1 + a_1 z^{-1} + \cdots + a_n z^{-n})y(k) = (b_1 z^{-1} + \cdots + b_n z^{-n})u(k) \tag{8.5.2}$$

or

$$y(k) + a_1 y(k-1) + \cdots + a_n y(k-n) = b_1 u(k-1) + b_2 u(k-2) + \cdots + b_n u(k-n)$$

in which

$$z^{-i}y(k) = y(k - i) \tag{8.5.3}$$

The state variables can now be introduced by a suitable choice; reasons for selecting the one described will hopefully become apparent in Section 8.6. Let the state variables be defined as:

$$
\begin{aligned}
x_1(k+1) &= x_2(k) \\
x_2(k+1) &= x_3(k) \\
&\vdots \qquad\qquad \vdots \\
x_{n-1}(k+1) &= x_n(k) \\
x_n(k+1) &= u(k) - a_1 x_n(k) - a_2 x_{n-1}(k) - \cdots - a_n x_1(k)
\end{aligned}
\tag{8.5.4}
$$

with the output equation:

$$y(k) = b_n x_1(k) + b_{n-1} x_2(k) + \cdots + b_1 x_n(k) \tag{8.5.5}$$

In matrix form, the state equations can thus be written as:

$$\underline{x}(k+1) = A\underline{x}(k) + Bu(k)$$
$$y(k) = C\underline{x}(k) \tag{8.5.6}$$

in which:

$$\underline{x}^{\mathrm{T}}(k) = [x_1(k), x_2(k), \ldots, x_n(k)]$$

$$A = \begin{bmatrix} 0 & 1 & 0 & \cdot & 0 \\ \cdot & & \cdot & & \cdot \\ \cdot & & & \cdot & 0 \\ 0 & \cdot & & \cdot & 1 \\ -a_n & -a_{n-1} & \cdot & -a_2 & -a_1 \end{bmatrix}; \quad B = \begin{bmatrix} 0 \\ \cdot \\ \cdot \\ 0 \\ 1 \end{bmatrix} \tag{8.5.7}$$

and

$$C = [b_n, b_{n-1}, \ldots, b_1]$$

These equations therefore take on a similar form to the phase variable continuous time state equations described in Chapter 6, the only difference in the A and C matrices being the method of defining the a_i and b_i parameters, i.e. here they are for convenience a_i and b_i: $i = 1, \ldots, n$ whereas in Chapter 6 they are defined as a_i and b_i: $i = n - 1, \ldots, 0$. One of the important properties of this particular state representation is that the coefficients occurring in the transfer function (8.5.1) also occur, with no modification, in the A and C matrices, i.e. the discrete time state–space representation can be written directly from the discrete time transfer function. This is also true for the state vector itself (8.5.4), the definition of which was based on the initial transfer function parameters.

8.5.2 Solution of state equations

The discrete time state equation can be solved, such that given any A, B and C matrices, the values of scalar state coefficients in the state vector can be found. Remembering the first of these state equations from (8.5.6) to be:

$$\underline{x}(k+1) = A\underline{x}(k) + Bu(k)$$

Then

$$z\underline{x}(k) - A\underline{x}(k) = Bu(k) \tag{8.5.8}$$

or

$$\underline{x}(k) = (zI - A)^{-1} Bu(k) \tag{8.5.9}$$

in which it is assumed that the initial state vector, denoted by $\underline{x}(0)$, contains only zero elements.

Substituting for (8.5.9) into the second equation of (8.5.6), namely:

$$y(k) = C\underline{x}(k)$$

the following expression is obtained:

$$y(k) = C(zI - A)^{-1}Bu(k) \qquad (8.5.10)$$

But from (8.5.1) it can be seen that

$$G(z) = C(zI - A)^{-1}B \qquad (8.5.11)$$

or

$$G(z) = \frac{C\{\mathrm{adj}(zI - A)\}B}{|zI - A|} \qquad (8.5.12)$$

where the characteristic equation is given by the denominator of $G(z)$ as:

$$|zI - A| = z^n + a_1 z^{n-1} + \cdots + a_n = 0 \qquad (8.5.13)$$

8.5.3 Worked example

Consider the system defined by:

$$y(k) = \frac{z + 0.5}{z^2 - 1.2z + 0.35} u(k)$$

This can also be written in the form:

$$y(k) = 1.2y(k-1) - 0.35y(k-2) + u(k-1) + 0.5u(k-2)$$

But by using the state vector definitions introduced in (8.5.4), the elements in this vector are:

$$x_1(k + 1) = x_2(k)$$
$$x_2(k + 1) = u(k) + 1.2x_2(k) - 0.35x_1(k)$$

or

$$\begin{bmatrix} x_1(k+1) \\ x_2(k+1) \end{bmatrix} = \begin{bmatrix} 0 & 1 \\ -0.35 & 1.2 \end{bmatrix} \begin{bmatrix} x_1(k) \\ x_2(k) \end{bmatrix} + \begin{bmatrix} 0 \\ 1 \end{bmatrix} u(k)$$

with an output equation:

$$y(k) = [0.5 \quad 1] \begin{bmatrix} x_1(k) \\ x_2(k) \end{bmatrix}$$

To check that this discrete time state–space representation is a true one, it will be transformed back to $G(z)$ by means of (8.5.12). Firstly:

$$|zI - A| = \begin{vmatrix} z & -1 \\ 0.35 & z - 1.2 \end{vmatrix} = z^2 - 1.2z + 0.35$$

which therefore equates the denominator and hence the characteristic polynomial of the two representations.

The numerator of $G(z)$ should be equal to:

$$C\{\text{adj}(zI - A)\}B = [0.5 \quad 1] \begin{bmatrix} z - 1.2 & 1 \\ -0.35 & z \end{bmatrix} \begin{bmatrix} 0 \\ 1 \end{bmatrix}$$

$$= [0.5 \quad 1] \begin{bmatrix} 1 \\ z \end{bmatrix}$$

Hence:

$$C\{\text{adj}(zI - A)\}B = 0.5 + z$$

which is certainly equal to the numerator of the transfer function, $G(z)$.

8.6   Pole placement design

The poles of a system transfer function are the factors of its denominator polynomial or roots of its characteristic equation. The control criterion based on pole placement is such that given an open-loop system transfer function the feedback provided will cause the closed-loop poles to lie in previously specified positions. It is only possible to do this when the nth order system in question is completely controllable; in this section it is therefore assumed that this is the case. Pole placement for discrete time systems will be considered firstly by means of state feedback and secondly by straightforward polynomial controller design.

8.6.1   State feedback pole placement

Assume the linear system is described by the state equations:

$$\underline{x}(k + 1) = A\underline{x}(k) + Bu(k)$$

and   (8.6.1)

$$y(k) = C\underline{x}(k)$$

Let the control input at time instant k be:

$$u(k) = -F\underline{x}(k) + v(k) \tag{8.6.2}$$

where $v(k)$ is the reference input signal at time instant k and F is the $1 \times n$ dimensional feedback matrix of the form:

$$F = [f_n, f_{n-1}, ..., f_1] \tag{8.6.3}$$

By substituting for (8.6.2) in the first equation of (8.6.1), the following is obtained:

$$\underline{x}(k+1) = A\underline{x}(k) - BF\underline{x}(k) + Bv(k) \qquad (8.6.4)$$

or

$$\underline{x}(k)[zI - \bar{F}] = Bv(k)$$

in which $\bar{F} = A - BF$. Thus:

$$\underline{x}(k) = [zI - \bar{F}]^{-1}Bv(k) \qquad (8.6.5)$$

and when this is substituted into the second state equation of (8.6.1), i.e. the output equation, the final closed-loop form relating reference input to system output is:

$$y(k) = C[zI - \bar{F}]^{-1}Bv(k) \qquad (8.6.6)$$

In pole placement design it is only the poles, and hence the denominator of the closed-loop equation (8.6.6), on which our interest is centered.

If a choice or selection of the n closed-loop poles has previously been made, by multiplying out these factors, the desired closed-loop pole polynomial can be expressed as:

$$T(z) = z^n + t_1 z^{n-1} + t_2 z^{n-2} + \cdots + t_n \qquad (8.6.7)$$

in which t_i; $i = 1, 2, \ldots, n$ are previously defined and hence known values. The closed-loop denominator from (8.6.6) is given by:

$$|zI - \bar{F}| = |zI - (A - BF)| \qquad (8.6.8)$$

It follows then that $f_1, f_2, \ldots, f_n$ must be selected such that the coefficients obtained from (8.6.8) are equal to those defined in the pole polynomial (8.6.7).

Using the A, B and C matrices employed in Section 8.5 and defined in (8.5.7), it can be seen that for an n-dimensional system:

$$BF = \begin{bmatrix} 0 \\ \cdot \\ \cdot \\ \cdot \\ 0 \\ 1 \end{bmatrix} [f_n, f_{n-1}, \ldots, f_1] = \begin{bmatrix} 0 & \cdots & 0 \\ \cdot & & \cdot \\ \cdot & & \cdot \\ 0 & \cdots & 0 \\ f_n, & f_{n-1}, \cdots & f_1 \end{bmatrix} \qquad (8.6.9)$$

and hence:

$$\bar{F} = A - BF = \begin{bmatrix} 0 & 1 & 0 & \cdots & & 0 \\ \cdot & & \cdot & & & \\ \cdot & & & \cdot & & 0 \\ 0 & & 0 & & & 1 \\ -f_n - a_n, & -f_{n-1} - a_{n-1}, & \cdots & -f_2 - a_2, & -f_1 - a_1 \end{bmatrix} \qquad (8.6.10)$$

Therefore

$$
| zI - \bar{F} | =
\begin{bmatrix}
z & -1 & 0 & \cdots & & 0 \\
0 & z & -1 & 0 & & 0 \\
\vdots & & & \ddots & & 0 \\
0 & & & 0 & z & -1 \\
a_n + f_n, & a_{n-1} + f_{n-1}, & & \cdots, & a_2 + f_2, & z + a_1 + f_1
\end{bmatrix}
\quad (8.6.11)
$$

Thus

$$
| zI - \bar{F} | = z^n + (a_1 + f_1)z^{n-1} + (a_2 + f_2)z^{n-2} + \cdots + (a_{n-1} + f_{n-1})z + (a_n + f_n)
$$
$$(8.6.12)$$

The coefficients of (8.6.12) can now be equated with those of the specified pole polynomial (8.6.7), from which it follows that for complete pole placement:

$$
\begin{aligned}
f_1 &= t_1 - a_1 \\
f_2 &= t_2 - a_2 \\
&\vdots \\
f_n &= t_n - a_n
\end{aligned}
\qquad (8.6.13)
$$

8.6.2 Worked example using state feedback

When using the phase variable state–space system description as was done in Section 8.6.1, it was shown that the coefficients of the transfer function numerator polynomial, b_i, did not appear in (8.6.13). In general, where the state–space description is not in this particular phase variable form the b_i coefficients will indeed be present in the state feedback parameter, f_i, calculations thus leading to a more difficult problem. It will be considered here, however, that the state equations are in the form of (8.5.7), hence let $A(z) = z^2 - 0.9z + 0.2$, and $T(z) = z^2 - 1.3z + 0.42$. These polynomials can also be written in factored form such that:

$$A(z) = (z - 0.4)(z - 0.5)$$

and

$$T(z) = (z - 0.6)(z - 0.7)$$

in which $A(z)$ is the system denominator whereas $T(z)$ is the desired closed-loop denominator. The problem can therefore be regarded as one in which the open-loop poles situated at $z = 0.4$ and $z = 0.5$ must be acted upon, by means of the controller, such that in the closed-loop transfer function the poles are placed at $z = 0.6$ and $z = 0.7$.

From (8.5.7) the state representation, excluding output equation, can be written as:

$$
\begin{bmatrix} x_1(k+1) \\ x_2(k+1) \end{bmatrix} =
\begin{bmatrix} 0 & 1 \\ -0.2 & 0.9 \end{bmatrix}
\begin{bmatrix} x_1(k) \\ x_2(k) \end{bmatrix} +
\begin{bmatrix} 0 \\ 1 \end{bmatrix} u(k)
$$

The state feedback parameters can be found from (8.6.13) to be:

$$-f_1 = a_1 - t_1 = -0.9 + 1.3 = 0.4$$
$$-f_2 = a_2 - t_2 = 0.2 - 0.42 = -0.22$$

The control input at time instant k is therefore

$$u(k) = -0.22x_1(k) + 0.4x_2(k) + v(k)$$

where $v(k)$ is the reference input signal. Also

$$A + BF = \begin{bmatrix} 0 & 1 \\ -0.42 & 1.3 \end{bmatrix} = \bar{F} = \begin{bmatrix} 0 & 1 \\ -t_2 & -t_1 \end{bmatrix}$$

Thus

$$|zI - \bar{F}| = \begin{vmatrix} z & -1 \\ 0.42 & z - 1.3 \end{vmatrix}$$

Hence

$$|zI - \bar{F}| = z^2 - 1.3z + 0.42 = T(z)$$

8.6.3   Polynomial pole placement

The system is defined as in (8.5.1), i.e.

$$y(k) = G(z)u(k) = \frac{B(z)}{A(z)} u(k) \qquad (8.6.14)$$

with the pole polynomial specified by the operator and defined in (8.6.7) as $T(z)$.

The control input, $u(k)$, can take on one of several forms to provide feedback of past and present output values and feedback of past input values. However, general feedback polynomials $D(z)$ and $F(z)$ are defined here to be:

$$D(z) = d_0 z^{n-1} + d_1 z^{n-2} + \cdots + d_{n-1}$$

and

$$F(z) = z^{n-1} + f_1 z^{n-2} + \cdots + f_{n-1}$$

These polynomials are applied in the form of:

$$u(k) = \frac{D(z)}{F(z)} y(k) + \frac{z^{n-1}}{F(z)} v(k) \qquad (8.6.15)$$

or

$$F(z)u(k) = D(z)y(k) + v(k + n - 1)$$

where the coefficients of $D(z)$ and $F(z)$ are calculated to provide a closed-loop pole polynomial equal to $T(z)$. This can be seen if (8.6.15) is substituted into (8.6.14) to give:

$$y(k) = \frac{B(z)D(z)}{A(z)F(z)} \, y(k) + \frac{B(z)z^{n-1}}{A(z)F(z)} \, v(k)$$

or

$$[A(z)F(z) - B(z)D(z)]y(k) = B(z)z^{n-1}v(k) \tag{8.6.16}$$

Then the coefficients of $F(z)$ and $D(z)$ are calculated such that:

$$[A(z)F(z) - B(z)D(z)] = T(z)z^{n-1} \tag{8.6.17}$$

in which the z^{n-1} term is required to cancel with that on the top line of (8.6.15), acting on $v(k)$. It follows that the closed-loop equation must be represented, with $T(z)$ as its denominator by:

$$y(k) = \frac{B(z)}{T(z)} \, v(k) \tag{8.6.18}$$

It is noticeable that although the open-loop denominator, $A(z)$, has now been modified to the desired polynomial, $T(z)$, in the closed loop; the open-loop numerator $B(z)$ remains intact as the closed-loop numerator. The same example as that used in the previous section with a $B(z)$ polynomial included this time, will now be employed to show the mode of operation of the polynomial pole placement controller.

8.6.4 Worked example showing polynomial pole placement

Consider the system in which $B(z) = z + 0.5$, the polynomials $A(z)$ and $T(z)$ are those defined in the worked example of Section 8.6.2. Then, from the definitions of the controller polynomials, these will be of the form:

$$D(z) = d_0z + d_1$$

and

$$F(z) = z + f_1$$

Hence equation (8.6.17) can be written in terms of the available coefficients as:

$$[(z^2 - 0.9z + 0.2)(z + f_1) - (z + 0.5)(d_0z + d_1)] = (z^2 - 1.3z + 0.42)z$$

In matrix form, this is:

$$\begin{bmatrix} 1 & -1 & 0 \\ -0.9 & -0.5 & -1 \\ 0.2 & 0 & -0.5 \end{bmatrix} \begin{bmatrix} f_1 \\ d_0 \\ d_1 \end{bmatrix} = \begin{bmatrix} -1.3 + 0.9 \\ 0.42 - 0.2 \\ 0 \end{bmatrix} = \begin{bmatrix} -0.4 \\ 0.22 \\ 0 \end{bmatrix}$$

From this, the coefficients are calculated as:

$$f_1 = -0.2333, \qquad d_0 = 0.1667$$

and

$$d_1 = -0.0933$$

Hence the control input is provided by utilizing equation (8.6.15), that is:

$$u(k) = \frac{(0.1667z - 0.0933)y(k)}{(z - 0.2333)} + \frac{zv(k)}{(z - 0.2333)}$$

On multiplying both sides of the above equation by $(z - 0.2333)z^{-1}$, this can be implemented by obtaining the control input at instant k from:

$$u(k) = 0.2333u(k - 1) + 0.1667y(k) - 0.0933y(k - 1) + v(k)$$

where $v(k)$ is the reference input at instant k and $u(k - 1)$, $y(k)$, $y(k - 1)$ are known values.

8.7 Summary

Digital control, especially in the form of computer control, constitutes a continually larger slice of the control systems cake, i.e. a larger proportion of controllers are now based on digital control design philosophies. Microprocessors have more recently continued this trend and it is envisaged that special purpose devices or machines based on VLSI chips will retain its impetus.

This chapter has merely acted as an introduction to the enthralling world of digital control which includes controllers whose parameters can vary in line with any system alterations while the controller is still in operation. The controller can therefore adapt or tune itself to variations in the plant under control and can lead to more robust controllers with far less necessary operator supervision.

This book will, however, content itself with looking at one or two more special forms of digital control, as described in the remaining chapters.

Problems

8.1 Sketch the output response of the following open-loop systems, following a unit step input:

(a) $G(z) = \dfrac{0.5}{z - 0.5} = \dfrac{0.5z^{-1}}{1 - 0.5z^{-1}}$

(b) $G(z) = \dfrac{1.5}{z + 0.5}$

(c) $G(z) = \dfrac{0.1}{z - 0.9}$

(d) $G(z) = \dfrac{0.9}{z - 0.1}$

(e) $G(z) = \dfrac{10}{z + 0.1}$

(f) $G(z) = \dfrac{10(z + 0.5)}{(z + 0.1)}$

(g) $G(z) = \dfrac{10(z - 0.5)}{(z + 0.1)}$

8.2   Investigate the effect on the stability of a system, when the z-transform (+ zero order hold) is taken. Use

(a) $G(s) = \dfrac{1}{s + 1}$

(b) $G(s) = \dfrac{1}{s - 1}$

as starting points.

8.3   Find the digital compensator of the form $D(z) = K(z + b)/(z + a)$, such that when cascaded with a system $G(z) = 0.5(z + 0.7)/(z - 1)(z - 0.5)$, within a unity feedback loop, the closed-loop denominator is given by $z^3 - 0.8z^2 - 0.25z + 0.2$
Note: It is assumed that the function $G(z)$ already includes the effect of a zero order hold.

8.4   Find the output time response, following a unit step input, for the closed-loop system shown in Fig. 8.7, where $G(s)$ consists of a zero order hold in cascade with the open-loop system

$$G_o(s) = \dfrac{5}{s(s + 6)}.$$

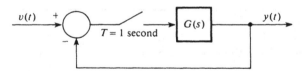

Fig. 8.7

8.5   For the closed-loop system of Problem 8.4, find the output time response when

$$T = 0.2 \text{ seconds and } G_o(s) = \dfrac{3}{s + 2}.$$

8.6   Consider the system shown in Fig. 8.8 where $G(z)$, which includes the effect of a

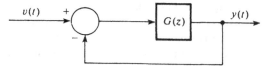

Fig. 8.8

zero order hold, is given by

$$G(z) = \frac{K(z - 0.5)}{(z - 0.9)(z + 0.3)}$$

Show how the closed-loop root positions move around in the z-plane as the gain K changes, i.e. plot the root loci for this system.

8.7 For the closed-loop system of Problem 8.6, plot the z-plane root loci for changes in gain K when

$$G(z) = \frac{K(z + 0.6)}{(z - 1)(z - 0.2)}$$

8.8 For the closed-loop system of Problem 8.6, find the range of values for K, over which the closed-loop system is stable, when

$$G(z) = \frac{K(z + 0.3)}{(z - 1)(z + 0.1)}$$

8.9 Consider the closed-loop system of Problem 8.4, with $T = 0.5$ seconds and

$$G_o(s) = \frac{4}{(s + 2)^2}$$

Find the open-loop system transfer function $G(z)$, noting that a zero order hold must be included.

A digital controller of the form $D(z) = K[(z - K')/(z - 1)]$ is cascaded with $G(z)$, within the unity feedback loop. Plot the root loci for this system for variations in K when $K' = 0.5$.

8.10 Consider the discrete-time system in state–space form

$$\underline{x}(k + 1) = A\underline{x}(k) + Bu(k)$$
$$y(k) = C\underline{x}(k)$$

where

$$A = \begin{bmatrix} 0 & 1 & 0 \\ 0 & 0 & 1 \\ -12 & -19 & -8 \end{bmatrix}, \qquad B = \begin{bmatrix} 0 \\ 0 \\ 1 \end{bmatrix} \quad \text{and} \quad C^T = \begin{bmatrix} 2 \\ 3 \\ 1 \end{bmatrix}$$

Show that this system is unobservable. Find the system transfer function and hence obtain a state–space representation which is observable.

8.11   Consider the system shown in Fig. 8.9 where $G_1(z)$ is the multiple of a zero order hold and the z-transform of s^{-1}. Also, the sampling period $T = 0.5$ seconds, and $G_2(z)$ is the z-transform of $(s + 2)^{-1}$.

   If $D(z) = K$, find the maximum value of K for which the closed-loop system is stable, by plotting the root loci for varying $\infty > K > 0$. What happens if an integral action controller is employed instead? i.e.

$$D(z) = \frac{K}{1 - z^{-1}}$$

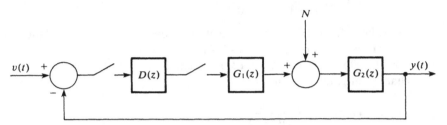

Fig. 8.9

8.12   For the system of Problem 8.11, with $D(z) = K$, set K to equal half its value at the limit of stability. If N is a unit step disturbance input, by calculating several output signal values with respect to time, find the effect of the disturbance on the system output.

   If $D(z) = K/(1 - z^{-1})$ and K is half its value at the limit of stability, calculate the output signal response to the same disturbance.

8.13   For the discrete-time state equation

$$\underline{x}(k + 1) = \begin{bmatrix} 1 & 1 \\ 0 & 1 \end{bmatrix} \underline{x}(k) + \begin{bmatrix} 0.5 \\ 1 \end{bmatrix} u(k)$$

find a transform matrix to place the state equation in canonical form.

   By employing state feedback $u(k) = -F\underline{x}(k)$, obtain a vector F which produces a characteristic polynomial with roots at $z = 0.8 \pm j0.1$.

8.14   For the discrete-time state–space system description

$$\underline{x}(k + 1) = \begin{bmatrix} 0 & 0 & -0.06 \\ 1 & 0 & -0.52 \\ 0 & 1 & -1.3 \end{bmatrix} \underline{x}(k) + \begin{bmatrix} 1 \\ 2 \\ 3 \end{bmatrix} u(k)$$

$$y(k) = [0 \quad 0 \quad 1]\underline{x}(k)$$

If $u(k) = -F\underline{x}(k) + v(t)$, find a vector F such that the closed-loop pole polynomial is:

$$z^3 + 0.6z^2 - 0.19z - 0.024$$

8.15   For the system diagram shown in Problem 8.11, $G_1(z)$ is merely a zero order hold, $G_2(z)$ is the z-transform of s^{-2} and the sampling period $T = 2$ seconds. What is the system output response to a unit ramp input signal $v(t)$, and what steady-state error is caused when N is a unit step disturbance input?

Further reading

Astrom, K. J. and Wittenmark, B., *Computer controlled systems*, Prentice-Hall, 1984.

Bishop, A. B., *Introduction to discrete linear controls*, Academic Press, 1975.

Cadzow, J. A. and Martens, H. R., *Discrete time and computer control systems*, Prentice-Hall, 1970.

Cadzow, J. A., *Discrete-time systems*, Prentice-Hall, 1973.

Chen, G. P., Malik, O. P. and Hope, G. S., 'Generalised discrete control system design method with control limit considerations', *Proc. IEE*, Part D, **141**, pp. 39–47, 1994.

Dorf, R. C., *Time domain analysis and design of control systems*, Addison-Wesley, 1965.

Franklin, G. F. and Powell, J. D., *Digital control of dynamic systems*, Addison-Wesley, 1980.

Franklin, G., *Digital control systems*, 2nd ed., Addison-Wesley, 1991.

Isermann, R., *Digital control systems*, Springer-Verlag, New York, 1991.

Jury, E. I., *Sampled data control systems*, John Wiley and Sons Inc., New York, 1958.

Katz, P., *Digital control using microprocessors*, Prentice-Hall, 1981.

Kuo, B. C., *Digital control systems*, Holt, Rinehart and Winston, 1980.

Leigh, J. R., *Applied digital control*, Prentice-Hall, 1984.

Phillips, C. L. and Nagle, H. T., *Digital control system analysis and design*, Prentice-Hall, 1984.

Ragazzini, J. R. and Franklin, G. F., *Sampled-data control systems*, McGraw-Hill, 1958.

Smith, C. L., *Digital computer process control*, Intext, Scranton, 1972.

Soeterboek, R., *Predictive control*, Prentice-Hall, 1992.

Stoten, D. P. and Harrison, A. J. L., 'Generation of discrete and continuous time transfer function coefficients', *Int. J. Control*, **59**, pp.1159–1172, 1994.

Strejc, V., *State space theory of discrete linear control*, John Wiley and Sons, 1981.

Tou, J. T., *Digital and sampled-data control systems*, McGraw-Hill, 1959.

Tzafestas, S. G. (ed.), *Applied control*, Marcel Dekker, 1993.

Warwick, K. and Rees, D. (eds.), *Industrial digital control systems*, Peter Peregrinus Ltd, 1986.

9

Computing methods

9.1   Introduction

In Chapter 8, the use of a computer as a central element in a closed-loop control system is discussed, and advantages, when compared with conventional analog controllers, are considered. Computers have a much wider applicability in the field of control systems, however, rather than merely being a tool for the purpose of digital controller implementation. In this chapter various areas for the employment of computers are outlined, while still remaining within the field of control systems analysis.

In Section 9.2, system simulation by means of computers is covered. Simulating a real-world system with a computer-based model almost always results in a reduction in cost, and often also in size, when compared with the system itself. Further, with the simulation, tests can be carried out not only under all likely operating conditions but also under some alternative special conditions. Hence system performance can be investigated, in terms of the simulation, both within and outside the expected operation range, thus any problem areas or improvement possibilities can subsequently be taken care of as far as the actual system is concerned.

Several different aspects of computer-based control systems are considered in Section 9.3, by means of a discussion as to the various roles that a computer can take. In its simplest form it can be employed merely as a collection point for plant input/output information, and although excellent data display capabilities are usually proffered by the computer, it is only when the computer is connected within a feedback loop that it has an effect on system performance. This can mean however that the computer is employed simply as a digital control device, multiplying input/output signals with controller coefficients and sending the resultant control signal to the plant. At a higher level though, the computer can take on a more intelligent role and can be used as an expert system to make decisions concerning the type of controller to be employed or action to be taken when fault conditions occur.

When not within a feedback loop, computers can be used for far more than simply data logging or system simulation. In fact complete packages are available which can take as input plant input/output data. Included are facilities for signal processing, system modelling, controller design and system simulation, with the latter two elements being based on results obtained from the first two. Such computer-aided control system design packages invariably contain self-testing procedures in order to ensure the validity of the system models obtained and to check on the final overall controller performance by means of a simulation trial run.

9.2   System simulation

The design, construction and analysis of a complete control system, or even an element of a more complex process, can be both time-consuming and expensive. This is particularly true if it is found, in the analysis stage, that the system is not quite right and some modifications are required, leading to greater expense. An alternative approach is to model the required system in the form of a computer simulation, either in terms of an analog hardware design or as a digital software-based scheme. The simulated system can then be modified with very little effort and no further expense as far as rebuilding or restructuring components is concerned.

The simulation of systems by means of a computer is very useful when cost is restricted, in that the purchase of the actual system under investigation may well not be a possibility, e.g. we cannot all afford a space shuttle. Also, with a simulation it is possible to apply different signals and disturbances without worrying unduly over the system blowing up or causing physical harm. Indeed, modifications can be carried out on the simulation and when an idea for a new design is found to work successfully, it can then also be tried out on the actual system with a much greater degree of confidence.

Accuracy is an important factor for the simulated model, in that if the simulation does not represent the original system with a certain amount of accuracy, then results obtained from the simulation may well not hold for the original system. Nevertheless, the simulation can still be employed to get a feel for the characteristics of the actual system.

9.2.1   Analog models

Analog simulations are obtained by means of an analog computer, where the term analog is used to signify the fact that the voltage and current signals employed are analogous to the signals found in the original system. Electronic circuits are used as basic building blocks, forming links between signals, rather than the actual physical components in the case of the system itself. Although analog models and analog computers are still encountered, system simulations are now almost always realised using digital computers. Hence, no further discussion is therefore given here on analog models and readers are directed to Gordon (1978) for further details.

9.2.2   Software models

In the design and analysis of a complicated control system, a computer is an extremely important tool by which means the system can be simulated not only to check results already available, but also to develop improved controller performance.

The vast majority of modern-day simulation studies are carried out by means of software development on digital computers. The main reasons for this are perhaps improved accuracy and reliability, along with increased flexibility, allowing system models to be varied arbitrarily and for a large selection of input signals to be applied, without the need for amplitude scaling because of saturation effects. A wide variety of computers are available for programming and an ever-increasing choice of programming languages is apparent, although standardization within establishments and research groups is commonplace, allowing for multi-user access to specific system simulation packages.

As discussed in the previous chapter, a digital computer operates on discrete signal values. Hence, when simulation of a continuous-time system is to be carried out, approximations must be made in order to obtain discrete-time signals. The concept of a delay operator system description is discussed in Section 8.2.1 and the method of sampling a continuous-time signal by means of a zero order hold, shown in Fig. 8.4, a basic assumption being made that with a sampling period of T seconds, then the continuous-time signal remains constant during that T seconds. Unfortunately, this latter assumption is not in general true, as is shown in Fig. 8.4, leading to the conclusion that a digital computer simulation of a continuous-time system will be subject to modeling errors due to the discretization process. In fact further errors will also be present due to the use of a computer, caused by truncation of real numbers to fit the computer's finite storage length, i.e. only a finite number of bits are used to represent each real number. Therefore, although digital computer simulations are usually far more accurate than those obtained by analog computer, it must be remembered that errors do exist and that in some cases these errors can be critical.

When a system is represented in terms of a mathematical model, where that model is in the form of a continuous-time equation, such as a differential equation, then this must be converted into a difference equation, by means of appropriate approximations.

Example 9.2.1

Consider the RC system shown in Fig. 9.1, where the output voltage, in response to a unit step in input voltage, is described by the equation:

$$v_i(t) = i(t)R + v_o(t) : v_o(t_o) = 0$$

and as

$$i(t) = C\frac{dv_o(t)}{dt}$$

this means that

$$CR\frac{dv_o(t)}{dt} = v_i(t) - v_o(t) \tag{9.2.1}$$

is a differential equation which represents the output voltage with respect to time.

A difference equation approximation of (9.2.1) is required, with the assumption being made that the $v_i(t)$ and $v_o(t)$ voltages are constant in value over each inter sample interval lasting T seconds. This is certainly true for $v_i(t)$, as it is a unit step input applied at $t = t_0$.

The value of voltage $v_o(t)$ up to a particular time instant kT where k is an integer number of sample periods T, from the previous time instant $kT - T$, is assumed to be equal to its value when sampled at $kT - T$, i.e.

$$v_o(t) = v_o(kT - T) : (kT - 1) \leq t < kT \tag{9.2.2}$$

The derivatives of the output voltage can then be approximated in terms of the gradient between the output voltage value at $t = (kT - T)$ and the new value at $t = kT$, such that

$$\left.\frac{dv_o(t)}{dt}\right|_{t=kT} \simeq \frac{v_o(kT) - v_o(kT - T)}{T} \tag{9.2.3}$$

or, by defining the time index in terms of sample periods rather than as an absolute value:

$$\frac{dv_o(k)}{dt} = \frac{v_o(k) - v_o(k - 1)}{T} \tag{9.2.4}$$

A discrete-time approximation of the differential equation (9.2.1) is therefore given as:

$$\frac{CR}{T}[v_o(kT) - v_o(kT - T)] = v_i(kT - T) - v_o(kT - T)$$

or

$$v_o(k) = v_o(k - 1) + \frac{T}{CR}[1 - v_o(k - 1)] \qquad (9.2.5)$$

remembering that $v_i(kT - T) = 1 = v_i(k - 1) = v_i(k)$; etc.

The output voltage signal $v_o(t)$ is now described, at any discrete-time instant $t = kT$, in terms of its previous value, and given the initial value of $v_o(t_o)$, then equation (9.2.5) can be recursively evaluated in order to find the time response of the RC circuit for a particular choice of values T, C and R. A similar form of time response is considered in the worked example of Section 8.4.2 and therefore (9.2.5) is not laboriously recursively calculated here. The first four iterations are however shown, when $CR = 10$ and $T = 1$, then

$$v_o(k) = v_o(k - 1) + 0.1 \cdot [1 - v_o(k - 1)]$$

or

$$v_o(k) = 0.9v_o(k - 1) + 0.1$$

k	$v_o(k)$
0	0
1	0.1
2	0.19
3	0.271
4	0.3439

The exponential rise is thus already apparent, as expected from the original system in Fig. 9.1, with the output voltage heading asymptotically towards a final value equal to unity, the magnitude of step input.

A simulation of the simple RC system can therefore be carried out quite quickly by means of a digital computer once the initial program, based around the recursive equation (9.2.5), has been written. Any modifications to C, R or T can then easily be made and the program rerun, without the need for time consuming rewiring, which might be necessary with the actual system. Also, any arbitrary values can be selected for C, R and T - we are not restricted into using the handful of capacitors which are physically in stock.

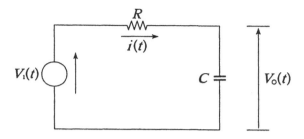

Fig. 9.1   RC circuit for Example 9.2.1

In the example given, a fairly straightforward first-order differential equation was approximated by a simple gradient technique. For more complex control systems and/or where improved accuracy is required, other approaches to the digital approximation problem can be considered; the text by Banks (1986) details many such schemes.

9.3   Computer control

In Section 9.2 the discussion is centered on the use of computers as a basis for control system simulations, i.e. merely an alternative means for system reproduction. In this section, various ways are considered in which a computer can be used either directly or indirectly in the control of a system. This can mean, in the simplest case, merely monitoring the behaviour of a system over a period of time, although it can also mean that a computer calculates the required control action by fulfilling the role of feedback element.

Using a computer for control purposes provides a great deal more flexibility than is otherwise obtainable. The computer can, for instance, be used merely as a "dumb" controller by evaluating a set of equations and giving out an appropriate actuator signal. The trend is, however, to use a computer as the basis for a more intelligent, active, controller which assesses and coordinates data from several sources and which takes a "common-sense", expert, approach to its calculations. By means of a computer therefore, not only can the type of control action and objective be modified due to certain occurrences, but also the role of other associated controllers, possibly themselves computer based, can be restructured.

9.3.1   Data logging

Probably the simplest means by which a computer can be made use of in a control system is for data logging purposes as shown in Fig. 9.2.

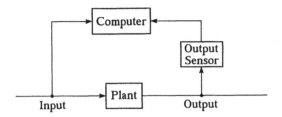

Fig. 9.2 A computer connected for data logging

When in this mode the computer takes no direct active part in the control of the plant. However, information obtained over a period of time may well be used as the basis of a control system reformulation. With such a system as shown in Fig. 9.2 it is very often the case that the computer will actually log data from several, possibly interconnected, systems and will provide detailed graphical output which is representative of the data collected. Where the plant itself is a fairly simple system it is likely that a low-cost microcomputer will be used as a local data logger, passing values periodically to a central, more powerful, machine.

Although remaining within the general classification of a data logger, as shown in Fig. 9.2, it is possible for certain actions to be taken within the computer, without a direct feedback link being provided to the system. One such case, which is of extreme importance, is the use of a computer as a *fault detection* device. By monitoring the output signal over a period of time, in response to a logged input signal, an output pattern, which has previously been denoted as that caused by a fault condition, can be detected and signalled by means of an alarm. Examples of simple fault conditions are, (a) increasingly large output oscillations and (b) output signal stuck for a long period of time at a high or low value, despite input signal fluctuations.

Remaining within the framework of Fig. 9.2, it is also possible for the computer to use the data collected in order to calculate a mathematical model of the system, i.e. it can use the data to estimate the coefficients in a structured model of the system. Any further data collected can then be used to assess the goodness of the model calculated and to either cause the model coefficients to be modified in order to improve the model, or for a new model to be calculated when it is eventually decided that the old model is no longer accurate enough.

Alternatively, the data collected from a system can be used to predict performance or output position, although this may well mean that a mathematical model of the system is also formulated and it is from this model that the prediction is made. When it is predicted that fault conditions will occur if the present course of action is continued, then an alarm can be given. However, if the computer is used as an aid to carrying out corrective action on the system input or the system itself, effectively this means that it is being employed within a feedback loop and is therefore no longer a simple data logger.

9.3.2 Real-time control

Initially, when computers were first employed for control purposes, their role was merely that of a supervisor, sounding an alarm under fault conditions. It was only in the 1960s that it became viable to replace hard-wired analog controllers with digital computer counterparts, often referred to as direct digital controllers (DDC). Although carrying out a similar function, the computer in a DDC scheme forms part of the control loop and has a direct controlling action on the plant.

If an analog controller has been working successfully on a plant, for some time, reasons for replacing the controller with a computer-based scheme must be convincing. This is particularly true as the initial cost of implementation of the computer will most likely be very high, when compared with the analog format. But the computer-based controller has several advantages over analog schemes, some of which are as follows:

1. Flexibility in control action: computer program modification rather than analog rewiring.
2. Reduced cost when further control loops are added: extra loops merely need to be connected up to the same computer, more analog controllers would otherwise be required.
3. Improved user interface: information can be displayed graphically on a VDU when required, rather than employing a large panel for the display of limited analog information.
4. Coupling of control loops: simply by linking different programs, several control loops can be coupled in any desired mode.
5. Increased controller complexity: versatility in programming allows for an infinite number of possible controllers, without limitations relating to physically realizable schemes, as in the analog case.
6. Adaptive control: the controller can be modified in real-time when operating conditions, e.g. ambient temperature, vary. Analog controller performance can often deteriorate considerably with time.

Although DDC schemes can readily be programmed in common high-level languages such as Pascal, Fortran, Basic, several special-purpose languages have been derived. By these means the operator is able to refer directly to input/output variables and important plant/controller features, without the need to know intricate programming details. It is then also possible for the operator to select an appropriate control algorithm from those available, along with a suitable sampling period for the plant under control.

Schematic diagrams of a computer-based control system are shown in Figs. 8.1 and 8.2; a further assumption must be made however that the computer is required to contain an on-board real-time clock, i.e. a clock which periodically gives an output pulse. The analog-to-digital converter can then be synchronized, such that the system output measurement is sampled with a period equal to an integer number of clock pulses. The time between two consecutive samples being taken is referred to as the *sampling period,* whereas the exact time at which a sample is taken is referred to as a *sampling instant.*

The main difference between a computer-based (DDC) control scheme and an analog controller is that the former employs a control action which is derived from discrete-time plant input and output signals, taken at the sampling instants. If the plant under control has a relatively fast response pattern then samples must be taken frequently, whereas when the plant's response pattern is slow, samples are required much less frequently. The sampling period must therefore be directly dependent on the response speed of the plant, and this presents a major problem for *real-time* control. In Section 9.3.3 the concept of a simulated application is discussed and in this case it is usual for the computer to employ its own time scaling, based essentially on the time required by the computer for calculation purposes. However, when a computer is connected on-line to a system in order to control that system, changes in control action must be carried out at a frequency which is found to be most suitable for the system, e.g. if a system responds with a time constant of 3 seconds to an input signal change, then there is little use in the computer providing a control action which is updated once every three days. The computer must therefore operate in real-time, that is the time frame of the plant, thus complex updating controller calculations can only be carried out if they can be completed within each sampling period.

Controller forms employed for real-time control are based on the z-operator description, as discussed in Chapter 8, and are not considered further here. See Astrom and Wittenmark (1984) for a more in-depth view.

9.3.3   Simulated application

Once a suitable computer-based control (DDC) algorithm has been selected, where on-line control is desired, the computer can be connected, via appropriate interfacing, to the system to be controlled. As soon as the operator is content that everything is in order, the controller can then be switched on-line and hence will take over system control. This does not in general mean though that the operator can immediately take a long vacation, because a supervisory role will normally be required in the first instance. Before the controller is actually switched on-line however, it is most likely the case that the plant manager will need to be convinced that the programmed controller will definitely work on the plant in question. As an aid to proof of controller viability it is most sensible to first carry out a simulated plant application (see Fig. 9.3).

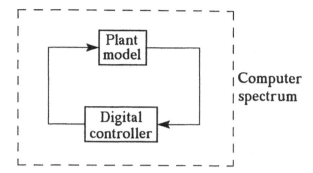

Fig. 9.3 Simulated application

With a simulated application, a mathematical model of the plant to be controlled is stored in the computer. The programmed control algorithm can then be run with the controller output. *Note:* This is also the plant input, being retained within the computer and merely fed to the plant model rather than to interface circuitry. Similarly the evaluated model output is immediately fed back into the controller program. The complete simulation process therefore takes place within the spectrum of the computer, with the computer being used to not only calculate the control signals but also to calculate the plant model response.

The accuracy and usefulness of a simulated application, when compared to the relevant real-world system is dependent largely on the accuracy with which the mathematical plant model mimics the real-world plant. It is often important therefore to include, where necessary, nonlinear effects, such as saturation and hysteresis, within the plant model. The plant itself will however, almost certainly, be a continuous-time system, whereas its computer-based model will be a discrete-time approximation.

All of the signals connected with the simulation therefore are discrete-time signals due to their remaining within the computer spectrum. The simulation speed is then only dependent on the speed of the computer, and is not limited by any real-time restrictions placed by the on-line control of a real-world system. So although the control of a real-world plant may take several days to analyze, due to the slow response time of that plant, a simulated application can take only a fraction of a second.

With a simulated application it is possible to try many different control schemes on a particular plant model, or to tailor one specific control scheme in order to improve performance. If a certain controller is found to work well, then it can later be applied to the actual system. Conversely if a controller is found to work very badly within the confines of the simulation, a lesson will have been learnt without adversely affecting the plant itself. So, if a controller causes a simulated plant to "blow up", i.e. for the plant output to continually increase in value to unacceptable levels, then this will only be a numerical "blow-up" rather than a completely destroyed real plant, and can be remedied by simply restarting the program with a different set of parameters. A simulated application is therefore a good means by which controller program bugs can be ironed out.

It is also possible to try out, via a simulation, one type of controller on many different plant models. By these means, one can deduce the general application areas of the controller, discover any likely problems and most importantly prove that the controller is, or is not, a viable proposition.

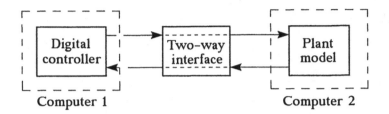

Fig. 9.4 Multi-user simulated application

There are many advantages to carrying out a simulated application, when compared with a real-world implementation, not the least of which is cost. It is generally much cheaper to simulate a plant with a computer-based plant model, and to apply the controller to the model, rather than to buy in the actual plant. This is especially true when one is not completely certain of the operating effectiveness and reliability of the controller. Also, while information from a simulation is retained within the computer, detailed displays can be obtained, either in graphical or numerical forms, hence providing an aid to further analysis.

To conclude this section, it is worth noting that although the simulated applications considered here are completely computer based, this is not necessarily always the case. It is also possible for two (or more) computers to be used, one containing the controller program and one containing a program representing the plant model. In order to carry out a simulated example, the two computers must then be connected together via appropriate interfacing, as shown in Fig. 9.4. As a follow-on from this, one final form of simulated application is that encountered when Computer 2 in Fig. 9.4 is replaced by an analog model representation of the real-world plant. This technique adds a certain amount of realism to the simulation, in that the interfacing circuitry must bridge the digital and analog worlds; however, a great deal is lost in terms of flexibility and ease of modifications.

9.3.4   Expert systems and artificial intelligence

In practice it is not usually possible to straightforwardly implement a DDC algorithm, no matter how complex it might be. The effect of operational problems such as nonlinearities, transients, spurious noise signals and manual/automatic switching must somehow also be taken into account. With a desire to operate a digital controller successfully within an overall plant control scheme, logical decisions must be made as to the action taken when a particular operational problem occurs. This decision making mechanism can, in its simplest form, be regarded merely as software jacketing around the digital control algorithm, e.g. if the plant output goes above a certain level then an alternative back-up controller must be switched in. But to be truly effective the logic should be based on heuristics resulting from both experience and common sense, as well as relevant mathematical inequality tests. Such a requirement makes the implementation of computer-based digital controllers amenable to *artificial intelligence* techniques, particularly because of the decisional logic which can be introduced via software.

In its widest sense artificial intelligence can be defined as "the science of making machines do things that require intelligence if done by people" (Marvin Minsky, MIT).* Relevant application areas are to be found in the fields of robotics and intelligent sensors, and more directly in software development for automatic programming and the expression of natural language. In the particular framework of this discussion however, artificial intelligence is referred to in the sense of a bank of knowledge, relating to a well-defined subject area, stored in a data base. The main purpose of this stored knowledge is to produce a response to certain stimuli which exactly mimics that which would be exhibited by a human expert. Such a decision making mechanism is called an *expert system*. The performance of an expert system is dependent on the size and quality of the data base, with a fallibility equal to that of an expert.

An expert system is made up of three main parts:

1. The knowledge of relevant facts;
2. Relationships between the facts;
3. A heuristic fact storage/retrieval procedure.

This means that in order to construct an expert system, suitable strategy, procedure and logic information must be extracted from one or more experts and stored efficiently in a data base. The person who constructs an expert system in such a way is referred to as a *knowledge engineer*, depicted in Fig. 9.5.

The knowledge engineer's job is to interrogate experts and to make use of case studies in order to obtain detailed answers and solutions. This information must then be translated into a data set which is suitable for storage, retrieval and operation.

Once an expert system has been constructed, it would be wrong to forget about the human expert, indeed there may well be cases in which it is preferable, for the time being at least, to retain a role for the human expert. Most cases will require occasional updating, improving and gap filling by human experts, particularly in the event of

---

* "Intelligence is the ability, which some people have, to make other people regard them as being intelligent" (Kevin Warwick).

changes in operating function or ambient conditions. Human experts retain several advantages over artificial (computer-based) experts in that they (a) are creative, (b) can apply common sense, (c) can take a broad view, (d) can adapt to different circumstances and (e) can use sensorial observations. However, artificial experts also have distinct advantages in that they (a) are consistent, (b) can be easily transferred, (c) are affordable, (d) are permanent and (e) are easily documentable.

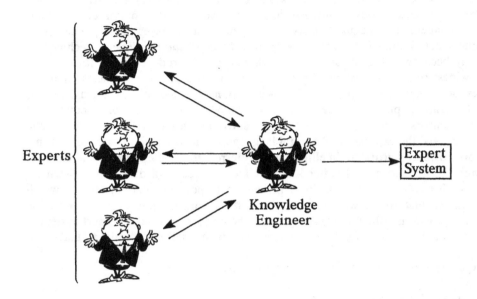

Fig. 9.5   Construction of an expert system

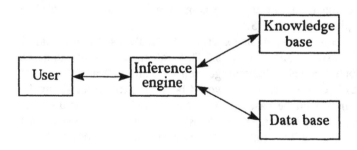

Fig. 9.6   Expert system organization

An expert system is usually organized in terms of two levels, as shown in Fig. 9.6;
1. The knowledge base/data base level;
2. The inference engine;

The *inference engine* provides an interface between the user (by user is meant the program which ''calls'' the expert system), and the knowledge/data base. The purpose of this is such that once a hypothesis has been selected, data can be compared with ''known'' values in order for a justifiable conclusion to be drawn. This conclusion can then be acted upon, again with reference to the knowledge base (see Fig. 9.6).

The most commonly encountered expert system rule is the ''*if ... then*'' structure, although this can take on one of several forms:

1. Deduction: *if* A *then* B;
2. Induction: *if* for a large number of cases, where A *then* B, so expect B *if* A;
3. Abduction: *if* B, when we have *if* A *then* B, there is support for A.

The different forms are expanded in Example 9.3.1.

Example 9.3.1

Consider the following example of an expert system in operation, Fig. 9.7 relates.

In the forward path one commences with a trigger, e.g. an alarm condition, a hypothesis is then suggested, e.g. a reason for the alarm is given. Subsequently a cause is pointed to which means that a certain fault is apparent. The fault can then be confirmed by a test and/or repaired by an appropriate action.

A reverse path also exists however, and this commences with a test which points to a number of faults, one of which is confirmed. By these means one can locate, and be prepared for, a particular trigger.

In the forward path the four rules applied are:

1. *If* <trigger> *then suggest* <hypothesis>;
2. <hypothesis> *caused by* <fault>;
3. <fault> *confirmed by* <test>;
4. *If* <fault> *then do* <action>.

In terms of the DDC algorithm, an expert system serves more in a supervisory/monitoring role and also modifies controller action, in what could be a drastic way, when certain triggering conditions apply. The usefulness of an expert system is dependent on the depth with which its knowledge base details the triggers which can possibly occur, and the action which needs to be taken when each trigger occurs. If a trigger occurs for which the expert system has no information, or a trigger occurs and the expert system suggests an incorrect action in response, then the operative quality of that system should be questioned.

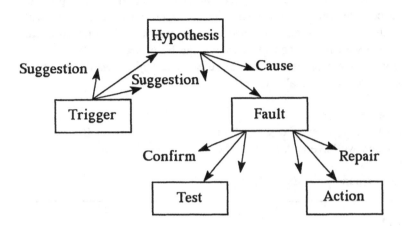

Fig. 9.7 Flow diagram for Example 9.3.1

9.3.5   Fuzzy logic controllers

Digital computers are based on standard logic; i.e. on or off, 1 or 0; whereas it is often the case when controlling a real-world system that we would like the temperature to be "fairly warm" or the height of liquid to be in the middle. These concepts led to the development of fuzzy logic as an alternative computer based approach to control.

The basis for fuzzy logic is the idea of a fuzzy set which contains all the possible values of a particular variable. So if F is the set and f is an individual value in F, so F is the set of all possible values of f, i.e. $F = \{f\}$. As an example consider the speed of a car with F being the set of all fast speeds for that car. The variable f could then range from, say 50 m.p.h to 120 m.p.h, although this is dependent on (a) the car, (b) the person involved, and (c) the problem domain. All of these factors are in fact very important in general fuzzy controller design.

Normal logic would simply state whether a variable was or was not a member of a set, e.g. 55 m.p.h is either a fast speed or it is not! However a fuzzy set has a grade of membership associated with it such that each value in a fuzzy set, F, is associated with a real number in the range [0,1]. 55 m.p.h could then be assigned a value, say 0.12, which means that it is not such a fast speed as 100 m.p.h. which might have a value of 0.83.

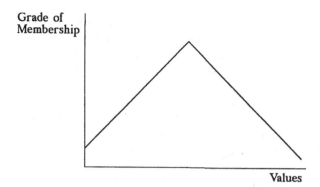

Fig. 9.8 Triangular membership function

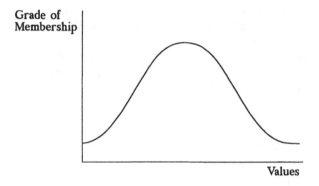

Fig. 9.9 Continuous membership function

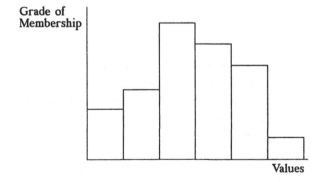

Fig. 9.10 Quantised membership function

Dependent on the problem encountered, so different types of functions, representing the grades of membership in a fuzzy set, are required. Three typical functions are shown in Figs. 9.8 to 9.10. The triangular membership function, shown in Fig. 9.8, is perhaps the most commonly encountered version, this being completely specified by only 3 values.

The continuous membership function, shown in Fig. 9.9, gives a more direct mathematical relationship, however it can be extremely difficult to decide on the exact shape of the function and the on-line calculation of a values' grade of membership can be quite time consuming. It does have the advantage though of including a wide range of values with a low grade of membership.

The quantised membership function of Fig. 9.10 is relatively easy to define by splitting the overall range of values considered into a number of sub-ranges. This method has the advantage that for on-line implementation it is very fast, often the function being realised simply by means of a look-up table.

Variables associated with real-world control situations are measurable values, e.g. 63 m.p.h, 15 cm, whereas fuzzy systems are based on ideas such as fast speed, medium temperature, etc. In order to apply a fuzzy controller therefore, it is necessary to firstly convert the real-world value of interest into its fuzzy equivalent, this process is called "Fuzzification". A fuzzy control procedure, as will shortly be described, can then be applied, with the resultant fuzzy controller output being converted back to a real-world value via a process termed "Defuzzification".

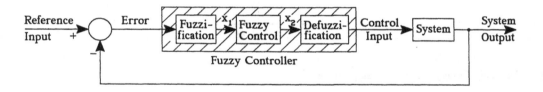

Fig. 9.11 Fuzzy logic controller

The overall implementation of a fuzzy logic controller is shown in Fig. 9.11, where it must be remembered that in terms of a computer implementation the fuzzy controller input will be a sampled signal via an analog-to-digital converter, whereas the controller output will be fed through a digital-to-analog converter. The error and control input in Fig. 9.11 are thus real-world, measurable values, whereas x_1 and x_2 are fuzzy values; all of the shaded area being considered as within the computer.

In order to achieve fuzzification, a number of functions describing the grade of membership for a particular value are defined, such that the measured value in question is merely converted into its fuzzy equivalent. Figure 9.12 shows an example range of functions used to convert a measured speed value into its fuzzy equivalent. Using this figure, a real-world speed of 40 m.p.h would become ($gL = 0.2$, $gM = 0.67$, $gF = 0$)

where *gL = grade of membership of low speeds*, *gM = grade of membership of medium speeds*, and *gF = grade of membership of fast speeds*. Also a real-world speed of 55 m.p.h. would become ($gL = 0$, $gM = 0.92$, $gF = 0$) whereas a real-world speed of 100 m.p.h would become ($gL = 0$, $gM = 0.2$, $gF = 0.6$). Clearly the extent of overlap between functions and the range of values selected is very much problem dependent.

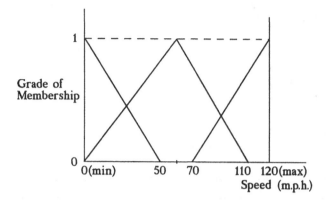

Fig. 9.12 Fuzzification

It is worth pointing out that normally a number of signals are actually passed through the fuzzification exercise, with different membership functions apparent for each signal. So, as well as speed we could have acceleration and distance provided as signals, i.e. derivative and integral forms.

Once a signal has been fuzzified it can be operated on by the fuzzy controller, the two most commonly encountered forms of which are (a) Rule-based control, and (b) Relational systems.

The rule-based system takes on the form described in Section 9.3.4, i.e. if *< trigger >* then suggest *< hypothesis >*. The trigger part of this mechanism is based on fuzzy logic operators, and when grouped together these can realise quite a complicated expression. The hypothesis part then assigns a qualitative value to the controller output in the form of an output reference set. Example 9.3.2 is used to indicate, by example, how a rule-based controller is applied.

Example 9.3.2

Consider a rule-based controller described by five rules (a) – (e), where *e = error*, *u = control input*,

(a) IF *e* is Large AND δe is Large THEN δu is Large.
(b) IF *e* is Medium THEN *u* is Medium.

(c)   IF *e* is Medium THEN δu is Medium.
(d)   IF δe is Medium THEN δu is Medium.
(e)   IF *e* is Small AND δe is Small THEN *u* is Small.

Assume, for a particular example, that fuzzification gives:

geS $= 0.3$ geM $= 0.6$ geL $= 0.2$
gδeS $= 0.1$ gδeM $= 0.4$ gδeL $= 0.5$

The overall degrees of truth obtained from the five rules are then:

(a) $\gamma_a = $ min $(0.2, 0.5) = 0.2$ that δu is Large;
(b) $\gamma_b = 0.6$ that *u* is Medium;
(c) $\gamma_c = 0.6$ that δu is Medium;
(d) $\gamma_d = 0.4$ that δu is Medium;
(e) $\gamma_e = $ min $(0.3, 0.1) = 0.1$ that *u* is Small;

which results in the output reference sets shown in Figs. 9.13 and 9.14

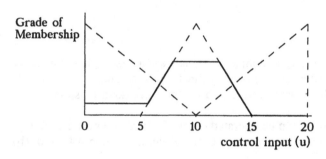

Fig. 9.13 Fuzzy control output reference set (u)

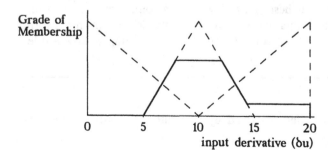

Fig. 9.14 Fuzzy control output reference set (δu)

Two key points arise from Example 9.3.2, firstly the AND operation carried out within the trigger part of the rules and secondly the resultant degree of truth of the hypothesis part. The latter point, the degree of truth, is merely giving an indication of how true a fact is dependent on how true the rule's trigger is. It is, therefore, a very similar measure to the grade of membership.

A number of fuzzy logic operators exist, not simply the AND statement found in Example 9.3.2, and these can be summarized as follows:

If $g_A(x)$ and $g_B(x)$ indicate the grade of membership of a variable x in functions A and B, then

(i) A is empty if $g_A(x) = 0$, for all x;
(ii) $g_A^1(x)$ is the complement of $g_A(x)$, such that $g_A^1(x) = 1 - g_A(x)$;
(iii) $g_A(x)$ is a subset of $g_B(x)$ if $g_A(x) \le g_B(x)$, for all x;
(iv) $g_C(x)$ is the intersection of $g_A(x)$ and $g_B(x)$ when

$$g_C(x) = \min(g_A(x), g_B(x),$$

this is the AND function;
(v) $g_C(x)$ is the union of $g_A(x)$ and $g_B(x)$ when

$$g_C(x) = \max(g_A(x), g_B(x))$$

this is the OR function.

One further point on rule-based fuzzy controllers is that if, as hypothesis support, two degrees of truth are realised for the same fact, the higher of these values should be employed. This is the case in Example 9.3.2 for the hypothesis that δu is Medium which receives $\gamma_C = 0.6$ and $\gamma_d = 0.4$, such that γ_c overrides γ_d in this case.

Rule-based fuzzy controllers suffer in the same way as expert systems in general, in that obtaining the rules can cause distinct problems, particularly if "experts" must be monitored or interviewed, in which case the results can be quite inaccurate and certainly dependent on which expert has been asked. It can also be a very expensive procedure if system down time is involved. Recent research in this area has therefore concentrated on self-organizing fuzzy controllers in which the rules modify themselves, particularly in terms of grades of membership and degrees of truth.

An alternative method for fuzzy controller implementation is the use of relational systems. Here qualitative inputs are related to qualitative outputs by means of an array, the relational array, containing degrees of truth associations. A big advantage of such an approach is that the relational array can be obtained through experimentation, i.e. finding input/output signal pairs. The method is also not tied down by the particular "experts" involved, as can be the case for rule-based fuzzy controllers.

The basic definition of a relational equation is:

$$u = R \circ e \tag{9.3.1}$$

in which u = array of degrees of truth of fuzzy controller output;
R = fuzzy relational array;
e = array of grades of membership of fuzzy controller input.

Once an input is provided, the relational array is employed to obtain:

$$\gamma_{u(i)} = \max_{j} \ [\min(g_R(i,j), g_e(j))] \tag{9.3.2}$$

where
$$\gamma_{u(i)} = \text{degree of truth in } i\text{th output set;}$$
$$g_e(j) = \text{grade of membership of } j\text{th input;}$$
$$g_R(i,j) = \text{relational array value relating } j\text{th input to } i\text{th output set.}$$

The concept of the relational array is perhaps best shown by means of an example, this being Example 9.3.3

Example 9.3.3

For a particular fuzzy controller, the relational array is given by:

$$R = \begin{bmatrix} 0.3 & 0.7 \\ 0.1 & 0.3 \end{bmatrix}$$

Given that the input to the fuzzy controller is

$$e = \begin{bmatrix} 0.2 \\ 0.6 \end{bmatrix}$$

Find the degree of truth of the output set:

$$u = \begin{bmatrix} \gamma_{u(1)} \\ \gamma_{u(2)} \end{bmatrix} = \begin{bmatrix} \max(\min(0.3, \ 0.2), \min(0.7, \ 0.6)) \\ \max(\min(0.1, \ 0.2), \min(0.3, \ 0.6)) \end{bmatrix} = \begin{bmatrix} 0.6 \\ 0.3 \end{bmatrix}$$

The final element of a fuzzy logic controller is that of defuzzification. The output from either a rule-based fuzzy controller or a relational array, is in terms of a number of reference sets, as shown in Figs. 9.13, 9.14. Each reference set must however be converted into a single figure to be used, as a crisp value, in the real-world control input. Three methods of defuzzification are most commonly encountered:

(a) Mean of Maxima Method.
    This is a fairly simple method in which a single maximum value, in the reference set, is employed as the defuzzified signal, subject to such a maximum existing. If several maximum points exist then the mean value of these points is found. Example 9.3.4 indicates a specific case. One negative point with this method is that small changes in fuzzy controller input can cause large changes in the defuzzified signal.
(b) Centre of Area Method.
    This method involves a bisection of the area under the overall reference set function. Although this solves the problem of large defuzzified signal swings, it does require a relatively high computational effort.
(c) Fuzzy Mean Method.
    This is very much of a compromise between the previous two methods, and is achieved by assigning each individual reference set a characteristic value, often this coincides with the largest reference set value. A weighted mean of the characteristic values is then taken, as shown in Example 9.3.4.

For the fuzzy mean method the controller output signal value is thus found from:

$$u = \frac{\sum\limits_{i=1}^{n} u_i \gamma_i}{\sum\limits_{i=1}^{n} \gamma_i} \qquad (9.3.3)$$

in which γ_i is the degree of truth of set i and u_i is the characteristic value of set i.

Example 9.3.4

Consider the overall reference set shown in Fig. 9.15, the defuzzified value for this set is found to be:

(a) Mean of Maxima:

Maximum $= 0.5$, from value $= 8.75$ to value $= 16.25$

$$\textit{Mean of Maxima} = \frac{16.25 + 8.75}{2} = 12.5$$

(b) Centre of Area:

Centre of area = 11   (see Problem 9.1).

(c) Fuzzy Mean Method:

Choose:

Characteristic value of function (i) = 7.5
Characteristic value of function (ii) = 12.5

these both being mid-range points.

The output value is then found from:

$$value = \frac{0.3 \cdot 7.5 + 0.5 \cdot 12.5}{0.3 + 0.5} = 10.625$$

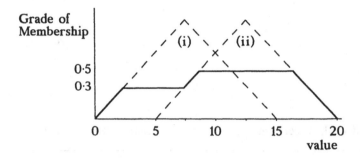

Fig. 9.15   Reference set for defuzzification

To conclude this section on fuzzy logic controllers, it is worth pointing out that although such controllers are now quite popular they are not so "exact" in their calculation or in the control action produced. Choice of reference sets is left up to the user/operator and yet such a choice is extremely important in that a poor choice will result in poor control, whilst there is no defined way of making a good choice. However, with most problems it is usually possible to select reference sets which will provide reasonable control.

9.3.6 Operational problems

When a computer-based digital controller is implemented on a plant, numerous operational problems will most likely appear. These can range from slight numerical inaccuracies in calculations to loss of control due to saturation effects. A range of the most common problems are discussed here and their relationship with an expert system structure solution, detailed in the previous section, is highlighted where appropriate.

Numerical problems are caused principally by the finite word length, not only of the computer data bus and memory, but also of the analog-to-digital and digital-to-analog conversion mechanisms. *Quantization* errors are caused by the representation of an analog signal by means of a digital base in terms of a finite number of digits. Their cause can be seen as either due to rounding and/or truncating the analog representation in order to convert it to its nearest digital form. Subsequently when two analog numbers are stored digitally, as soon as a calculation takes place in which those numbers are concerned so further errors will be introduced because of the *finite precision* of the *calculations*. Finally, the controller parameters, e.g. polynomial coefficients, will themselves be merely digital representations stored with finite precision, hence leading to further errors when calculations take place.

A different aspect, in terms of operational difficulties, is the problem caused by the time taken for calculations to take place, when the computer is controlling a real-world plant. It may well be necessary for one computer to service several control loops as well as operating a logical expert system testing sequence, on top of which the display of information is required and also communication with the operator. All of these factors combined could provide an impossible schedule between sampling instants, unless sequencing and priority conditions are applied. A natural solution to this problem is to employ parallel processing techniques both in the form of special-purpose hardware and concurrent programming. A certain amount can, however, usually be done within the control algorithm itself, by suitable modifications, in order to aid the parallel implementation.

With regard to the plant under control, this usually presents several operational problems in its own right, although very often these problems are due primarily to the signal actuators rather than the plant itself. Actuators are usually fairly limited in both their range and slew speed and sometimes can contain a dead zone, i.e. a small control signal change will not produce any corresponding actuator change. If such a signal occurs, this information should be fed back to the controller, possibly as an expert system trigger.

When a controller contains integral action, the PID controllers discussed in Chapter 10 are an example, then the controller will continually integrate the error between actual and desired output signals. Assume for a moment that the output signal actuator has saturated, then the resultant error will be continually integrated producing very large, increasing, controller values; this is called *integrator saturation* (or reset windup). One possible solution to this is to employ only the integrating effect of the controller, when the error is small, i.e. within defined minimum and maximum bounds. Such a choice of integrator on/off can be made within an expert system hierarchy.

The final operational problem considered here is the effect of manual/automatic transfer. When switching over from manual to automatic control of a plant, transient signal spikes can occur. Indeed such spikes can also be witnessed when controller coefficients are manipulated on-line. Unfortunately a solution to this problem is not simple, and involves the use of a state-space system description in order to match not only the manual and automatic control signals, but also the state vectors at the instant of transfer.

9.4   Computer-aided design

As the availability of computers became more widespread in the 1960s, so digital system simulations replaced analog simulations, because of the many apparent advantages. Programming was based around a block diagram formulation combined with a selection of fixed operations, although modeling in terms of mathematical equations was soon introduced.

For all types of system simulation work, it is important that model parameters can be altered easily, and this is certainly possible with a computer-based system. The most successful simulation schemes are in fact those implemented as an *interactive* package, whether they are run on a stand-alone personal computer or a multi-user computer with time-sharing facilities. The user is then able to interact directly on-line with the package such that any modification made is immediately reported on by the computer, whereby the user can take appropriate necessary action, at once.

The interactive capabilities can be presented to the user in one of a number of ways. One possibility is the presentation of a *menu* at each particular stage of simulation – the user can then choose one of the options listed in the menu. Once a selection has been made it may then be the case that the computer will take immediate action, or alternatively the presentation of a further menu may occur from which another selection must be made. The menu approach is extremely useful for inexperienced or nontechnical end users, on condition that the menu is not cluttered with too much irrelevant, confusing detail.

More flexibility can be obtained with a computer-based system simulation, if a *command* driven package is employed. In this case the user selects from a list of available commands, and although this leads to much greater efficiency in operation, it does necessitate a higher level of experience on the user's part. Often learning to work with a particular package can be a harrowing and time-consuming period. A good set of explanation facilities at all levels of operation is therefore of great importance within the package itself, in order to give a concise but comprehensive outline of the relevant stage such that users' problems can be dealt with. The *help* command is invariably employed for this purpose, and is generally also available for menu-driven packages. As an extension, and/or introduction, to the help facility, back-up documentation is also an important feature, something which was sadly neglected in earlier packages and which can still be variable in quality when a new package is introduced (until initial problems with the package have been found and removed or avoided).

Presentation of simulation package results is usually available in several modes, ranging from specific numerical values dredged from the depths of an algorithm, to colorful graphic displays of time-varying signals. Interactive graphical facilities allow for a diagram, or graph, to vary with time as different stages of the simulation are reached, thus presenting an up-to-date picture of the study being carried out.

However, it is not merely simulation work for which a computer is useful. Due to the numerical and graphical properties associated with a computer-based facility, almost all of the control system studies and controller design procedures described in this book, both continuous-time and discrete-time, can be carried out by means of an appropriate package. Although for continuous-time designs, e.g. a Nyquist diagram, this can mean that certain discrete approximations are necessary, it is usually the case that a time saving is obtained, particularly when interactive capabilities are available.

9.4.1 Modeling and identification

Of primary importance in the study of system performance and controller design is the construction of a mathematical model which adequately describes the behaviour of that system. Computer-based procedures are more of a necessity nowadays where system modeling is concerned, because of the complexity and variability in model structure which can be represented.

Although a plant model can by found be investigating the internal plant structure, it is usually the case that a complete picture cannot be obtained in this way, due to unknown or immeasurable factors. It is in fact common practice to use the multi-experiment approach, in which various input signals are applied to the plant and the corresponding output signals recorded. To be representative of the system under consideration, it is important that these input/output data pairs are taken within the normal operating range of the system. However, it is unfortunately the case that the data obtained from a real-world system will invariably be corrupted by noise emanating from random fluctuations both in the plant itself and in the measuring devices employed. Making use of the collected input/output data, despite any associated noise, in order to estimate coefficients within a plant model is termed *system identification.*

System identification is the procedure of modeling systems mathematically, with due consideration being given to problems such as noisy signals, model form, data analysis and model testing. In general, statistical tests are applied to a set of plant input-output data, in order firstly to choose a suitable model structure and secondly to select appropriate coefficients in a model with the chosen structure. As many computer packages are available for system identification, the main problem for the user is often achieving sufficient real-world data for the package.

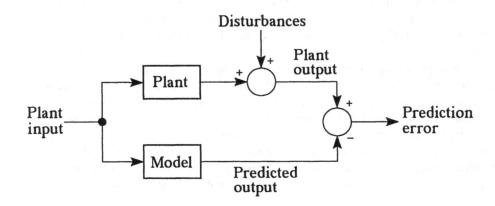

Fig. 9.16 Prediction error generation

Prediction error models are perhaps the most useful for identification purposes, whereby the same input signal is applied to the plant itself as is applied to a mathematical model of the plant, see Fig. 9.16. If the outputs from the plant and model are then compared, the error produced is dependent on how close the model is to representing the actual plant. If the structure of the model is wrong, however, i.e. not enough coefficients are employed, then no matter how much time is spent finding values for the coefficients, a large prediction error will occur. Conversely, because of the disturbance effects, complex models, with many coefficients, are not necessarily of use because variations in some of the coefficients may well have less of an effect on the prediction error than will the noise signal.

The vast majority of identification procedures are based around the idea of making the magnitude of the prediction error as near to zero as possible, hence the square of the prediction error is in general used. Essentially, a number of samples, N, of the prediction error signal squared are summed, and this sum of squares is employed as a function to be minimized by the selection of model coefficients. This process has the advantage that noise effects on the model are reduced considerably.

Many procedures exist for the minimization of the sum of squares function, the most commonly encountered method being that of *linear least squares*. Such techniques are however, ideally suited to the computer, which is merely required to operate on a given data set in order to produce a representative model.

9.4.2 Controller design

System evaluation and design can be split up into roughly four areas: (a) signal processing, (b) modeling and identification, (c) controller design and (d) system simulation; as shown in Fig. 9.17. A complete computer-aided control system design

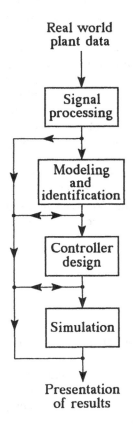

Fig. 9.17 Requirements for a CAD package

package then includes facilities for all four of these areas, although many schemes merely concentrate on one or two of the topics.

So, a controller design package is required to be not only operative in its own right, but also to fit in with identification and simulation routines. It is therefore necessary for the model structures, employed as a basis for the controller design stage, to be of a compatible type with those employed for identification procedures. Hence, as part of the input to the controller design process, a plant model must be entered, whether this model is from a known, reliable source, or whether it is directly obtained as the output from an identification procedure.

Note: As shown in the diagram, it might be the case that data output from the signal processing stage is to be combined with a known plant model, for controller design purposes.

Due to the large number of possible controller actions, the controller design stage can attempt to offer an extremely wide variety of choice, although very often only a few different techniques are available (indeed in many cases only one scheme is on offer). It is usually the case that for any particular controller technique selected, e.g. assume that a PID controller (described in Chapter 10) has been chosen, then the coefficients of that controller can be found in one of a number of possible ways, e.g. pole placement or Ziegler–Nichols. Also, once the controller coefficients have been finalized, jacketing and compensation terms must be sorted out. Hence, the effect of time delays, nonlinear behaviour, disturbances and loop interactions can all be dealt with in this stage of the design procedure. Selectable actuator gains and interfacing specifications can also be decided upon in the control design stage, some by the user and some within the package program, in order that the overall control system response is manipulated to satisfy real-world engineering constraints.

Once the controller has been finalized, it can either be employed within a simulation study, from which useful operational details, such as performance and problem areas, can be ascertained, or coefficient information can be output from the CAD package such that the controller can be used directly in an on-line, most likely computer-based, control system.

9.5   Summary

In this chapter it has been shown how computers can be used as efficient tools in many aspects of control systems study and design.

Three important features have been looked at in detail, the first of these being the use of computers, whether analog or digital, to simulate the behaviour of systems. This is an extremely important issue in that large-scale plant, when represented by a computer model, can be investigated and modified by means of a flexible bench-top exercise. It also allows for the study of certain plants, which for cost and/or security reasons would normally be unavailable for analysis.

The second aspect of computer usage considered in this chapter was the employment of computer-based control systems, in which the computer itself makes up part of the overall system. Present trends are for more intelligent, faster controllers with the ability to adapt to changes in the environment. Hence, as well as discussing operational details of real-time on-line computer control systems, intelligent systems based on a knowledge store employed within an expert system framework were also considered.

Finally, various aspects of computer-aided control system design were outlined. Initially the use of CAD for system simulation work was highlighted, although system modeling and controller design packages were also covered. The diversity of CAD tools available to the control engineer makes it impossible to do other than give a flavour of the techniques on offer; the prospective user should however beware of the number of user-unfriendly packages which require specialist knowledge before they can be operated.

Problems

9.1   Verify that in Example 9.3.4 the centre of area method realises a value of 11.

9.2   Find defuzzified values for the control input, u, in Fig. 9.13, using the three methods described.

9.3   Find defuzzified values for the change in control input, δu, in Fig. 9.14, using the three methods described.

9.4   A system has been modelled using a relational equation with the output and input described by two fuzzy reference sets. The resultant relational array produced by identification is:

$$R = \begin{bmatrix} 0.3 & 0.6 \\ 0 & 0.4 \end{bmatrix}$$

Given that the input to the system, described in terms of its reference sets, is e = [0.8, 0.2], what is the output?

9.5   A fuzzy logic controller is to be constructed for the control of an electric motor. Outline an initial design for such a controller and give examples of rules to be applied, if a rule-based system is employed.

9.6   For a fuzzy controller based on a relational model, the controller output, in terms of degrees of truth of output sets, is given for two different inputs as:

(a) $u = \begin{bmatrix} 0.4 \\ 0.2 \end{bmatrix}$, and (b) $u = \begin{bmatrix} 0.3 \\ 0.3 \end{bmatrix}$

If the array of degrees of truth of input sets is:

(a) e = [0.4, 0.2], and (b) e = [0.3, 0.4]

suggest an appropriate fuzzy relational array.

9.7   Fuzzy logic control is perhaps better suited to certain control problems and less well suited to others. Give examples of where its advantages and disadvantages lie.

9.8   A fuzzy controller based on a relational equation, has a relational array

$$R = \begin{bmatrix} 0.1 & 0.2 & 0.3 \\ 0.6 & 0.5 & 0.4 \\ 0.7 & 0.2 & 0.4 \end{bmatrix}$$

Given that the input applied to the controller is:

$$e = [0.4 \quad 0.3 \quad 0.5]$$

what is the controller output?

9.9   What advantages are apparent if a computer control scheme is employed in place of a straightforward digital (noncomputer) control implementation? What are the disadvantages?

9.10   Consider the state-space system matrix

$$A = \begin{bmatrix} 2.1 & 6.1 \\ 1.0 & 2.9 \end{bmatrix}$$

which is nonsingular, and therefore invertible. However, if the matrix elements are measured values which are stored within computer memory, each element will have errors, e.g. quantization, rounding, associated with it. For the example given, show how these errors can result in a singular, noninvertible matrix.

9.11   In the construction of an expert system for a home heating system, what important rules should be included?

9.12   In the construction of an expert system for an automated vehicle driver, with regard merely to vehicle guidance along a well defined roadway, what factors must be accounted for?
Having constructed such an expert system, are there any occurrences that have not been taken into account but which a human expert could easily cope with, e.g. adverse weather conditions?

9.13   An important aspect of many complex control systems is to ensure that even if one system element fails, the overall system remains in operation. Apart from alarm sounding, how can a computer control scheme provide fault tolerant control?

Further reading

Aleksander, I. and Morton, H., 'Artificial intelligence: an engineering perspective', *Proc. IEE*, **134**, Pt.D, 1987, pp. 218–23.

Artwick, B., *Microcomputer interfacing*, Prentice-Hall, 1980.

Astrom, K. and Wittenmark, B., *Computer controlled systems*, Prentice-Hall, 1984.

Banks, S. P., *Control systems engineering*, Prentice-Hall Int., 1986.

Becker, R. G., Heunis, A. J. and Mayne, D. Q., 'Computer aided design of control systems via optimisation', *Proc IEE*, **126**, 1979, pp. 573–84.

Bennett, S. and Linkens, D. A. (eds.), *Real time computer control*, Peter Peregrinus Ltd, 1984.

Boullart, L., 'An introduction to artificial intelligence and expert systems', *Proc. Int. Seminar on Expert Systems and Artificial Intelligence in Industry*, Antwerp, 1986.

Bratley, P., Fox, B. L. and Schrage, L. E., *A guide to simulation*, Springer-Verlag, 1983.

Brdys, M. A. and Malinowski, K., *Computer aided control system design*, World Scientific Publishing Co., 1993.

Clayton, G. B., *Operational amplifiers*, Butterworths, 1979.

Driankov, D., Hellendoom, H. and Reinfrank, M., *An introduction to fuzzy control*, Springer-Verlag, 1993.

Efstathiou, J., 'The chocolate biscuit factory', *Journal A*, **27**, No. 2, 1986.

Fisher, D. G. and Seborg, D. E., *Multivariable computer control*, North-Holland, 1976.

Gordon, G., *System simulation*, Prentice-Hall, 1978.

Hartley, M. G. (ed.), *Digital simulation methods*, Peter Peregrinus Ltd, 1975.

Jacoby, S. L. S. and Kowalik, J. S., *Mathematical modelling with computers*, Prentice-Hall, 1980.

Millman, J. and Halkias, C. C., *Integrated electronics*, McGraw-Hill, Kogakusha, 1972.

Minsky, M., *The psychology of computer vision*, McGraw-Hill, New York, 1975, pp. 211–77.

Ng, T. S., 'An expert system for shape diagnosis in a plate mill', *IEEE Trans. on Industry Applications*, pp. 1057–1064, 1990.

Norton, J. P., *An introduction to identification*, Academic Press, 1986.

Pedrycz, W., *Fuzzy control and fuzzy systems*, 2nd ed., Research Studies Press, 1992.

Peel, D., 'Adaptive control and artificial intelligence', *Journal A*, **28**, No. 3, 1987, pp. 143–7.

Pressman, R. S. and Williams, J. E., *Numerical control and computer-aided manufacturing*, John Wiley, 1977.

Rosenbrock, H. H., *Computer aided control system design*, Academic Press, 1974.

Sandoz, D. J., 'CAD for the design and evaluation of industrial control systems', *Proc. IEE*, **131**, No. 4, 1984.

Shoureshi, R. and De Silva, C.W. (eds.), *Intelligent control systems*, American Society of Mechanical Engineers, 1992.

Stout, T. M., 'Economic justification of computer control systems', *Automatica*, **9**, 1973, pp. 9–19.

Tzafestas, S. G. (ed.), *Applied control,* Marcel Dekker, 1993.

Warwick, K. (ed.), *Implementation of self-tuning controllers*, Peter Peregrinus Ltd, 1988.

Warwick, K. (ed.), *Applied artificial intelligence*, Peter Peregrinus Ltd, 1991.

Warwick, K., Karny, M. and Halouskova, A. (eds.), *Advanced methods in adaptive control for industrial applications*, Springer-Verlag, 1991.

Williams, T. J., 'Computer control technology – past, present and probable future', *Trans. Inst. MC.*, **5**, No. 1, 1983.

Winston, P. H., *Artificial intelligence*, Addison-Wesley, 1984.

Zanker, P. M. and La Pensee, S., 'Design of a computer controlled electrohydraulic servo-system', *Proc. IEE Conference on Control and its Applications*, 1981, pp. 96–101.

10

Three-term controllers

10.1   Introduction

When first attempting to control a particular plant, in all but the simplest cases, it is very soon appreciated that a controller which consists solely of a gain, or amplification, value is not versatile enough to cope with even the most basic requirements made for overall system performance. On the other hand there are a large number of elaborate controllers which satisfy a particular mathematical function, require high accuracy and may well take several days to calculate on the most expensive computers. Clearly there is room for a fairly straightforward controller type which although it is relatively simple and cheap to operate, can also deal successfully and reliably with the essential control requirements. To be acceptable this controller must be able to cope adequately with the vast majority of plants to which it is applied and prove its worth through many years of service, often in an environment for which it is not particularly designed and to which it is not previously accustomed. Such a controller is the three-term or PID controller which has been for many years by far the best known and most widely used control structure for industrial applications. It is applicable in either continuous- or discrete-time modes and is the most likely form of control system that a practising engineer will encounter.

In a sense, three-term control is just one form of compensation which can be compared directly with the lead-lag compensation described in Chapters 4 and 5, or with more specific schemes such as optimal, deadbeat or pole placement control. However, once the basic structure has been decided upon as consisting of three-terms, i.e. Proportional + Integral + Derivative (PID), the selection of gains associated with each of these terms is still to be found, and although in many cases methods such as root locus, Bode or Nyquist could be used to meet the desired control objective, it may well be that the controller parameters must be chosen to satisfy an optimal or pole placement criterion. It must be pointed out that in many practical situations an exact mathematical description of the plant is not available and this may result in the P + I + D gains being modified on-line until satisfactory performance is achieved.

In this chapter the advantages and disadvantages of the three-term control structure are investigated in a bottom-up approach. Firstly, an attempt is made to apply simply a proportional controller, and it is shown how this can be seen as positive as far as closed-loop stability is concerned, but very little help in terms of steady-state output following or transient response tailoring. It is then shown how the inclusion of both integral and derivative action can deal with these requirements, without the need for more elaborate schemes. Indeed, if an elaborate controller is to be used in practice, it should first be shown that the controller performance is superior to that of a three-term controller in at least one important aspect.

10.2   Proportional control

There are many ways in which a controller can be connected to a system in order to provide the required compensation. One common element between the different approaches is that the error between the actual system output and the desired reference value for the output needs to be available for manipulation. Perhaps the simplest and most commonly encountered method is that of cascade or series compensation, and this is shown in Fig. 10.1.

It is apparent in the cascade controller that the error $e(t)$, between the actual output, $y(t)$, and the desired reference, $v(t)$, is obtained directly and used as the input to the compensator/controller, D, whose output, $u(t)$, is then the input to the system/plant to be controlled. This chapter is concerned primarily with the nature of D, some of the forms it can take, and its effect on the system, G.

As a slight aside, an alternative but nevertheless popular form of compensation connection to that proposed in Fig. 10.1 is the minor-loop feedback controller shown in Fig. 10.2, in which the unity negative feedback loop is preserved. The controller,

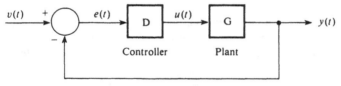

Fig. 10.1   Cascade compensation

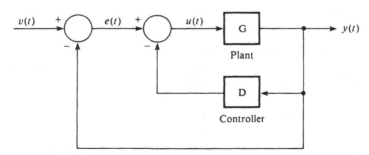

Fig. 10.2   Minor-loop feedback compensation

however, loses a lot of the intuitive appeal provided by the cascade connection, and will only be of indirect relevance here, as the controller is no longer a simple operator on the difference between what we want and what we have.

With reference to the cascade controller of Fig. 10.1, a fairly obvious and primitive attempt at control of the system would be to take into account the error at time t, $e(t)$, then to provide a large corrective action when $e(t)$ is large and a small corrective action when $e(t)$ is small. This is a nice and simple procedure because it only involves a controller D which consists entirely of a multiplying constant.

10.2.1 Continuous-time proportional controller

By connecting up the proportional controller in its cascade form, the controller input is the error, $e(t)$. Its output, $u(t)$, is made up of a single term, which is simply the signal $e(t)$ multiplied by the controller gain, K_p, as in equation (10.2.1)

$$u(t) = K_p e(t) \tag{10.2.1}$$

The closed-loop equation relating $v(t)$ to $y(t)$ is, when the control input of (10.2.1) is applied, given by

$$y(t) = \frac{GK_p}{1 + GK_p} v(t) \tag{10.2.2}$$

or in terms of its s-domain representation

$$Y(s) = \frac{G(s)K_p}{1 + G(s)K_p} V(s) \tag{10.2.3}$$

So a signal has been applied in an attempt to control the system, such that the larger the output error, the larger the control signal – but does this simple form of control provide a good result at the end of the day, i.e. does the output signal, $y(t)$, follow the input signal, $v(t)$? Consider firstly just what happens in the steady-state when all transients have settled down, i.e. when $t \to \infty$. It is obvious that for the simplest of systems, when G is just a straightforward gain value, e.g. $G = K$, then

$$y(t) = \frac{KK_p}{1 + KK_p} v(t) \tag{10.2.4}$$

Therefore as long as $KK_p \gg 1$, whatever type of input signal is applied it follows that the output will be just about equal to the input. In fact it is fairly irrelevant just how large K is, we can 'in theory' make K_p as large as we like, i.e. use a very 'high gain' indeed, in order to achieve the desired results. So as well as being a simple control action this scheme has other hidden benefits in that if the system is not quite how we had thought, i.e. if the actual system gain K is a little bit different from the designed value, possibly due to ageing or a modification, then it does not matter much as long as K_p is a very high value.

The 'high gain' results obtained for a simple system in fact extend to much more complicated cases; as an example consider a system of the form:

$$G(s) = \frac{Ke^{-s\tau}}{(s\tau_1 + 1)(s\tau_2 + 1)}$$

(10.2.5)

in which the term $\exp(-s\tau)$ is a pure time delay, i.e. a period of time elapses before any event or change at the output will be witnessed in response to an event or change at the input. The system $G(s)$ is typical of the transfer function obtained for a heating system with a measured temperature output value.

If a step input of magnitude 3 units is applied to such a system (10.2.5), then the steady-state output signal, which we wish to be equal to 3 units, is found from:

$$\lim_{t \to \infty} y(t) = \lim_{s \to 0} sY(s) = s \left. \frac{G(s)K_\mathrm{p}}{1 + G(s)K_\mathrm{p}} \frac{3}{s} \right|_{s \to 0}$$

$$= \left. \frac{3G(s)K_\mathrm{p}}{1 + G(s)K_\mathrm{p}} \right|_{s \to 0}$$

but

$$\lim_{s \to 0} G(s) = K$$

so

$$\lim_{t \to \infty} y(t) = \frac{3KK_\mathrm{p}}{1 + KK_\mathrm{p}}$$

(10.2.6)

which gives a similar result to (10.2.4) when $v(t) = 3$ units.

The results have thus far pointed to the benefits of proportional feedback, but what are the drawbacks? A first point to take into account is of a practical nature in that the gain K_p will most likely be a simple amplifier — it may therefore not be possible to obtain the high gain required before saturation occurs. For many systems we might have that $K = 0.001$, such that in order to obtain $KK_\mathrm{p} \gg 1$ it would be required that $K_\mathrm{p} > 10^6$ which is quite a tall order for a simple amplifier. So, if it is not possible to use a high gain K_p then an offset between desired and actual output values could occur in the steady-state. As an example, consider (10.2.6) with $K = 0.2$ and $K_\mathrm{p} = 10$, it then follows that the output would be 2 units in the steady-state and not the desired 3 units. In practical terms therefore, K_p is not a suitable control selection to be used for steady-state following of a reference signal.

A second point to note is the effect that the choice of K_p has on the transient response and closed-loop stability of the system. With reference to the closed-loop equation (10.2.3) and the example system (10.2.5), then if the time delay $\tau = 0$ for simplicity, the denominator is given by:

$$(\tau_1 s + 1)(\tau_2 s + 1) + KK_\mathrm{p}$$

or in terms of the characteristic equation:

$$\tau_1\tau_2 s^2 + (\tau_1 + \tau_2)s + (1 + KK_\mathrm{p}) = 0$$

(10.2.7)

For system stability it is easy to see that as all coefficients of the denominator polynomial must be positive, it is required that:

$$KK_p > -1$$

which would certainly be satisfied for a high gain system with $KK_p > 1$. Hence on condition that $\tau_1, \tau_2 > 0$ the closed-loop system will be stable as long as $KK_p > -1$. But what of the transient performance of the system? This is best seen by making reference to the transient response characteristics shown in Chapter 3, Section 3.3.3, in which the second-order denominator is considered to be of the form:

$$s^2 + 2\zeta\omega_n s + \omega_n^2 \tag{10.2.8}$$

For any given natural frequency, ω_n, as ζ tends to zero the response to a step input becomes more oscillatory and conversely as ζ increases the response becomes more heavily damped producing less overshoot. Comparing (10.2.7) with (10.2.8), we have that:

$$\omega_n = \sqrt{\frac{(1 + KK_p)}{\tau_1\tau_2}} \quad \text{and} \quad \zeta = \frac{(\tau_1 + \tau_2)}{2\tau_1\tau_2} \cdot \frac{1}{\omega_n} \tag{10.2.9}$$

and for a given system τ_1 and τ_2 will be constant. This means that an increase in controller gain K_p will increase ω_n, i.e. the frequency of oscillations in response to a step input will increase. The increase in ω_n will however result in a decrease in ζ which means that less damping of the oscillations will occur. In particular a high gain K_p, required for steady-state following, would result in a very low value for ζ and therefore the closed-loop system would perform with oscillations which occurred more frequently and were only slightly damped, i.e. the oscillations would go on for a long time.

In summary, by employing only a proportional control scheme, we appear to fall between two stools in that both steady-state and transient conditions need to be catered for, and these requirements cannot both be satisfied with K_p only. If K_p is chosen large to provide approximate steady-state following, this results in very poor transient control and steady-state conditions will only actually be reached in the distant future.

In Section 10.3 a slightly more complicated controller will be investigated, with an extra design feature employed to cope with the steady-state requirements. The digital version of a proportional controller will however be considered briefly before going on to Section 10.3.

10.2.2 Discrete-time proportional controller

With the controller, $D(z)$, connected up in the mode depicted in Fig. 8.6 in which $G(z)$ is the discrete-time version of the system to be controlled assuming a zero-order hold is present, then the closed-loop equation is given by:

$$y(t) = \frac{D(z)G(z)}{1 + D(z)G(z)} v(t) \tag{10.2.10}$$

as shown in (8.4.6).

The discrete version of a proportional controller is in fact just a discretized version of the continuous-time equation (10.2.1), in that now the error signal $e(t)$ will contain a measured output signal $y(t)$ which retains a steady value between samples. So we have once again that

$$u(t) = K_p e(t)$$

where now

$$u(t) = D(z)e(t) \tag{10.2.11}$$

which is identical to

$$u(k) = D(z)e(k) \tag{10.2.12}$$

In order to try and find out the effect of a particular gain selection K_p, consider as an example the discrete-time system transfer function

$$G(z) = \frac{0.5(z - 0.6)}{(z - 0.2)(z + 0.3)} \tag{10.2.13}$$

which is assumed to include the effect of a zero-order hold.

Firstly let us consider steady-state conditions which, for a three-unit step input $v(t)$, can be found by setting z to unity in $G(z)$ such that by means of (10.2.10):

$$\lim_{t \to \infty} y(t) = \frac{D(1)G(1)}{1 + D(1)G(1)} \cdot 3$$

noting that $D(1) = K_p = D(z)$.

Now

$$G(1) = \frac{0.5(1 - 0.6)}{(1 - 0.2)(1 + 0.3)} = \frac{1}{5.2}$$

and hence

$$\frac{D(1)G(1)}{1 + D(1)G(1)} \cdot 3 = \frac{K_p \cdot 3}{5.2 + K_p}$$

For steady-state following this value is required to be 3 units, in order to equal the stated reference input. However this will only be true if K_p is large or, as for the continuous-time case, the proportional controller must be 'high gain' in order to remove, as much as possible, any steady-state error. It is left as an exercise, see Problem 10.1, to show how the selection of K_p will affect closed-loop system stability, in terms of the unit circle, and also how a 'high gain' K_p will result in very lightly damped oscillations.

The results obtained from a discrete-time proportional controller correspond directly with those obtained from a continuous-time implementation and it is obvious that even for very simple systems, using a proportional controller alone will not provide sufficient flexibility such that the operator may specify closed-loop system performance. However, the operator can ensure, for certain systems, closed-loop stability by an appropriate choice.

10.3   Proportional + integral control

It was highlighted in the previous section that several fairly rudimentary requirements can be asked of a closed-loop system, namely stability, steady-state following and a desired transient response. It was shown however that by using proportional control alone these requirements could not all be satisfied.

Let us consider the steady-state following properties of a closed-loop system with proportional feedback alone. It was stated that in order to obtain no error between the desired and actual output values in terms of steady-state conditions, then the proportional feedback gain must be a very high value. This means that at any instant, if the error is small and positive it will be multiplied by a large K_p to produce a large and positive control input signal, whereas if the error is small and negative it will be multiplied by the large K_p to produce a large and negative control input signal. Obviously the result of this is very wild control input variations caused by only slight fluctuations in output error, not too much of a desirable feature when for instance the practical requirements of an input signal actuator are considered. It will most likely be the case that the actuator employed is not able to change instantaneously from a large negative value to a large positive one, further it will only have linear properties over a certain range, saturation occurring if the input signals are large enough.

So something more than a simple proportional controller is necessary in order to deal with steady-state offset, and in order to see just what properties are required let us consider the problem of heating up a room. If we desire the temperature of the room to be 20 °C, then on a typical British day the outside temperature will be a lot colder than this, requiring heat to be input to the room. Because of heat losses from the room it will be necessary to continually input heat in order to keep the temperature at 20 °C. So the requirement of our room heating controller is a constant heat input signal in order to obtain zero output error, i.e. for the actual room temperature to be equal to the desired value of 20 °C. But with a proportional controller alone, if the error is zero then so is the input signal, see (10.2.1) or, looked at in a different way, if a constant input signal is necessary there must be an output error exhibited. So if a straightforward proportional controller is employed, in order to provide a constant heat input, the actual room temperature will not be 20 °C, i.e. a steady-state error will exist. What is required therefore is a different type of control action which provides a constant control input when the error is zero and which changes its value when an error does exist. By this means, if the desired temperature value is altered to 25 °C it would be expected that the

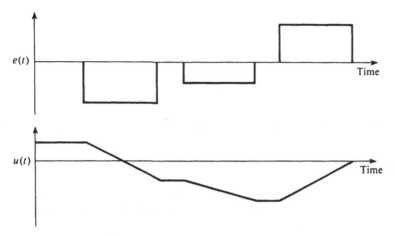

Fig. 10.3   Integrating action of a controller

necessary constant heat input would be greater than that necessary to keep the room at 20 °C. In fact an integrating action has just the desired property and achieves the type of control input $u(t)$ as shown in Fig. 10.3 for an error $e(t)$.

10.3.1 Continuous-time proportional + integral controller

The slope of the control input signal during an error period is dependent on the integrating gain, K_i, selected in Equation (10.3.1)

$$u(t) = K_i \int_0^{t_1} e(t)\,dt \tag{10.3.1}$$

So the integrating action, once started at time $t = 0$, holds a memory of all the error signals which occur up to and including that at time $t = t_1$. In terms of the Laplace transform the integrating controller can be restated as:

$$U(s) = \frac{K_i}{s} E(s) \tag{10.3.2}$$

such that when this forms the controller D in Fig. 10.1, the overall closed-loop equation is given by

$$Y(s) = \frac{G(s)K_i}{s + G(s)K_i} V(s) \tag{10.3.3}$$

The benefits of employing an integral action controller can be seen if steady-state conditions are investigated. Consider, as was done previously, the application of a step input having magnitude 3 units, i.e. $V(s) = 3/s$ in (10.3.3). It is required that the output signal will also be equal to 3 units, with no error, in the steady-state and this can be seen from

$$\lim_{t \to \infty} y(t) = \lim_{s \to 0} sY(s) = s \left. \frac{G(s)K_i}{s + G(s)K_i} \frac{3}{s} \right|_{s \to 0}$$

$$= \left. \frac{3G(s)K_i}{s + G(s)K_i} \right|_{s \to 0}$$

and thus the output will be equal to 3, regardless of the value of K_i or the complexity of $G(s)$. So

$$\lim_{t \to \infty} y(t) = 3 = \text{the desired output value}$$

Hence steady-state following of the desired output value is obtained by the use of an integrating controller. But there were other basic requirements made on the closed-loop system, namely stability and transient performance. Consider the first-order system represented by

$$G(s) = \frac{K}{(s\tau_1 + 1)} \tag{10.3.4}$$

When this system is cascaded with an integrating controller (no proportional action), the closed-loop characteristic equation becomes

$$s^2\tau_1 + s + KK_i = 0 \tag{10.3.5}$$

where K_i is the integral gain.

In order to obtain a stable closed-loop system it is directly apparent that all the coefficients of the closed-loop system must be positive and that this is satisfied as long as firstly $\tau_1 > 0$, which is true for a stable open-loop system (10.3.4), and secondly $KK_i > 0$, or $K_i > 0$ for $K > 0$. So using integral feedback alone does not necessarily mean that stability is lost. However, the transient performance of the closed-loop system that will result is defined by

$$\omega_n = \sqrt{\frac{KK_i}{\tau_1}} \quad \text{and} \quad \zeta = \frac{1}{2\omega_n} \tag{10.3.6}$$

and now there are problems. Selecting K_i to be positive and small in order to obtain a low frequency of oscillations will result in a high damping factor (i.e. it could take several years for a room to be heated up to 20 °C). However, if K_i is selected to be positive and large in order to obtain low damping, this will result in a high frequency of oscillations (i.e. room temperature will swing wildly, reaching possibly 30 °C and 10 °C).

Clearly, using integral control alone results in nice steady-state following properties but throws caution to the wind as far as transient conditions are concerned. Further, although for the simple system (10.3.4) closed-loop stability could be easily obtained, this is not generally the case and hence the usual procedure is to use a two-term controller, which consists of a proportional + integral control action, as shown in Fig. 10.4.

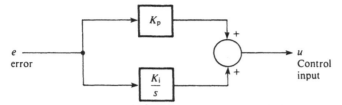

Fig. 10.4   Proportional + integral controller

The equation for this two-term controller is then

$$u(t) = \left[K_p e(t) + K_i \int_0^{t_1} e(t) \, dt \right]$$

(10.3.7)

or

$$u(t) = K_p \left[e(t) + \frac{1}{T_i} \int_0^{t_1} e(t) \, dt \right]$$

in which $T_i = K_p/K_i$ is the integral time constant. In terms of the Laplace transform this can also be written as

$$U(s) = \frac{K_p}{T_i} \left[\frac{1 + sT_i}{s} \right] E(s)$$

(10.3.8)

where it can be seen that the integral action, disclosed by the common s factor in the denominator, is still present; however, the transfer function differs, when compared to the solely integrating controller, by the inclusion of a zero in the numerator of (10.3.8). By applying this controller to the first-order system in equation (10.3.4), the closed-loop characteristic equation is found to be:

$$s^2 \tau_1 + (KK_p + 1)s + KK_i = 0$$

(10.3.9)

and on condition that $KK_i > 0$ and $KK_p > -1$ for stability, an extra controller term is now present for transient conditioning.

The overall philosophy of P + I control can be viewed in the following way. The integral term is included to remove any steady-state error, the time constant T_i then specifies how quickly it is attempted to achieve zero error — T_i large means a long time elapses before the error reduces to zero whereas T_i small means that the reduction to zero error occurs swiftly. The proportional gain K_p is then selected in order to achieve a stable closed-loop system with properties of relative stability, i.e. extremely stable or marginally stable. Although transient performance in the closed-loop can also be specified for many simple systems, as was shown in the example, see (10.3.9), this is not generally the case; however it must be remarked that the vast majority of process controllers, especially those for which transient conditioning is not important, are in practice P + I controllers.

It is fairly straightforward to implement integral action by means of an active network with reasonable accuracy, i.e. an operational amplifier integrating circuit; however, the

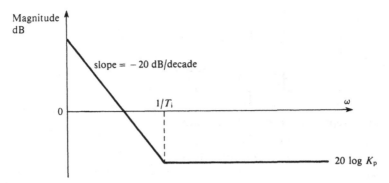

Fig. 10.5   Bode plot for a P + I controller

same cannot be said for the derivative action considered in the next section. It must be remembered, however, that such an integrating implementation has a limited operating range, dictated by the magnitude of the power supply, and as quite large magnitude inputs may be required in order to obtain relatively small output fluctuations, a certain amount of signal conditioning may also be required.

One final point worth taking into account as far as the P + I continuous-time controller is concerned is its Bode plot which, for the equation (10.3.8), is shown in Fig. 10.5.

The best way to describe such a response is perhaps as a high-frequency attenuator and this can be understood directly to mean that high-frequency modes of the closed-loop system are given a low weighting by this controller, whereas low-frequency modes are given a high weighting. This is consistent with the main aim of employing the integral action, which was to deal with the steady-state system response, i.e. the low-frequency end of the spectrum.

10.3.2 Discrete-time proportional + integral controller

The implementation of a discrete-time proportional + integral controller is by no means an exact procedure and it is perhaps best to take the proportional and integral elements separately, adding up the results. The proportional part is simply a repeat of that considered in Section 10.2.2, whereas the integral part is a lot more complicated in that in terms of its continuous-time realization the process of integration is one of finding the total area under the error, $e(t)$, curve from time $t = 0$ to $t = t_1$. The discrete-time representation of the same continuous-time error curve is, though, one in which the error value at a particular sampling instant is held until the next sampling instant, a predefined sampling period T later, when it is replaced by the new value. The effects of sampling a continuous-time signal are in fact shown in Fig. 8.4. The best we can hope to do therefore is to approximate the integral term by finding an approximation to the total area under the error curve from manipulations on the discrete error values that are at our disposal. There are quite a number of possible methods for carrying out this approximation, greater accuracy being obtained generally at the cost of greater complexity. Although the trapezoidal method is often used, the simpler technique of rectangular integration is perhaps the most popular and hence this is described directly.

The sampled error signal is shown in Fig. 10.6 and integration is carried out over the period $t = 0$ to $t = t_1 = n$, where n is equal to an integer multiple of the sampling period T.

If it is assumed that we are only interested in the integral from the initial time $t = 0$ onwards, then any error prior to that instant can be ignored and hence the initial summation of error values will start at zero. The integral is then

$$\phi(n) = \int_0^{t_1} e(t)\, dt = \int_0^n e(t)\, dt$$

and by approximating this with the error terms available, we have

$$\phi(n) \simeq T \sum_{kT=T}^n e(kT) \tag{10.3.10}$$

where k is an integer.

From (10.3.10) it follows directly that:

$$\phi(n) \simeq \phi(n-1) + Te(n) \tag{10.3.11}$$

Before proceeding it must be remarked that the method of approximation carried out in order to find $\phi(n)$ in (10.3.10) assumes that the error at time instant kT has in fact been the error over the period $(k-1)T$ to kT. This is in direct contrast to the sample and hold techniques described in Chapter 8, in which the value of continuous-time signal when sampled at time instant kT is held and therefore considered to be the true value until time $(k+1)T$. The two items should in no way be seen to be contradictory, the method just described for integration purposes merely being a way of approximating a particular signal over a given time, and, as has been stated earlier, many other approximations based on different philosophies are possible.

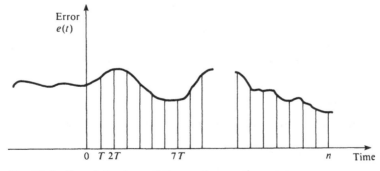

Fig. 10.6 Sampled values of the continuous-time error

Returning to the P + I controller, if the proportional element obtained in Section 10.2.2 is now combined with the integral part given by (10.3.11), this results in

$$u(n) = K_p e(n) + K_i[\phi(n-1) + Te(n)] \tag{10.3.12}$$

where $u(n)$ is the control input to be applied at time instant n and K_i is the integral gain. We have also that the previous control input was

$$u(n-1) = K_p e(n-1) + K_i[\phi(n-2) + Te(n-1)]$$

and by combining this latter equation with (10.3.11) and subtracting the result from (10.3.12) it follows that

$$u(n) - u(n-1) = K_p[e(n) - e(n-1)] + K_i Te(n)$$

which in terms of a general time instant of k sample periods, can also be written as

$$u(k)[1 - z^{-1}] = K_p\left\{\left[1 + \frac{T}{T_i}\right]e(k) - e(k-1)\right\} \tag{10.3.13}$$

where T_i is the integral time constant, and $z^{-1}u(k) = u(k-1) =$ control input at the previous sampling instant.

Consider now the example previously investigated for the discrete-time controller of Section 10.2.2, where the control input is given by (10.2.11) as

$$u(t) = D(z)e(t) \quad \text{or} \quad u(k) = D(z)e(k)$$

in which the P + I controller of (10.3.13) means that

$$D(z) = \frac{\bar{K}z - K_p}{z - 1}$$

where

$$\bar{K} = K_p\left[1 + \frac{T}{T_i}\right]$$

and if the system $G(z)$ of (10.2.12) is introduced, the final closed-loop expression is then

$$y(t) = \frac{0.5(\bar{K}z - K_p)(z - 0.6)}{(z - 1)(z - 0.2)(z + 0.3) + 0.5(\bar{K}z - K_p)(z - 0.6)} v(t)$$

Steady-state conditions can simply be investigated by setting z to unity in the above transfer function; however, it is only really necessary to consider the $(z - 1)$ term in the denominator, which will become zero. This means that in the steady-state

$$\lim_{t \to \infty} y(t) = \lim_{t \to \infty} v(t)$$

and thus steady-state following is achieved directly, as was hoped for with the inclusion of an integrating term.

The denominator has however become of third order and hence a simple analysis in terms of natural frequency and damping is not possible. It is sufficient to note that as the system is third order, it will have three roots, and we have only two controller parameters, $\bar{K}$ and K_p, which can be selected in order to place those roots in arbitrary positions in order to achieve a desired transient response. The conclusion then is that by the introduction of integral feedback, to link with the proportional feedback, it was found to be straightforward to achieve steady-state following of the desired output value without placing restrictions on the type of system response. However, in order to obtain greater flexibility in the type of transient response obtainable, and this includes closed-loop system stability, a further term is required in the controller and this is the subject of the next section.

10.4   PID control

So far in this chapter the discussion has concentrated primarily on the steady-state following capabilities of control systems, and hence an integral type of controller has been of importance. It may be the case however that the system to be controlled has itself an inherent integral action and thus, for a straightforward step reference input, the actual output signal level will automatically follow in the steady-state. This property is shown for a simple proportional controller on the position servomechanism of Fig. 10.7.

The open-loop transfer function for the servomechanism is

$$Y(s) = \frac{K}{s(1 + s\tau_1)} \, U(s)$$

which means that this is a type 1 system, because of the common s^1 factor in the denominator.

The closed-loop transfer function of Fig. 10.7 is however

$$Y(s) = \frac{KK_p}{s(1 + s\tau_1) + KK_p} \, V(s) \qquad (10.4.1)$$

and it can be seen directly that if $V(s)$ is a 2-unit step input, then for steady-state conditions

$$\lim_{t \to \infty} y(t) = \lim_{s \to 0} sY(s) = s \, \frac{KK_p}{s(1 + s\tau_1) + KK_p} \, \frac{2}{s} \, \bigg|_{s \to 0}$$

so

$$\lim_{t \to \infty} y(t) = 2 = \lim_{t \to \infty} v(t)$$

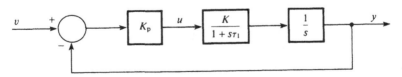

Fig. 10.7   Position servomechanism with proportional control

Hence steady-state following of the desired output value is achieved for a step input on a type 1 system, i.e. a system which has its own in-built integrator, without the need for an integrating type of controller.

It is worth pointing out here though, that a further important feature arises from the closed-loop equation (10.4.1) in that although by a correct choice of K_p it can be ensured that the closed-loop system is stable, it is not possible with only the one controller parameter, K_p, to define the type of transient response which we might like. It is in fact the case that with one further controller parameter, incorporated in the mode described in the next section, the transient response of the system (10.4.1) can be characterized exactly as desired.

10.4.1 Proportional + derivative control

There are instances in many subjects when different terminologies are used to describe things which are exactly the same, and this only serves to confuse someone learning about a topic. Such is the case with the terms derivative control and velocity feedback, because as far as control systems analysis of position control systems is concerned, they mean exactly the same thing. Having said that, in the previous section an attempt was made to employ integral control alone without incorporating any proportional element. If the same attempt is made with derivative action alone then the result is very much dependent on the system type. With solely derivative action applied to the type 1 system of Fig. 10.7, in place of the proportional term, the effect is merely a cancellation of the inherent integrating action of the system, thus producing problems of steady-state offset. If the system to which derivative action is applied is type 0 however, i.e. no common denominator s factor, then no cancellation of s-terms will occur and hence an s factor due to the derivative action will appear in the closed-loop numerator. The output position will in this case be dependent not on the actual desired position value but rather on its rate of change – not generally a good feature. To summarize then, use of derivative feedback alone is not generally a sensible arrangement and hence in the first instance it will be considered that the derivative term is coupled with a proportional element, as shown in Fig. 10.8, with K_d as the derivative gain.

The closed-loop transfer function of this system with proportional and derivative (P + D) feedback is then

$$Y(s) = \frac{K(K_p + K_d s)}{s(1 + s\tau_1) + K(K_p + K_d s)} \, V(s) \tag{10.4.2}$$

or

$$Y(s) = \frac{KK_p + KK_d s}{s^2 \tau_1 + s(1 + KK_d) + KK_p} \, V(s)$$

The effect of the derivative term is two-fold, firstly it has introduced a zero into the closed-loop system at a point $s = -K_p/K_d$, i.e. a point selected by the two controller

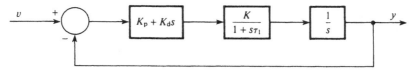

Fig. 10.8 Position servomechanism with proportional + derivative control

parameters. Secondly, and this is really of much greater importance, it has given an extra degree of freedom as far as choice of the closed-loop denominator coefficients is concerned, and hence allows for selection of the closed-loop transient response as desired. It must be noticed also that inclusion of the derivative action, along with the proportional term, has not altered the steady-state following characteristics of the closed-loop system to a step input, i.e. the integrating property of a type 1 system has not been open-loop cancelled.

As an example of how the design of a P + D controller could be very simply carried out, consider that we have a position servomechanism in which $K = 5$ and $\tau_1 = 0.5$. The design requirements are then to select K_p and K_d such that the closed-loop system has a response with natural frequency $\omega_n = 1.0$ and damping factor $\zeta = 0.7$.

Now the closed-loop characteristic equation can be written as:

$$s^2 + s \frac{(1 + KK_d)}{\tau_1} + \frac{KK_p}{\tau_1} = 0$$

where

$$\omega_n = \sqrt{\frac{KK_p}{\tau_1}} \quad \text{and} \quad \zeta = \frac{(1 + KK_d)}{2\sqrt{KK_p\tau_1}}$$

So it is a straightforward process to select K_p in order to achieve the required ω_n, and then to select K_d in order to satisfy the damping factor requirements. In this example:

$$\omega_n = 1.0 = \sqrt{10} \cdot \sqrt{K_p}, \quad \text{so} \quad K_p = 0.1$$

and $\zeta = 0.7 = (1 + 5K_d)$, so

$$K_d = -0.06$$

Hence the controller becomes

$$K_p + K_d s = 0.1[1 - 0.6s]$$

Previously with the P + I controller it was found advantageous to consider the controller's Bode plot, as this gave a further insight into performance – see Fig. 10.5. If this is done with the P + D controller, as is shown in Fig. 10.9, then it can be seen that for this type of controller, while low-frequency components are attenuated, high-frequency components are considered to be of importance and are given greater emphasis by the controller.

The P + D controller, which is usually written in the form

$$U(s) = K_p[1 + T_d s]E(s) \tag{10.4.3}$$

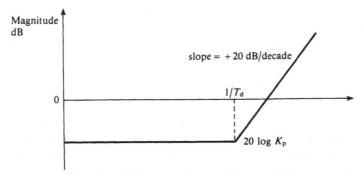

Fig. 10.9   Bode plot for P + D controller

where the derivative time constant, $T_d = K_d/K_p$, then ensures that the higher the frequency of a signal, the more it is amplified. Two notes must be made from this: firstly the gain of the P + D controller tends to infinity as the frequency tends to a high value, and realistically it is not possible to achieve this with standard active devices; secondly the system input/output signals will generally be of relatively low frequency, whereas noise signals present will almost certainly have fairly substantial high-frequency components. The inclusion of a derivative term in the controller therefore will serve to highlight any noise present while at the same time attenuating the actual system signals. This in fact is the converse of the P + I controller which, because of its low-pass filtering action, tends to remove the effects of noise.

It has been shown that when P + D control is applied to a system it is possible to characterize, exactly as desired, the closed-loop transient response of that system. However, unless the system is of a certain type, steady-state position following properties may not be present in the closed loop. For this reason, the general form of Proportional + Integral + Derivative (P + I + D) control will now be considered.

10.4.2 Proportional + integral + derivative control

In this chapter so far the benefits of the individual proportional, integral and derivative control elements have been shown and major drawbacks of certain combinations have been pointed out. The derivative term is based on a multiple of the rate of change of error and therefore can have little effect on steady-state values, whereas the integral term is based on the value of error over a relatively long period of time and therefore can have little effect on transient values. Each element has a purpose for which it is particularly useful, and these cover the basic requirements for almost all linear control systems, no matter how complicated the system might be. A block schematic of the P + I + D controller is shown in Fig. 10.10, and in this scheme the benefits of all three types of control action can be obtained, without the drawbacks exhibited by the individual terms.

By combining the derivative term of the P + D controller (10.4.3) with the P + I

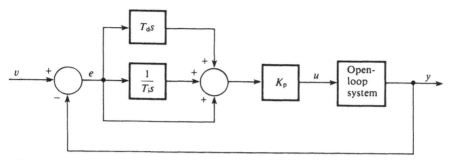

Fig. 10.10   P + I + D controller

controller (10.3.8) it can be seen that, as given in Fig. 10.10, the overall three-term P + I + D controller has the error to control input equation

$$U(s) = K_p\left[1 + \frac{1}{T_i s} + T_d s\right] E(s) \tag{10.4.4}$$

or

$$U(s) = \left[K_p + \frac{K_i}{s} + K_d s\right] E(s) \tag{10.4.5}$$

where in the latter form of the three-term equation, (10.4.5), the three separate coefficient gains can be seen.

It can be noted that the three-term controller equation is to be found in various forms, for example in terms of its time-domain representation

$$u(t) = K_p\left[e(t) + \frac{1}{T_i}\int_0^t e(t)\,\mathrm{d}t + T_d\frac{\mathrm{d}e(t)}{\mathrm{d}t}\right] \tag{10.4.6}$$

or with a common 'integrating' denominator s factor

$$U(s) = K_p\left[\frac{s^2 + s/T_d + 1/T_i T_d}{s}\right] E(s) \tag{10.4.7}$$

The overall controller now has three coefficients, K_p, T_i and T_d, which must be selected by the operator to meet some desired criterion or objective. An example could be given to show how for one specific plant, a set of coefficient values can be obtained, indeed this was the approach taken for both the P + I and P + D controllers previously described. Here however, it is wished to make a further point in that it is often the case that an exact representation of the plant to be controlled is not available. This might be due to one of a number of reasons, for example actually obtaining a good plant description could prove to be too expensive or the plant may well have deteriorated due to ageing. Whatever the reason, carrying out an elaborate exact controller calculation by using inaccurate plant parameters seems rather rash, as the accuracy of the controller can be no better than the accuracy of our plant model. In fact many approximate methods for the selection of proportional, integral and derivative coefficients are to be found, two

Table 10.1 Ziegler–Nichols three-term
controller – coefficient values

	Proportional controller	P + I controller	P + I + D controller
K_p	$0.5K_c$	$0.45K_c$	αK_c $(0.6 \leqslant \alpha \leqslant 1)$
T_i	—	$0.83T_c$	$0.5T_c$
T_d	—	—	$0.125T_c$

essential properties being that they are both relatively simple and fairly inexpensive to implement.

Perhaps the most widely employed method for the setting of P, P + I or P + I + D controllers, certainly for process control, is that detailed by Ziegler and Nichols. The method is based largely on, and has been proved to work effectively by, much experimental evidence, hence very little background theoretical detail is available, and none is given here. In this section the procedure used in order to implement the Ziegler–Nichols method is described, and the information necessary to calculate each of the controller coefficients is tabulated.

The system is set up in the form of a simple cascade compensator, Fig. 10.1, with the controller consisting simply of a proportional gain term, as in equation (10.2.1). The Ziegler–Nichols procedure then requires that the gain K_p be increased until the closed-loop system just oscillates, i.e. the system output will just begin to oscillate. The critical value of gain K_p at which oscillations just start/stop is then denoted as K_c. Also the frequency of the oscillations which occur is measured as ω_c with a period $T_c = 2\pi/\omega_c$. In terms of the values K_c and T_c, the 'best' values for proportional-integral-derivative terms are shown in Table 10.1, it must be noted however that these are not necessarily the best of all possible coefficient values.

Unfortunately the method described for controller coefficient calculation has one or two drawbacks when applied to some plants. In practice it can be extremely difficult to measure exactly the point at which the onset of oscillations occurs; this is particularly true when there is an amount of noise affecting the system and/or measurements. A further consideration which must be made is that for certain plants it can be extremely dangerous to cause the output to oscillate, in that this may well result in physical plant problems, such as mechanical breakdown.

Rather than use a closed-loop technique for Ziegler–Nichols tuning, where necessary it is possible to employ an open-loop alternative, which requires solely that a plant step response (also called a reaction curve) is obtained, see Fig. 10.11. Note that no controller parameters are varied or modified in order to achieve the step response, it is merely found as the system output in response to a step input of sensible magnitude.

The final steady-state value of the step response is of no importance as far as the controller parameters are concerned. What is of importance is (a) the lag time and (b) the

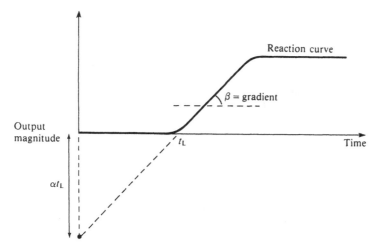

Fig. 10.11 System response to a step input – showing lag time, t_L, and gradient, β

gradient, also called the reaction rate, the latter of these being dependent on the magnitude of the step input signal.

The two values K_c and T_c required in order to calculate the controller parameters can be found by the Ziegler–Nichols formulae, from the measured β and t_L values, shown in Fig. 10.11, as

$$K_c = \frac{2\gamma}{\beta t_L}, \qquad T_c = 4t_L \qquad\qquad (10.4.8)$$

where γ is the step input magnitude.

10.4.3 Discrete-time P + I + D controller

The discrete-time version of a P + I controller was considered previously in Section 10.3.2, where it was shown how approximations were necessarily made to the integration procedure in order to incorporate such a technique in a discrete-time format. In this section it therefore suffices to consider merely the discretization of the derivative term, and to subsequently append this to the P + I controller.

An approximation is required for the derivative term, although this is rather more straightforward than the integral approximation – especially so if the idea of finding a differential or difference is associated directly with the derivative.

By referring to Fig. 10.12, the derivative of the error at time instant $kT = n$, where k is an integer number of sample periods T, is simply the gradient between the nearest measurement $e(kT)$, and the previous measurements, $e(kT - T)$, i.e. the derivative is

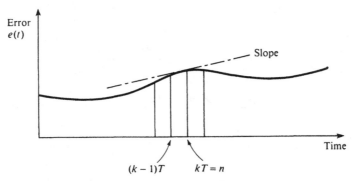

Fig. 10.12 Backward differencing operation

taken to be the difference between the two values divided by the time, T, between them:

$$\left.\frac{de(t)}{dt}\right|_{t=n} \simeq \frac{e(kT) - e(kT - T)}{T} \qquad (10.4.9)$$

or alternatively

$$\frac{de(n)}{dt} = \frac{e(n) - e(n - 1)}{T} \qquad (10.4.10)$$

In order to obtain the complete three-term controller, it remains only to append the derivative term when multiplied by the derivative gain K_d, to the control input equation at time instant n, found previously for the P + I controller as equation (10.3.12). The control input therefore becomes

$$u(n) = K_p e(n) + K_i[\phi(n - 1) + Te(n)] + K_d \frac{[e(n) - e(n - 1)]}{T} \qquad (10.4.11)$$

where $\phi(n - 1)$ is the integral summation of errors at time instant $(n - 1)$, defined in (10.3.10).

We have also that

$$u(n - 1) = K_p e(n - 1) + K_i[\phi(n - 2) + Te(n - 1)] + \frac{K_d[e(n - 1) - e(n - 2)]}{T}$$

and on subtracting this equation for $u(n - 1)$ from (10.4.11) it follows that

$$u(n) - u(n - 1) = e(n)\left\{K_p + K_i T + \frac{K_d}{T}\right\} - e(n - 1)\left\{K_p + 2\frac{K_d}{T}\right\}$$

$$+ e(n - 2)\frac{K_d}{T}$$

remembering that (10.3.11) holds, i.e. $\phi(n) \simeq \Phi(n - 1) + Te(n)$.

By rearrangement of this difference equation, the general P + I + D equation at time instant t, for discrete-time control, can be written as

$$u(k)[1 - z^{-1}] = K_p\left[e(k)\left\{1 + \frac{T}{T_i} + \frac{T_d}{T}\right\} - e(k - 1)\left\{1 + 2\frac{T_d}{T}\right\}\right.$$

$$\left. + e(k - 2)\left\{\frac{T_d}{T}\right\}\right] \qquad (10.4.12)$$

in which the integral time constant $T_i = K_p/K_i$ and the derivative time constant $T_d = K_d/K_p$. Remembering that $z^{-1}u(k) = u(k - 1) =$ the value of control input one sampling instant prior to time t, similarly $z^{-2}u(k) = u(k - 2)$, etc.

The implementation of the discrete-time P + I + D equation of (10.4.12) is fairly simple to carry out in that it is only necessary to store an error signal value for two sample periods before discarding. Also the terms multiplying the error values would usually remain fixed once an initial selection has been made, and thus the total computation per iteration only amounts to three multiplications and three additions/subtractions as can be seen in Fig. 10.13.

The multiplying factors K_1, K_2 and K_3 in Fig. 10.13 are given as:

$$K_1 = K_p\left\{1 + \frac{T}{T_i} + \frac{T_d}{T}\right\}$$

$$K_2 = K_p\left\{1 + 2\frac{T_d}{T}\right\} \qquad (10.4.13)$$

$$K_3 = K_p\frac{T_d}{T} = \frac{K_d}{T}$$

which means that the three-term controller equation can be written in the form,

$$u(k) = u(k - 1) + K_1e(k) - K_2e(k - 1) + K_3e(k - 2)$$

Although fairly straightforward to develop, it must be remarked that the three-term controller discussed here does exhibit one or two drawbacks. Firstly the problem of noise is highlighted by the fact that the derivative effect acts directly on the error term which

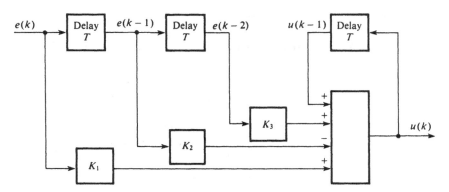

Fig. 10.13 Schematic of the discrete-time three-term controller ($t = kT$)

consists of the difference between a noise-free reference signal and a noise-corrupted measured value. A sensible way round this problem would be to filter the measured value before feeding the signal back, thereby attenuating the high-frequency noise components which would be accentuated directly by the derivative action. A second problem is the production of a large error derivative value every time a change in desired output, set point, value occurs. This temporary, but potentially dangerous, action is known as 'derivative kick' and although it is common to both continuous and discrete-time controllers, it is particularly problematic in the latter. The 'derivative kick' effect can be reduced by filtering the set-point input or by employing an alternative derivative approximation scheme, however in many cases the derivative action is applied solely to the fed-back measured value rather than the error and indeed three-term controllers can also be found where the same is true of the proportional action.

10.4.4 Worked example

An example is given for the design of a continuous-time three-term controller and is based on a problem given in Kuo for the control of the air/fuel ratio in an engine. As shown in Fig. 10.14, the three-term controller acts on the error between the desired and actual air/fuel ratios and is required to achieve a predefined response pattern.

The engine transfer function is given by

$$G(s) = \frac{10e^{-0.2s}}{s + 10}$$

where 0.2 seconds is a pure transport delay.

However, by taking an approximation to the transport delay, by means of a truncated power series expansion, it will be assumed that

$$G(s) = \frac{50}{(s + 5)(s + 10)} \tag{10.4.14}$$

It is required that a three-term controller of the form shown in (10.4.5), i.e.

$$D(s) = K_p + \frac{K_i}{s} + K_d s \tag{10.4.15}$$

be employed, in order to place the roots of the closed-loop system denominator at $s = -100$, $-10 + j10$, $-10 - j10$.

By cascading the controller of (10.4.15) with the engine equation (10.4.14) and applying a unity feedback loop, as in Fig. 10.14, the closed-loop equation

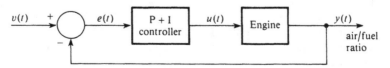

Fig. 10.14 Engine air/fuel ratio controller

relating desired to actual air/fuel ratio is found to be

$$Y(s) = \frac{50(K_d s^2 + K_p s + K_i)}{s(s+5)(s+10) + 50(K_d s^2 + K_p s + K_i)} \ V(s) \qquad (10.4.16)$$

But it is only the denominator of this equation which specifically concerns us for design purposes, and thus the three controller parameters must be found in order to satisfy the equation:

$$s(s^2 + 15s + 50) + 50(K_d s^2 + K_p s + K_i) = (s + 100)(s + 20s + 200)$$

or

$$s^3 + (15 + 50K_d)s^2 + 50(1 + K_p)s + 50K_i = s^3 + 120s^2 + 2,200s + 20,000$$

Therefore our design requirements are satisfied if, for this example, the three controller coefficients are:

$$K_i = 400, \qquad K_p = 43 \qquad \text{and} \qquad K_d = 2.1$$

It is useful to note that the derivative coefficient is much smaller in magnitude; this is often found to be the case in practice, particularly in process control systems, resulting in P + I controllers being sufficient in many cases. It is fairly easy to check, by employing the final value theorem, that if a step input is applied to the open-loop system (10.4.14), a steady-state error will result, and yet if the same step input is applied to the closed-loop system of (10.4.16) no steady-state error will appear; this is a direct result of the use of an integral action controller.

10.5 Summary

Despite the availability of many different control design techniques, by far the most widely encountered procedure is that of three-term or PID control. A complete chapter has been designated here specifically for the introduction and description of both continuous-time and discrete-time three-term controllers, and simple examples have been given to show how a typical design can be carried out. The build-up has been such in this chapter as to attempt to get away with as simple a controller as possible, hence initially an attempt was made to use solely a proportional controller. For many systems this is complex enough. However, it can be the case that a steady-state error exists between the actual and desired output values, and the inclusion of an integrator solves this problem. Finally, in order to achieve greater flexibility in the transient design procedure, a derivative term is also introduced.

Various problems such as 'derivative kick' and noise were mentioned as far as PID controller implementation is concerned. These conditions are both caused by the derivative action, however other problems can occur due to the other terms, one example being integral 'wind-up' which is apparent when an offset exists persistently between the actual and desired output values. While an offset is present the integral part of the controller output will increase and therefore lengthen considerably the time taken for the closed-loop system to settle down.

Finally, it must be remembered that the extensive use of a PID structure is due only partly to its simplicity, when compared with alternative controller frameworks, and much of the reason behind its employment must lie in the fact that satisfactory performance is invariably achieved along with reliability and robustness.

Problems

10.1   Consider the discrete-time proportional controller discussed in Section 10.2.2, where $u(t) = K_p e(t)$.

For the system:

$$G(z) = \frac{0.5(z - 0.6)}{(z - 0.2)(z + 0.3)}$$

with $D(z)$ and $G(z)$ cascaded within a unity feedback loop, find the resultant closed-loop denominator.

Hence, show how the choice of K_p will affect closed-loop stability, and how a choice of K_p very large will result in lightly damped output oscillations.

10.2   Consider a continuous-time P + I + D controller cascaded with the system

$$G(s) = \frac{5}{s(s + 3)}$$

within a unity feedback loop.

It is required that the closed-loop poles be placed at $s = -7$ and $s = -3 \pm j1$. Find K_p, K_i and K_d such that this requirement is satisfied.

10.3   Proportional + derivative action can be applied either to the error between the desired and actual output signals or directly to the actual output signal alone, as shown in the diagrams (Fig. 10.15).

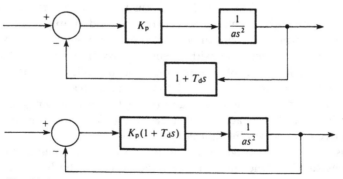

Fig. 10.15

Consider the effect of the different controller connections, in terms of steady-state errors, firstly when a unit step input is applied, and secondly when a unit ramp input is applied.

10.4 Consider a $P + I$ controller $D(s) = K_p(1 + 1/s)$ which is cascaded with the open-loop system:

$$G(s) = \frac{25}{(s + 2)(s + 5)}$$

within a unity feedback closed-loop system. Calculate the value K_p which will ensure a steady-state velocity error of 10%.

10.5 A three-term controller is of the form:

$$D(s) = K_p\left[1 + \frac{1}{sT_i} + T_d s\right]$$

Show that if $T_i = 4T_d$ the frequency response of the controller has the form

$$D(j\omega) = \frac{K_p[1 + j\omega T]^2}{j2\omega T}$$

where

$$T = 2T_d$$

10.6 For the closed-loop system shown in Fig. 10.16, find the time solution of the transient ramp error for the case in which $K_p = 14$ and $K_i = 8$.

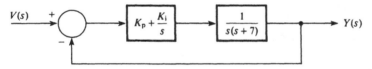

Fig. 10.16

10.7 For the closed-loop system shown in Fig. 10.17, find values for K_p and T_d to provide ideal transient performance and a 1% ramp error.

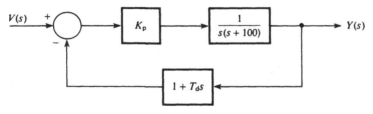

Fig. 10.17

10.8   Consider the use of a three-term controller $D(s)$ in the system shown in Fig. 10.18. If it is required to track disturbance steps with zero steady-state error and also to exhibit overall closed-loop system stability, show that all three controller coefficients are required. Find the range of values for the proportional gain K_p, such that the closed-loop system is stable.
Note: $D(s)$ is identical to that in Problem 10.5.

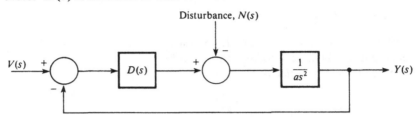

Fig. 10.18

10.9   A P + I controller is employed in cascade with the open-loop system which has a transfer function:

$$G(s) = \frac{3}{s^2 + 4s + 8}$$

Find the integral gain K_i necessary to ensure that the steady-state ramp error for the closed-loop unity feedback system is 5% at most, also find the range of values for the proportional gain K_p over which the closed-loop is stable.

10.10   A continuous-time P + I + D controller is cascaded with the system

$$G(s) = \frac{10}{(s + 1)(s + 2)}$$

within a unity feedback loop.
Find K_p, K_i and K_d such that the closed-loop poles are placed at $s = -50$ and $s = -4 \pm j5$.

10.11   A discrete-time P + I + D controller, of the form

$$[1 - z^{-1}]u(k) = [K_1 - K_2 z^{-1} + K_3 z^{-2}]e(k)$$

is cascaded with the open-loop input $u(t)$ to output $y(t)$ transfer function

$$G(z) = \frac{3z^{-1}}{(1 - 2z^{-1})(1 - 0.5z^{-1})}$$

with $e(t) = v(t) - y(t)$, and the cascaded functions are connected within a unity feedback loop. $v(t)$ is a reference input. Find K_p, K_i and K_d such that the closed-loop poles are given by $z = -0.5$, $+0.5$ and -0.9; see (10.4.13). Is there any steady-state offset in the output signal following a unit step input?

10.12   Consider the open-loop system described by the discrete-time equation:

$$G(z) = \frac{10z^{-1}}{(1 - 3z^{-1})(1 - 0.1z^{-1})}$$

By using the general P + I + D control equation (10.4.12), show that if $K_i = 0$, i.e. only a P + D control is employed, then $K_2 = K_1 + K_3$. What problems does the employment of a P + D controller therefore raise if an attempt is made to place the closed-loop poles of a unity feedback system by the selection of K_1, K_2 and K_3? If a P + D controller is cascaded with $G(z)$, can the closed-loop poles be placed at $z = -0.5$, $+0.5$ and -0.9?

10.13 Show that if it is desired to place the closed-loop poles of a system in specified locations, this can only be accomplished by a digital P + I + D controller when cascaded with a system $G(z)$, within a unity feedback loop, if the system is no more complicated than:

$$G(z) = \frac{b_1 z^{-1}}{1 + a_1 z^{-1} + a_2 z^{-2}}$$

i.e. any further numerator or denominator terms will mean that closed-loop pole positions cannot be arbitrarily specified.

10.14 Consider the system shown in Fig. 10.19 with $D(s) = K_p + K_i/s$, i.e. a P + I controller is applied. Also, the disturbance $N(s)$ can in this case be regarded as a ramp of slope $= 0.5$.

If it is required that the steady-state value of $Y(s)$ is equal to 100 with 2% error, find values for the step input magnitude and the proportional and integral gains when a closed-loop damping ratio $= 0.7071$ is desired.

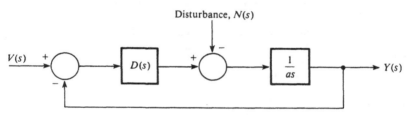

Fig. 10.19

10.15 Consider the system shown in Fig. 10.20 with $a = 2$ and $D(s) = K_p + K_i/s$.

If $V(s)$ is a ramp input of slope $= 0.01$, find the steady-state ramp error between desired $V(s)$ and actual $Y(s)$ outputs when:

(a) $K_p = 20$, $K_i = 0$;
(b) $K_p = 10$, $K_i = 0.1$;
(c) $K_p = 1$, $K_i = 1$.

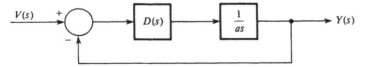

Fig. 10.20

Further reading

Astrom, K. J. and Hagglund, T., 'Automatic tuning of simple reactors with specifications on phase and amplitude margins', *Automatica*, 20, No. 5, 1984, pp. 645–51.

Doebelin, E. O., *Control system principles and design*, John Wiley and Sons, 1985.

Fensome, D. A., 'Understanding 3-term controllers', *Electronics and Power*, **647**, 1983.

Fertik, H. A., 'Tuning controllers for noisy processes', *ISA Trans.*, **14**, No. 4, 1975, pp. 292–304.

Franklin, G. *Digital control systems*, 2nd ed., Addison-Wesley, 1991.

Gawthrop, P. J., 'Self-tuning PID controllers: algorithms and implementation', *IEEE Trans. on Automatic Control*, **AC-31**, No. 3, 1986, pp. 201–9.

Gawthrop, P. J. and Nomikos, P., 'Automatic tuning of commercial PID controllers', *IEEE Control Systems Magazine*, pp. 34–41, 1990.

Kuo, B. C., *Automatic control systems*, Prentice-Hall, 1982.

Lopez, A. M., Smith C. L. and Murrill, P. W., 'Tuning PI and PID digital controllers', *Instruments and control systems*, **42**, No. 2, 1969, p. 89.

McAlpine, S. M., 'Applications of self-tuning control in the chemical industry', *Proc. IEE Conference 'Control 85'* 1985, pp. 382–6.

McMillan, G. K., *Tuning and control loop performance*, ISA, Research Triangle Park, NC, USA, 1983.

Nishikawa, Y. and Sannomiya, N., 'A method for auto-tuning of PID control parameters', *Proc. IFAC Congress* (1981), Kyoto, Japan.

Ortega, R. and Kelly, R., 'PID self-tuners: some theoretical and practical aspects', *IEEE Trans. on Industrial Electronics*, **IE-31**, No. 4, 1984, pp. 332–8.

Pemberton, T. J., 'PID: the logical control algorithm', *Control Engineering*, 1972, pp. 64–7 (May) and pp. 64–7 (July).

Pessen, D. W., 'Optimum three-mode controller settings', *ASME Paper* 52-A-58, 1952.

Porter, B., Jones, A. H. and McKeown, C. B., 'Real-time expert tuners for PI controllers', *Proc. IEE*, **134**, Pt.D, 1987, pp. 260–3.

Roberts, P. D., 'Simple tuning of discrete PI and PID controllers', *Trans. Inst. MC*, **9**, 1976, pp. 227–34.

Shinskey, F. G., *Process control systems*, McGraw-Hill, 1979.

Smith, C. L., Corripio, A. B. and Martin Jr., J., 'Controller tuning from simple process models', *Instr. Tech.*, 1975, pp. 39–44.

Thomas, H. W., Sandoz, D. J. and Thompson, M., 'New desaturation strategy for digital PID controllers', *Proc IEE*, **130**, Pt.D, No. 4, 1983, pp. 184–92.

Warwick, K. and Rees, D. (eds.), *Industrial digital control systems*, Peter Peregrinus Ltd, 1986.

Warwick, K., 'PID methods in self-tuning control', *Proc. 4th IEE Workshop on Self-Tuning and Adaptive Control*, Digest No. 1987/33, 1987, pp. 311–314.

Zhuang, M. and Atherton, D. P., 'PID controller design for a TITO system', *Proc. IEE*, Part D, **141**, 1994, 111–120.

Ziegler, J. G. and Nichols, N. B., 'Optimum settings for automatic controllers', *Trans. ASME*, **4**, 1942, pp. 759–68.

Appendices

Appendix 1   Matrix algebra

In this appendix, some of the basic rules of matrix algebra are covered briefly, as an aid to problem solving.

A1.1   Basic notations

Consider the three simultaneous equations

$$a_{11}x_1 + a_{12}x_2 + a_{13}x_3 = y_1$$
$$a_{21}x_1 + a_{22}x_2 + a_{23}x_3 = y_2 \qquad\qquad\qquad (A1.1)$$
$$a_{31}x_1 + a_{32}x_2 + a_{33}x_3 = y_3$$

for which a solution is required for the three unknowns x_1, x_2, x_3.

These equations can also be written in an alternative, matrix, form as:

$$Ax = y \qquad\qquad\qquad (A1.2)$$

where

$$A = \begin{bmatrix} a_{11} & a_{12} & a_{13} \\ a_{21} & a_{22} & a_{23} \\ a_{31} & a_{32} & a_{33} \end{bmatrix}, \qquad x = \begin{bmatrix} x_1 \\ x_2 \\ x_3 \end{bmatrix} \qquad \text{and} \qquad y = \begin{bmatrix} y_1 \\ y_2 \\ y_3 \end{bmatrix}$$

The array A is called a *matrix* and in more general terms, e.g. by considering m simultaneous equations with n unknowns, it can be written as:

$$A = \begin{bmatrix} a_{11} & a_{12} & \dots & a_{1n} \\ a_{21} & a_{22} & \dots & a_{2n} \\ \vdots & \vdots & & \vdots \\ a_{m1} & a_{m2} & \dots & a_{mn} \end{bmatrix} \qquad\qquad\qquad (A1.3)$$

in which the coefficients a_{ij} are termed the matrix *elements*, with the indices i and j denoting the row and column respectively, in which that element lies. A, as shown in (A1.3), is called an $m \times n$ matrix, such that when $m = n$, as was the case in (A1.2), it is called a *square* matrix.

When the index $n = 1$, the $m \times 1$ matrix is called a *column vector* (or column matrix); examples being both x and y in (A1.2), whereas when $m = 1$, the $1 \times n$ matrix is called a *row vector* (or row matrix) as shown in (A1.4):

$$w = [w_1, w_2, ..., w_n] \tag{A1.4}$$

A special case of a square matrix exists when $a_{ij} = a_{ji}$, in which case the matrix is called a *symmetrical* matrix, an example being given in (A1.5).

$$A = \begin{bmatrix} 3 & 2 & 4 & 8 \\ 2 & 6 & -5 & 17 \\ 4 & -5 & 0 & -1 \\ 8 & 17 & -1 & 14 \end{bmatrix} \tag{A1.5}$$

This is taken a stage further if the matrix coefficients $a_{ij} = a_{ji} = 0$ for all $i \neq j$, leaving only those elements of A on the principal diagonal. Such a matrix is called a *diagonal* matrix, as shown in (A1.6).

$$A = \begin{bmatrix} a_{11} & 0 & & & 0 \\ 0 & a_{22} & \cdot & & 0 \\ \cdot & & 0 & \diagdown & \cdot \\ \cdot & & & \diagdown & \cdot \\ \cdot & & & & 0 \\ 0 & 0 & ... & 0 & a_{nn} \end{bmatrix} \tag{A1.6}$$

When all the principal diagonal elements of a diagonal matrix are equal to unity, i.e. $a_{ii} = 1$, the matrix is termed an *identity* matrix, and is usually denoted by the letter I.

$$I = \begin{bmatrix} 1 & 0 & \cdot & \cdot & \cdot & 0 \\ 0 & 1 & 0 & & & 0 \\ \cdot & & 0 & \diagdown & & \cdot \\ \cdot & & & \diagdown & & \cdot \\ \cdot & & & & 0 \\ 0 & 0 & \cdot & \cdot & 0 & 1 \end{bmatrix} \tag{A1.7}$$

Finally, when all the elements of a matrix are zero, i.e. $a_{ij} = 0$, for all i and j, then the matrix is called a *null* matrix, and is denoted by the symbol 0.

A1.2   Matrix operations

The *transpose* of a matrix A, denoted by A^T, is found by interchanging the rows and columns of the matrix, i.e. $a_{ij} \rightarrow a_{ji}$. As an example consider the matrix

$$A = \begin{bmatrix} 3 & -2 & 8 \\ 1 & 4 & 3 \\ 6 & 0 & -5 \end{bmatrix}$$

then

$$A^T = \begin{bmatrix} 3 & 1 & 6 \\ -2 & 4 & 0 \\ 8 & 3 & -5 \end{bmatrix}$$

Also, the transpose of a row vector is a column vector and the transpose of a column vector is a row vector, e.g. if the vector

$$x = \begin{bmatrix} x_1 \\ x_2 \\ x_3 \end{bmatrix}$$

then $x^T = [x_1 x_2 x_3]$.

Note: Taking the transpose of a transpose produces the original matrix, or vector, as a result, i.e.

$$(A^T)^T = A \tag{A1.8}$$

The *trace* of a square matrix is the sum of the elements on the principal diagonal, i.e. for the $n \times n$ matrix

$$\text{trace } A = a_{11} + a_{22} + \cdots + a_{nn} \tag{A1.9}$$

For the matrix

$$A = \begin{bmatrix} 3 & -2 & 8 \\ 1 & 4 & 3 \\ 6 & 0 & -5 \end{bmatrix}$$

$$\text{trace } A = 3 + 4 - 5 = 2$$

The *determinant* of a matrix is a scalar and is only applicable to square matrices. The determinant of matrix A is written as $|A|$. When the determinant of a matrix is equal to zero, that matrix is said to be *singular*, whereas when the determinant of a matrix is not equal to zero, that matrix is said to be *nonsingular*.

For an $n \times n$ matrix, its determinant is found by means of cofactors and minors. The cofactor of an element a_{ij}, in the matrix A, is denoted by A_{ij}, where

$$A_{ij} = (-1)^{i+j} M_{ij} \tag{A1.10}$$

M_{ij} is the minor of a_{ij} and is the determinant of the square array which remains when the ith row and jth column of the original array are removed, e.g. if

$$|A| = \begin{bmatrix} a_{11} & a_{12} & a_{13} \\ a_{21} & a_{22} & a_{23} \\ a_{31} & a_{32} & a_{33} \end{bmatrix}$$

then

$$A_{32} (-1)^5 M_{32} = - \begin{vmatrix} a_{11} & a_{13} \\ a_{21} & a_{23} \end{vmatrix}$$

The determinant of a general 2×2 matrix is

$$\begin{vmatrix} a_{11} & a_{12} \\ a_{21} & a_{22} \end{vmatrix} = a_{11}a_{22} - a_{12}a_{21} \qquad \text{(A1.11)}$$

such that the determinant of an $n \times n$ matrix A is given by either

$$|A| = \sum_{i=1}^{n} a_{ij}A_{ij}, \text{ for any one selected column } j$$

or

$$|A| = \sum_{j=1}^{n} a_{ij}A_{ij}, \text{ for any one selected row } i$$

Example A1.1

The determinant of matrix

$$A = \begin{bmatrix} 3 & -2 & 8 \\ 1 & 4 & 3 \\ 6 & 0 & 5 \end{bmatrix}$$

can be found from $|A| = a_{11}A_{11} + a_{12}A_{12} + a_{13}A_{13}$ by selecting the first row $(i = 1)$. So

$$|A| = 3\begin{vmatrix} 4 & 3 \\ 0 & 5 \end{vmatrix} + 2\begin{vmatrix} 1 & 3 \\ 6 & 5 \end{vmatrix} + 8\begin{vmatrix} 1 & 4 \\ 6 & 0 \end{vmatrix}$$

$$= 3(20) + 2(5 - 18) + 8(-24)$$

Therefore $|A| = -158$, or $|A| = a_{21}A_{21} + a_{22}A_{22} + a_{23}A_{23}$ by selecting the second column $(j = 2)$. So

$$|A| = 2\begin{vmatrix} 1 & 3 \\ 6 & 5 \end{vmatrix} + 4\begin{vmatrix} 3 & 8 \\ 6 & 5 \end{vmatrix} - 0\begin{vmatrix} 3 & 8 \\ 1 & 3 \end{vmatrix}$$

$$= 2(5 - 18) + 4(15 - 48)$$

Therefore $|A| = -158$, which confirms the previous answer.

The *adjoint* of a square matrix A is found by firstly replacing each element by its cofactor and secondly by transposing, i.e. if

$$A = \begin{bmatrix} a_{11} & a_{12} & a_{13} \\ a_{21} & a_{22} & a_{23} \\ a_{31} & a_{32} & a_{33} \end{bmatrix}$$

then

$$\text{adj } A = \begin{bmatrix} A_{11} & A_{21} & A_{31} \\ A_{12} & A_{22} & A_{32} \\ A_{13} & A_{23} & A_{33} \end{bmatrix}$$

A1.3 Matrix manipulation

Addition and subtraction of matrices, which are of the same dimensions, are carried out by simply adding (or subtracting) the corresponding elements. So

$$C = A + B = B + A \tag{A1.12}$$

where the elements of C are $c_{ij} = a_{ij} + b_{ij}$. As an example, consider

$$C = \begin{bmatrix} 3 & -6 & 0 \\ 8 & 2 & 4 \\ 1 & 7 & 5 \end{bmatrix} + \begin{bmatrix} -1 & 4 & 2 \\ 0 & 3 & 7 \\ 6 & -7 & 5 \end{bmatrix} = \begin{bmatrix} 2 & -2 & 2 \\ 8 & 5 & 11 \\ 7 & 0 & 10 \end{bmatrix}$$

For subtraction however, $c_{ij} = a_{ij} - b_{ij}$, such that

$$C = \begin{bmatrix} 3 & -6 & 0 \\ 8 & 2 & 4 \\ 1 & 7 & 5 \end{bmatrix} - \begin{bmatrix} -1 & 4 & 2 \\ 0 & 3 & 7 \\ 6 & -7 & 5 \end{bmatrix} = \begin{bmatrix} 4 & -10 & -2 \\ 8 & -1 & -3 \\ -5 & 14 & 0 \end{bmatrix}$$

Multiplication of two matrices requires that the number of columns of the first matrix is equal to the number of rows of the second, the two matrices do not therefore necessarily have to be of the same dimension. So

$$C = AB \tag{A1.13}$$

where the elements of C are $c_{ij} = \sum_{k=1}^{n} a_{ik} b_{kj}$, i.e. the element c_{ij} is found by multiplying the elements of the ith row of A and the jth column of B and summing the results. This means that if A is of dimension $m \times n$ and B is of dimension $n \times p$, then C will be of dimension $m \times p$. As an example, consider

$$C = \begin{bmatrix} 3 & -6 & 1 \\ 4 & 7 & 2 \end{bmatrix} \begin{bmatrix} 1 & 0 \\ 3 & 2 \\ 0 & -5 \end{bmatrix} = \begin{bmatrix} -15 & -17 \\ 25 & 4 \end{bmatrix}$$

Note: By the definitions of multiplication given, it follows that in many cases

$$AB \neq BA \tag{A1.14}$$

This is easily verified by a simple example, e.g. try

$$A = \begin{bmatrix} 2 & 1 \\ 4 & 2 \end{bmatrix}, \qquad B = \begin{bmatrix} 6 & 1 \\ 3 & 1 \end{bmatrix}$$

However, (A1.14) is certainly not a general rule, consider the case when $B = I$ for example.

Multiplication of a matrix by a scalar results in all the elements of the matrix being multiplied by the scalar, so, if

$$C = \lambda A \qquad\qquad (A1.15)$$

where λ is a scalar, then $c_{ij} = \lambda a_{ij}$. As an example, consider

$$C = 3 \begin{bmatrix} 2 & -1 \\ 4 & 0 \end{bmatrix} = \begin{bmatrix} 6 & -3 \\ 12 & 0 \end{bmatrix}$$

Multiplication of vectors can take on two forms. Firstly, when a row vector premultiplies a column vector, in which case the number of elements in each vector must be identical, e.g. consider the $n \times 1$ column vector x, then

$$x^T x = [x_1 \quad x_2 \quad \cdots \quad x_n] \begin{bmatrix} x_1 \\ x_2 \\ \vdots \\ x_n \end{bmatrix} = x_1^2 + x_2^2 + \cdots + x_n^2$$

However, when a column vector premultiplies a row vector the number of elements in each vector need not be identical, e.g. consider the $n \times 1$ column vector x and the $1 \times m$ row vector y, then

$$xy = \begin{bmatrix} x_1 \\ x_2 \\ \vdots \\ x_n \end{bmatrix} [y_1 \quad y_2 \quad \cdots \quad y_m] = \begin{bmatrix} x_1 y_1 & x_1 y_2 & \cdots & x_1 y_m \\ x_2 y_1 & x_2 y_2 & \cdots & x_2 y_m \\ \vdots & \vdots & & \vdots \\ x_n y_1 & x_n y_2 & \cdots & x_n y_m \end{bmatrix}$$

i.e. an $n \times m$ matrix results.

The product of two transposed matrices is equal to the transpose of the product of the matrices when commuted, i.e.

$$A^T B^T = (BA)^T \qquad\qquad (A1.16)$$

This can easily be verified with a simple example.

The *inverse* of a nonsingular square matrix A is defined as:

$$A^{-1} = \frac{\text{adj } A}{|A|} \qquad\qquad (A1.17)$$

and has the property that:

$$AA^{-1} = A^{-1}A = I \qquad\qquad (A1.18)$$

Example A1.2

Find the inverse of the matrix

$$A = \begin{bmatrix} 3 & -2 & 8 \\ 1 & 4 & 3 \\ 6 & 0 & 5 \end{bmatrix}$$

The determinant of A was found in Example A1.1 to be $|A| = -158$. The adjoint of A is however

$$\text{adj } A = \begin{bmatrix} 20 & 10 & -38 \\ 13 & -33 & -1 \\ -24 & -12 & 14 \end{bmatrix}$$

The inverse of A is therefore

$$A^{-1} = -\frac{1}{158} \begin{bmatrix} 20 & 10 & -38 \\ 13 & -33 & -1 \\ -24 & -12 & 14 \end{bmatrix}$$

A1.4   Simultaneous equations solution

The concept of a matrix description was introduced in terms of writing a set of simultaneous equations (A1.1) in an alternative form (A1.2), namely

$$Ax = y$$

On the assumption that the number of equations n is equal to the number of unknowns in the vector x, A will be a square matrix of dimension $n \times n$. As long as A is nonsingular, a solution will then exist to the set of simultaneous equations, and A^{-1} exists. It also follows that

$$A^{-1}(Ax) = A^{-1}y$$

i.e.

$$x = A^{-1}y \tag{A1.19}$$

which gives a unique solution for x.

Example A1.3

Find a solution to the set of simultaneous equations given by:

$$Ax = y$$

where

$$A = \begin{bmatrix} 6 & -3 & 4 \\ -2 & 5 & 1 \\ 1 & -1 & 1 \end{bmatrix}, \qquad x = \begin{bmatrix} x_1 \\ x_2 \\ x_3 \end{bmatrix}, \qquad \text{and} \qquad y = \begin{bmatrix} -1 \\ 10 \\ -2 \end{bmatrix}$$

So

$$|A| = 6\begin{vmatrix} 5 & 1 \\ -1 & 1 \end{vmatrix} + 3\begin{vmatrix} -2 & 1 \\ 1 & 1 \end{vmatrix} + 4\begin{vmatrix} -2 & 5 \\ 1 & -1 \end{vmatrix} = 15$$

and

$$\text{adj } A = \begin{bmatrix} 6 & -1 & -23 \\ 3 & 2 & -14 \\ -3 & 3 & 24 \end{bmatrix}$$

which means that

$$A^{-1} = \frac{1}{15} \begin{bmatrix} 6 & -1 & -23 \\ 3 & 2 & -14 \\ -3 & 3 & 24 \end{bmatrix}$$

and hence

$$x_1 = \tfrac{1}{15}(6 \times -1 \quad -1 \times 10 \quad -23 \times -2) = \tfrac{30}{15}$$
$$x_2 = \tfrac{1}{15}(3 \times -1 \quad +2 \times 10 \quad -14 \times -2) = \tfrac{45}{15}$$
$$x_3 = \tfrac{1}{15}(-3 \times -1 \quad +3 \times 10 \quad +24 \times -2) = -\tfrac{15}{15}$$

The solution is therefore:

$$x_1 = 2, \qquad x_2 = 3 \qquad \text{and} \qquad x_3 = -1$$

It is assumed, to ensure that a solution exists to the equation (A1.2), that the square matrix A is nonsingular. It can also be said that a nonsingular matrix is of full *rank*, i.e. rank = n for an $n \times n$ nonsingular matrix. If, however, such a matrix is singular, i.e. its determinant is zero, then its rank is equal to the row (or column) dimension of the largest (in terms of dimension) minor within that matrix. So, if the matrix A possesses a nonzero minor of order r, whilst every minor of higher order is zero, then A has rank r.

By this means, the matrix

$$A = \begin{bmatrix} 3 & -1 & -7 \\ 1 & 4 & 2 \\ 0 & 1 & 1 \end{bmatrix}$$

has rank = 2.

Finally in this section, two useful matrix relationships are:

$$|A||B| = |AB| \tag{A1.20}$$

and

$$AI = IA = A \tag{A.1.21}$$

Further reading

Kailath, T., *Linear systems*, Prentice-Hall, Inc., 1980.
Tropper, A. M., *Linear algebra*, Nelson, 1978.

Appendix 2   Tables of common transforms

Table A2.1 Laplace transforms

Function, $f(t)$ [for $t \geqslant 0$]	Laplace transform, $F(s)$ i.e. $\mathscr{L}\{f(t)\}$
$\delta(t)$, unit impulse at $t = 0$	1 or s^0
1, unit step	s^{-1}
t, unit ramp	s^{-2}
$t^{(n-1)}/(n-1)!$	s^{-n}
$\exp(-\alpha t)$	$(s+\alpha)^{-1}$
$\dfrac{t^{(n-1)}}{(n-1)!}\exp(-\alpha t)$	$(s+\alpha)^{-n}$
$1 - \exp(-\alpha t)$	$\dfrac{\alpha}{s(s+\alpha)}$
$\exp(-\alpha t) - \exp(-\beta t)$	$\dfrac{\beta-\alpha}{(s+\alpha)(s+\beta)}$
$\{\gamma-\alpha\}\exp(-\alpha t)$ $\quad - \{\gamma-\beta\}\exp(-\beta t)$	$\dfrac{(\beta-\alpha)(s+\gamma)}{(s+\alpha)(s+\beta)}$
$\alpha\{1-\exp(-\beta t)\}$ $\quad - \beta\{1-\exp(-\alpha t)\}$	$\dfrac{\alpha\beta(a-\beta)}{s(s+\alpha)(s+\beta)}$
$\gamma\{\alpha-\beta\} + \beta\{\gamma-\alpha\}\exp(-\alpha t)$ $\quad - \alpha\{\gamma-\beta\}\exp(-\beta t)$	$\dfrac{\alpha\beta(a-\beta)(s+\gamma)}{s(s+\alpha)(s+\beta)}$
$\{\alpha-\delta\}\exp(-\beta t) + \{\delta-\beta\}\exp(-\alpha t)$ $\quad + \{\beta-\alpha\}\exp(-\delta t)$	$\dfrac{(\beta-\delta)(\delta-\alpha)(\alpha-\beta)}{(s+\alpha)(s+\beta)(s+\delta)}$
$\{\gamma-\beta\}\{\alpha-\delta\}\exp(-\beta t)$ $\quad + \{\gamma-\alpha\}\{\delta-\beta\}\exp(-\alpha t)$ $\quad + \{\gamma-\delta\}\{\beta-\alpha\}\exp(-\delta t)$	$\dfrac{(\beta-\delta)(\delta-\alpha)(\alpha-\beta)(s+\gamma)}{(s+\alpha)(s+\beta)(s+\delta)}$
$\cos \omega t$	$\dfrac{s}{s^2+\omega^2}$
$\sin \omega t$	$\dfrac{\omega}{s^2+\omega^2}$
$\cos(\omega t + \phi)$	$\dfrac{s\cos\phi - \omega\sin\phi}{s^2+\omega^2}$
$\sin(\omega t + \phi)$	$\dfrac{s\sin\phi + \omega\cos\phi}{s^2+\omega^2}$
$\cosh \omega t$	$\dfrac{s}{s^2-\omega^2}$
$\sinh \omega t$	$\dfrac{\omega}{s^2-\omega^2}$
$\exp(-\alpha t)\cos \omega t$	$\dfrac{(s+\alpha)}{(s+\alpha)^2+\omega^2}$
$\exp(-\alpha t)\sin \omega t$	$\dfrac{\omega}{(s+\alpha)^2+\omega^2}$
$\exp(-\alpha t)\sin(\omega t + \phi)$	$\dfrac{\omega}{[\omega^2+(\gamma-\alpha)^2]^{1/2}} \cdot \dfrac{(s+\gamma)}{(s+\alpha)^2+\omega^2}$ where $\gamma = \alpha + \omega/\tan\phi$

Table A2.1 (contd.)

Function, $f(t)$ [for $t \geqslant 0$]	Laplace transform, $F(s)$ i.e. $\mathscr{L}\{f(t)\}$
$\dfrac{1}{\beta^2}\left[1 + \dfrac{\beta}{\omega}\exp(-\alpha t)\sin(\omega t - \phi)\right]$ $\omega = -\alpha\tan\phi$ and $\beta = (\alpha^2 + \omega^2)^{1/2}$	$\dfrac{1}{s[(s+\alpha)^2 + \omega^2]}$
$\dfrac{1}{\beta^2}\left[\gamma + \dfrac{\beta\delta}{\omega}\exp(-\alpha t)\sin(\omega t + \phi)\right]$ $\delta = [(\lambda - \alpha)^2 + \omega^2]^{1/2}$ $\beta = (\alpha^2 + \omega^2)^{1/2}$ $\phi = \tan^{-1}\left[\dfrac{\omega}{\gamma - \alpha}\right] - \tan^{-1}\left[\dfrac{\omega}{-\alpha}\right]$	$\dfrac{(s+\gamma)}{s[(s+\alpha)^2 + \omega^2]}$
$[\exp(-\beta t)/\delta]$ $+ \dfrac{1}{\omega}[\exp(-\alpha t)\sin(\omega t + \phi)]$ $\tan\phi = \dfrac{\omega}{\beta - \alpha}$	$\dfrac{\delta}{(s+\beta)[(s+\alpha)^2 + \omega^2]}$ $\delta = [(\beta - \alpha)^2 + \omega^2]^{1/2}$
$\exp(-\alpha t)\exp j(\omega t + \phi)$	$\dfrac{1}{(s+\alpha)^2 + \omega^2}[(s+\alpha)\cos\phi - \omega\sin\phi$ $\qquad + j(s+\alpha)\sin\phi + \omega\cos\phi]$
$\dfrac{\omega_n}{\alpha}\exp(-\zeta\omega_n t)\sin(\alpha\omega_n t)$ $\alpha^2 + \zeta^2 = 1$	$\dfrac{\omega_n^2}{s^2 + 2\zeta\omega_n s + \omega_n^2}$ $\zeta < 1$
$1 - \dfrac{1}{\alpha}\exp(-\zeta\omega_n t)\sin(\alpha\omega_n t + \phi)$ $\alpha^2 + \zeta^2 = 1$ $\zeta = \cos\phi$	$\dfrac{\omega_n^2}{s(s^2 + 2\zeta\omega_n s + \omega_n^2)}$ $\zeta < 1$

Example A2.1

For an example of transform with zero order hold present, consider the original system transfer function

$$G_0(s) = \frac{1}{s + 3}$$

which is to be cascaded with a zero order hold, subject to a sample period of T seconds. The overall transfer function, with zero order hold included is

$$G(s) = \frac{[1 - \exp(-sT)]}{s(s + 3)}$$

Table A2.2 z-transforms

Note: In the following table of z-transforms, it is assumed that no zero order hold is included in the required transform. Therefore if it is the case that a zero order hold is present, the overall transform should be obtained as described in Chapter 8, with particular reference to Example 8.3.3. Example A2.1 is a further example.

Function, $f(t)$ [for $t \geqslant 0$]	Laplace transform $F(s)$	z-transform, $F(z)$ (T = sample period)
$\delta(t)$	1	1
$\delta(t - kT)$	$\exp(-kTs)$	z^{-k}
1	s^{-1}	$\dfrac{z}{z-1}$
t	s^{-2}	$\dfrac{Tz}{(z-1)^2}$
$\dfrac{t^2}{2}$	s^{-3}	$\dfrac{T^2 z(z+1)}{2(z-1)^3}$
$\exp(-\alpha t)$	$\dfrac{1}{(s+\alpha)}$	$\dfrac{z}{z - \exp(-\alpha T)}$
$t \cdot \exp(-\alpha t)$	$\dfrac{1}{(s+\alpha)^2}$	$\dfrac{Tz \exp(-\alpha T)}{[z - \exp(-\alpha T)]^2}$
$1 - \exp(-\alpha t)$	$\dfrac{\alpha}{s(s+\alpha)}$	$\dfrac{z[1 - \exp(-\alpha T)]}{(z-1)[z - \exp(-\alpha T)]}$
$t + \dfrac{1}{\alpha}[\exp(-\alpha t) - 1]$	$\dfrac{\alpha}{s^2(s+\alpha)}$	$\dfrac{Tz}{(z-1)^2}$ $-\dfrac{z[1 - \exp(-\alpha T)]}{\alpha(z-1)[z - \exp(-\alpha T)]}$
$\sin \omega t$	$\dfrac{\omega}{s^2 + \omega^2}$	$\dfrac{z \sin \omega T}{z^2 - 2z \cos \omega T + 1}$
$\cos \omega t$	$\dfrac{s}{s^2 + \omega^2}$	$\dfrac{z(z - \cos \omega T)}{z^2 - 2z \cos \omega T + 1}$
$\exp(-\alpha t)\sin \omega t$	$\dfrac{\omega}{(s+\alpha)^2 + \omega^2}$	$\dfrac{z \exp(-\alpha T)\sin \omega T}{z^2 - 2z \exp(-\alpha T)\cos \omega T + \exp(-2\alpha T)}$
$\exp(-\alpha t)\cos \omega t$	$\dfrac{s+\alpha}{(s+\alpha)^2 + \omega^2}$	$\dfrac{z^2 - z \exp(-\alpha T)\cos \omega T}{z^2 - 2z \exp(-\alpha T)\cos \omega T + \exp(-2\alpha T)}$

such that

$$G(z) = (1 - z^{-1}) \cdot Z\left\{\frac{1}{s(s + 3)}\right\}$$

The z-transform for the function $1/s(s + 3)$, is found, from the above table, to be:

$$\frac{z[1 - \exp(-3T)]}{3(z - 1)[z - \exp(-3T)]}$$

giving an overall transform of:

$$G(z) = \frac{1 - \exp(-3T)}{3[z - \exp(-3T)]}$$

which can be evaluated for any specified sample period T.

Appendix 3   Nonlinear systems

The study of control systems carried out in this book is concerned with those systems which can be regarded as *linear*, definitions of linearity being given in Chapter 2, Section 2.2.3. In practice the vast majority of systems to be controlled are, to an extent, nonlinear, i.e. they break at least one of the rules of linearity. Despite this, it is often sufficient to consider a system to be linear, to a good approximation, though this might mean that the range of operation is limited. However, a point is reached with some systems where nonlinearity effects are too large to ignore, and they must be taken into account. In this appendix a brief overview of nonlinear systems, and the methods employed for their control, is given.

Because there are many different ways in which a system can exhibit nonlinear behavior, it is difficult to build up a general theoretical background, other than by classifying particular types of nonlinearity. For any one nonlinear system however, the response of that system will depend on the type and magnitude of the input signal applied, and unfortunately it is not always possible to know exactly the range of values to which the system will be subjected.

Example A3.1

Consider the case of an amplifier cascaded with a linear system as shown in Fig. A3.1. The amplifier characteristics will very likely result in amplifier output saturation caused by the application of too large an input signal, see Fig. A3.2.

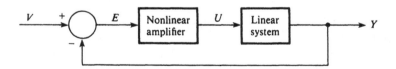

Fig. A3.1   Cascade nonlinear amplifier

While the signal $e(t)$ remains within the range $-e_1 \leqslant e(t) \leqslant +e_1$, it is apparent that the overall system in Fig. A3.1 will operate as a linear system, nonlinear effects only being witnessed when $|e(t)| > |e_1|$.

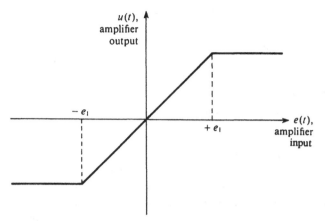

Fig. A3.2   Nonlinear amplifier characteristics (saturation)

Amplifier saturation, motor stiction, transducer errors and gear backlash are examples of *inherent* nonlinearities. These are nonlinear effects which are part of a system we wish

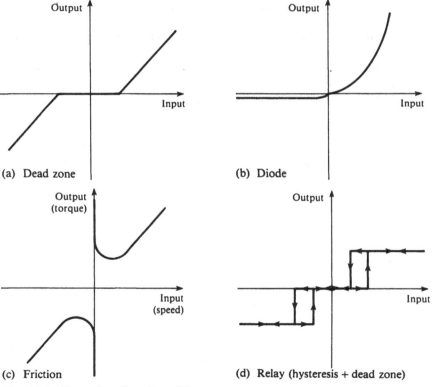

(a) Dead zone

(b) Diode

(c) Friction

(d) Relay (hysteresis + dead zone)

Fig. A3.3   Examples of nonlinearities

to control. In general, they are effects which can be represented by means of an input to output nonlinear system function, an example being that shown in Fig. A3.2 with further examples given in Fig. A3.3.

Another type of nonlinearity is encountered simply by multiplying together a number of linear system signals, this is termed an operational nonlinearity and is explained by means of the following example.

Example A3.2

As an example of an operational nonlinearity, consider the linear system $y_1(t) = u_1(t) \cdot v_1(t)$, in which both the input signals are variable. Similarly, therefore for two different input signals $y_2(t) = u_2(t) \cdot v_2(t)$. From the definition of a linear system, if an input $u_1(t)$ produces an output $y_1(t)$, and an input $u_2(t)$ produces an output $y_2(t)$, then an input $u_1(t) + u_2(t)$ must produce an output $y_1(t) + y_2(t)$. However, we must also have that if $v_1(t)$ produces $y_1(t)$ and $v_2(t)$ produces $y_2(t)$, then $v_1(t) + v_2(t)$ must produce $y_1(t) + y_2(t)$.

Let $u_1(t) = 3$ and $v_1(t) = 4$, such that $y_1(t) = 12$. Also, let $u_2(t) = 2$ and $v_2(t) = 5$, such that $y_2(t) = 10$. So, $u_1(t) + u_2(t) = 5$ and $v_1(t) + v_2(t) = 9$, and $y_1(t) + y_2(t) = 22$ does *not* result from the multiplication of the sums.

Inherent nonlinearities are very often, in practice, effects which are neglected, especially when high accuracy is not a requirement. Indeed, the application of a feedback control loop can, in many cases, result in the linearization of an open-loop nonlinear system. A speed control system, as an example, may well have an input voltage to output speed relationship such as that in Fig. A3.4(a) when in open-loop. However, this relationship can be made to look like that in Fig. A3.4(b) simply by selecting a suitable proportional feedback gain.

The result of employing feedback, in this case, is to drastically reduce the effect of nonlinearities − the cost of such an 'improvement' being a significant reduction in system gain.

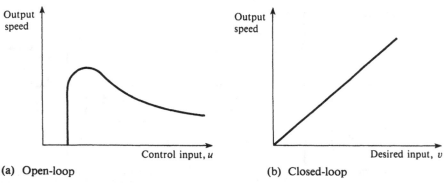

(a) Open-loop   (b) Closed-loop

Fig. A3.4 Speed control system characteristics

When designing a controller for certain systems, whether they be linear or nonlinear themselves, it can be useful to introduce an *intentional* nonlinearity, the aim being either to simplify system design or merely to meet specific performance requirements. An example of such a nonlinearity is a closed-loop system in which the control action is only applied when the output exceeds either an upper or lower limit, e.g. the simple on–off type of control applied with a domestic thermostat.

Analysis and design of nonlinear control systems can take on one of several forms dependent on the extent and type of nonlinearities present, overall specifications however still remain in terms of stability requirements, steady-state values and transient response tailoring. Where a description of a particular nonlinearity is quite complicated, possibly in terms of several differential equations, then it is probably best to employ a computer in order to find the equation solutions necessary. If the nonlinear effects are more straightforward though, it is often possible to linearize the description in a way which produces roughly equivalent results. Several approaches to equivalent linearization are now discussed.

A3.1   Analysis at the extremes

The system is investigated by looking only at the extreme values (highest and lowest) of each of the variable parameters. Consider the following example, in which the variable parameter is the nonlinear amplifier stage.

Example A3.3

Consider the system shown in Fig. A3.5, in which an amplifier is cascaded with the linear system

$$G(s) = \frac{K}{s + 1}$$

where the gain $K > 0$ must be selected to ensure that the steady-state output value is of magnitude 10 units $\pm 5\%$, following the application of a 10 unit step input.

The characteristics of the nonlinear amplifier are shown in Fig. A3.2, with $e_1 = 1$, and $u(t) = 1$ when $e_1 = 1 = e(t)$. Because $v(t)$ is a step input, from $v(t) = 0$ to $v(t) = 10$, only positive values need be considered for the amplifier

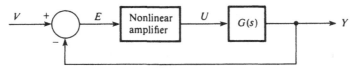

Fig. A3.5   Cascade nonlinear amplifier for Example A3.3

characteristics. Two extremes are then of interest:

(i) $0 < e(t_1) < 1$, $u(t_1) = e(t_1)$ from $t_1 = 0$ (when $y(t_1) = 9 : t_1 = 0$).
       Thus

$$Y = \frac{K}{s+1+K} \cdot \left[\frac{10}{s} + \frac{9}{K}\right] \Rightarrow y(t_1)$$

$$= \frac{10K}{1+K}\{1 - e^{-(1+K)t_1}\} + 9e^{-(1+K)t_1}$$

(ii) $e(t) > 1$, $u(t) = 1$ from $t = 0$ and $Y = GU$, i.e.

$$Y = \frac{K}{s+1} \cdot \frac{1}{s} \Rightarrow y(t) = K[1 - e^{-t}]$$

Immediately following the application of the 10 unit step input, the output
signal $y(t)$ will start to increase from zero as in (ii), when the error $e(t)$ will be
greater than unity. Once $y(t)$ has reached a value of 9 units, the error $e(t)$ will
be less than 1 and the output signal will increase from 9 to 10 units following the
equation found in (i). The steady-state output value will therefore be

$$y(t)\Big|_{t_1 \to \infty} = \frac{10K}{1+K}$$

and if it is required that $9.5 < 10K/(1 + K)$, a value of $K = 25$ will suffice.

With this value of K, the overall response of the system shown in Fig. A3.5,
following a 10 unit step input, is shown in Fig. A3.6, with

(i) $y(t_1) = \frac{250}{26}\{1 - e^{-26t_1}\} + 9e^{-26t_1}$
(ii) $y(t) = 25\{1 - e^{-t}\}$

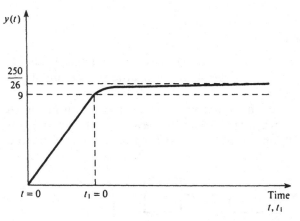

Fig. A3.6   Time response of the system in Example A3.3

A3.2   Analysis about a fixed operating point

Linearization of a nonlinear function about a fixed operating point is a commonly encountered procedure in practice, a basic assumption being made that the system under consideration is more or less linear at the operating point and any nonlinear effects are negligible.

Consider the effect of operational nonlinearities, such as those encountered about the fixed operating point $Y = U \cdot V$.

Let both U and V vary by small amounts $\bar{u}$ and $\bar{v}$ respectively from the operating point, such that

$$Y + \bar{y} = (U + \bar{u})(V + \bar{v}) \qquad \text{where} \qquad U \gg \bar{u} \qquad \text{and} \qquad V \gg \bar{v}$$

so

$$Y + \bar{y} = UV + \bar{u}V + U\bar{v} + \bar{u}\bar{v}$$

But the product $\bar{u}\bar{v}$ can readily be neglected, leaving

$$Y = UV$$

with an error

$$\bar{y} = \bar{u}V + U\bar{v}$$

and this error $\bar{y}$ is the error incurred by assuming that the operational nonlinearity is approximately linear.

A3.3   Describing functions

In the describing function method the isolated nonlinear element, in the system under consideration, is approximated by assuming that only the first harmonic (fundamental) component of its output response is significant (employing Fourier series analysis). Consider the effect of applying a sinusoidal input to a nonlinear amplifier with saturation, for example see Fig. A3.7. Typical amplifier characteristics are shown in Fig. A3.2.

The describing function of a nonlinear system element is the complex ratio of the first harmonic components. In the amplifier considered in Fig. A3.7, the describing function is

$$N = \frac{U_1}{E} \, \big/\!\underline{\phi_1}$$

where ϕ_1 is the phase shift exhibited by the first harmonic component of $u(t)$.

In general, the describing function of a nonlinear element can be found by obtaining the first harmonic component of the Fourier series which represents the nonlinear

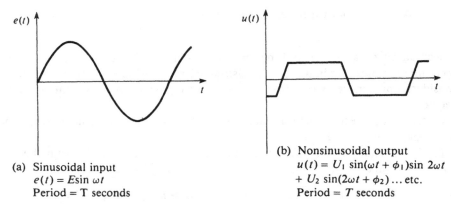

(a) Sinusoidal input
   $e(t) = E\sin \omega t$
   Period = T seconds

(b) Nonsinusoidal output
   $u(t) = U_1 \sin(\omega t + \phi_1)\sin 2\omega t$
   $+ U_2 \sin(2\omega t + \phi_2) \dots$ etc.
   Period = T seconds

Fig. A3.7 Input and output signals for a saturating amplifier

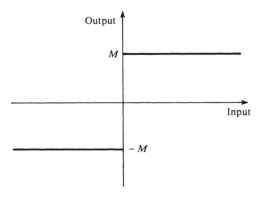

Fig. A3.8 Ideal relay characteristics

elements' output response, the describing function itself then is the ratio of the output (first harmonic) to input sine wave. As a further example, the describing function of the ideal relay shown in Fig. A3.8 is

$$N = \frac{4M}{\pi E}$$

where E is the peak magnitude of the input applied to the relay.

Where several nonlinear elements are present in a system, the describing function can be calculated by accounting for the first harmonic due to each. A unity feedback control system which contains one or more nonlinear elements can then be represented as shown in Fig. A3.9.

As long as harmonics of second order and above are sufficiently attenuated, the closed-loop system transfer function is given by

$$\frac{Y(j\omega)}{V(j\omega)} = \frac{NG}{1 + NG} \tag{A3.1}$$

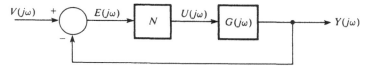

Fig. A3.9 Unity feedback system with nonlinear element

where

$$G(j\omega) = K \frac{B(j\omega)}{A(j\omega)} \qquad\qquad (A3.2)$$

and $K > 0$ is a scalar gain.

The characteristic equation of the closed-loop system (A3.1) is then

$$1 + NG = 0$$

which is

$$G = -\frac{1}{N} \qquad\qquad (A3.3)$$

or

$$[G(j\omega)]^{-1} = -N \qquad\qquad (A3.4)$$

A similar stability analysis to that carried out for Nyquist design (see Sections 5.4 and 5.5) can be employed to investigate the stability properties of a system defined by (A3.1), given an open-loop transfer function $G(j\omega)$ and a describing function N. It must be remembered that in general $G(j\omega)$ and N are characterized by both magnitude and phase components. Stability properties can then be considered either in terms of a normal Nyquist plot of $G(j\omega)$ over the range of frequencies ω, or by plotting $[G(j\omega)]^{-1}$ as an inverse Nyquist plot. However, rather than the point $-1 + j0$ being of importance, as is the case with the straightforward Nyquist plot, the locus of $-1/N$ or $-N$ respectively is of importance.

Stability criteria

1. If the complete $G(j\omega)$ locus does not enclose the $-1/N$ locus, then the unity feedback closed-loop system is stable.
2. If the complete $G(j\omega)$ locus does enclose the $-1/N$ locus, the unity feedback closed-loop system output will increase to some finite limiting value (possibly caused by saturation).
3. If the $G(j\omega)$ and N loci intersect, then the unity feedback closed-loop system output may well exhibit a *limit cycle*.

Note: A limit cycle is a sustained oscillation at a fixed frequency and with constant amplitude.

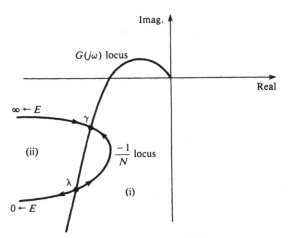

Fig. A3.10   Nyquist plot with $-1/N$ locus

Consider the plot in Fig. A3.10, where E is the peak magnitude of the signal input to the nonlinear element with

$$N = \frac{4M}{\pi E}$$

At the point λ, the $G(j\omega)$ and $-1/N$ loci intersect. If the signal E is slightly to the left of that at λ then this is within the stable region (ii), and E will continue to decrease. However, if E is slightly to the right of that at λ then this is within the unstable region (i), and E will continue to increase along the $-1/N$ loci towards the point γ. The limit cycle at the point λ is called an *unstable limit cycle*, any variation in E from its value at that point being sufficient to cause further movement along the $-1/N$ locus away from λ.

At the point γ, the $G(j\omega)$ and $-1/N$ loci intersect. If the signal E is slightly to the left of that at γ then this is within the stable region (ii), and E will tend to decrease back towards its value at γ. However, if E is slightly to the right of that at γ then this is within the unstable region (i), and E will tend to increase towards its value at γ. The limit cycle at the point γ is called a *stable limit cycle*, any variation in E from its value at that point resulting in a tendency for E to return to its value at γ.

In a similar fashion to the studies carried out for Nyquist design procedure in Section 5.5, system performance can be modified to meet specifications made by reshaping the $G(j\omega)$ loci with a suitable choice of linear compensator.

Example A3.4

Consider the open-loop transfer function:

$$G(j\omega) = \frac{10K}{j\omega(j\omega + 2)}$$

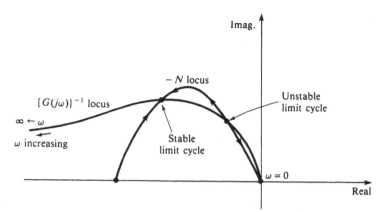

Fig. A3.11 Inverse Nyquist plot for Example A3.4

Then

$$[G(j\omega)]^{-1} = \frac{j\omega(j\omega + 2)}{10} \quad \text{for} \quad K = 1$$

The frequency response of the inverse function is shown plotted in Fig. A3.11 along with the frequency response of a nonlinear element N. It can be seen that two limit cycles exist, one stable and one unstable, and that simply by decreasing the gain K by a suitable amount, the magnitude of $[G(j\omega)]^{-1}$ is increased over all frequencies, resulting in a stable unity feedback closed-loop system when the $[G(j\omega)]^{-1}$ locus no longer intersects with the $-N$ locus.

Appendix 4   Unity feedback loops

A large number of tests employed to study the stability of a closed-loop system require that the system contains a unity feedback loop. This is true of the continuous-time tests described in Chapters 4 and 5, namely (a) root loci, (b) Bode, (c) Nyquist and (d) Nichols. For each of these it is necessary to consider initially the open-loop transfer function, the methods then give stability results for a unity feedback closed-loop system. Consider the open-loop transfer function to be G_0, a function of s or jw as appropriate, such than when connected within a unity feedback loop, as shown in Fig. A4.1, the

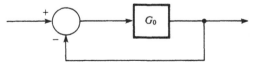

Fig. A4.1 Unity feedback loop

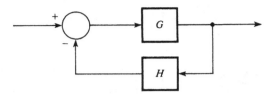

Fig. A4.2   General closed-loop system

closed-loop transfer function is given as G_c where

$$G_c = \frac{G_0}{1 + G_0} \tag{A4.1}$$

In practice however, the closed-loop system may well take on the more general feedback form shown in Fig. A4.2.

The closed-loop transfer function in the more general case is then:

$$G_c = \frac{G}{1 + GH} \tag{A4.2}$$

It was shown in Chapter 3 (Section 3.4.1) how the general form of Fig. A4.2 can be reconsidered in terms of a unity feedback loop, by equating G_0 with G and H.

The relationship is found to be:

$$G_0 = \frac{G}{1 + G[H - 1]} \tag{A4.3}$$

As far as stability of the closed-loop system is concerned, it is the roots of the characteristic equation which are of importance. In the general case the roots are obtained from:

$$1 + GH = 0 \tag{A4.4}$$

Let

$$G = K\frac{A}{B} \quad \text{and} \quad H = \bar{K}\frac{\bar{A}}{\bar{B}}$$

where $A, B, \bar{A}$ and $\bar{B}$ are polynomials, K and $\bar{K}$ are constants. So

$$1 + GH = \frac{B\bar{B} + K\bar{K}A\bar{A}}{B\bar{B}} = 0$$

Stability of the general closed-loop system can then be investigated by considering the roots of

$$B\bar{B} + K\bar{K}A\bar{A} = 0 \tag{A4.5}$$

With G and H as defined, this means that the equivalent G_0 can be found from (A4.3) as:

$$G_0 = \frac{KA\bar{B}}{B\bar{B} + KA[\bar{K}\bar{A} - \bar{B}]}$$

So the stability of the equivalent unity feedback system can be investigated by considering the roots of

$$1 + G_0 = \frac{B\bar{B} + K\bar{K}A\bar{A}}{B\bar{B} + KA[\bar{K}\bar{A} - \bar{B}]} = 0$$

which are found from

$$B\bar{B} + K\bar{K}A\bar{A} = 0 \tag{A4.6}$$

The resultant characteristic equation is therefore the same, whether the general system characteristic equation $1 + GH = 0$ or the equivalent unity feedback characteristic equation $1 + G_0 = 0$ is employed, i.e. (A4.6) is identical to (A4.5). Hence stability results will be the same if GH is employed, as though it were in fact the open-loop transfer function itself. For the stability tests discussed in Chapters 4 and 5 it is therefore sufficient to plot the characteristics of GH, on the appropriate diagram, rather than to firstly find the equivalent G_0, and subsequently to plot the characteristics of this new function.

Index